# Springer Series in Optical Sciences

Volume 46

Edited by Theodor Tamir

# Springer Series in Optical Sciences

Volumes 1–41 are listed on the back inside cover

W. Schumann   J.-P. Zürcher
D. Cuche

# Holography
## and
# Deformation Analysis

With 78 Figures

Springer-Verlag
Berlin Heidelberg GmbH

Prof. Dr. WALTER SCHUMANN
Dr. JEAN-PIERRE ZÜRCHER
Dr. DENIS CUCHE

Laboratorium für Photoelastizität, ETH Zürich, Rämistr. 101
CH-8092 Zürich, Switzerland

ISBN 978-3-662-13559-4     ISBN 978-3-540-38981-1 (eBook)
DOI 10.1007/978-3-540-38981-1

Library of Congress Cataloging-in-Publication Data. Schumann, Walter, 1927- Holography and deformation analysis. (Springer series in optical sciences ; v. 46) Bibliography: p. Includes index. 1. Holography. 2. Holographic interferometry. 3. Deformations (Mechanics) I. Zürcher, J.-P. (Jean-Pierre), 1953-. II. Cuche, D. (Denis), 1954-. III. Title. TA1542.S38 1985 621.36'75 85-12563

© Springer-Verlag Berlin Heidelberg 1985
Originally published by Springer-Verlag Berlin Heidelberg New York Tokyo in 1985
Softcover reprint of the hardcover 1st edition 1985

Typesetting: Schwetzinger Verlagsdruckerei GmbH

2153/3130-5 4 3 2 1 0

*To Brigitte Schumann*
*Monique Zürcher*
*Erika Cuche*

# Preface

In this book series on Optical Sciences, holography has been the subject of three previous volumes. In particular, Vol. 16, written by one of us (W.S.) and Dr. M. Dubas, treated holographic interferometry of opaque bodies from the standpoint of deformation analysis. However, the fundamental principles of holography are developed there only briefly in preparation for a discussion of interference fringe modifications.

This new volume in the series is intended to consider in detail many topics which were previously omitted, such as the deformation or distortion of holographic images, the theory of volume holograms, composite or multiplex holography, holographic interferometry of transparent media, time dependent effects, holographic contouring, and applications of fringe modifications to the deformation of opaque bodies. In addition, these and other subjects will be treated with the same unifying concept developed in Vol. 16, but with an additional emphasis on those features that have their origins in classical optics, especially the small-wavelength approach, the coupled-wave theory, and the Seidel aberrations. Since the field of holography and its various applications is growing rapidly, it is impossible to be comprehensive in a single book. Every effort has been made to avoid unnecessary duplication of Vol. 16. For example, displacement and fringe localization problems are only briefly discussed, while some modification techniques (e.g., sandwich holography) are not included. When needed, however, the reader is directly referred to complementary publications.

Apart from the literature itself, stimulation for this work originates from personal communications, particularly with Prof. R. Dändliker and Dr. K. A. Stetson, as well as from some reviews of previous work. The authors are most grateful to those who have made this publication possible: Dr. H. Lotsch for his constructive help and support; Prof. T. Tamir for accepting it in this series; collaborators at Springer-Verlag, in particular Mrs. A. Rapp, for the editing and printing; Drs. P. Colberg and G. Hanselmann for conscientiously revising the English text; Mr. L. Pellegrinelli for carefully drawing the many detailed figures; and Mrs. V. Schaer for her careful typing of the manuscript which Mr. Ph. Tatasciore and Mr. J.-P. Koob read over again.

Zürich, June 1985

*W. Schumann*
*J. P. Zürcher*
*D. Cuche*

# Contents

# 1. Introduction

As the history of holography is admittedly well known, we will confine the first part of this introduction to a brief summary of the most significant events. The foundations of holography were laid in 1948 by *Gabor* [1.1, 2] who demonstrated that it was possible, in principle, to reconstruct the image of an object from a diffraction pattern created at the recording, when the object and reference waves interfered with one another on a thin, photographic plate termed the hologram. However, this technique became a reality only 14 years later with the advent of the laser, which provided the needed source of intense, coherent, monochromatic light. In 1962, *Leith* and *Upatnieks* [1.3] used laser beams and simultaneously varied the directions of both the object and reference waves in order to avoid their overlapping. In 1963, *Denisyuk* [1.4] and *van Heerden* [1.5] independently proposed recording in a three-dimensional medium, resulting in production of thick holograms, which have subsequently gained importance in information storage, in white light reconstruction, and other applications. About 1965, it was technically possible to cause holographically produced wavefields to interfere with other wavefields, holographically or non-holographically produced, which resulted in fringe patterns. The first contributions in the so-called fields of holographic interferometry were made by *Powell* and *Stetson* [1.6], *Burch* [1.7], *Brooks* et al. [1.8]. Most of their investigations were concerned with the deformation and vibration analysis of opaque bodies, on the one hand, and the determination of changes of indices of refraction in fluids (phase objects), on the other hand. These studies later found applications also in photoelasticity. Since discussion of this latter area will be omitted, interested readers may, for instance, consult Vol. 11 in this series. Shortly after the introduction of the laser, the basic principles of holography were rapidly elucidated. However, the more refined analysis of image formation and fringe patterns in general configurations, such as needed for industrial applications, required a much longer time to develop.

In holography, the case in which the reconstructed image is identical to the configuration of the recorded object represents the exception rather than the rule. Intentionally or unintentionally produced optical modifications are almost always perceived at the reconstruction as aberrations of the image points; these points become blurred and distortions of the whole apparent object image appear as a virtual deformation. Investigation of these effects is not trivial and sometimes they cannot even be avoided as, for instance, in conjugate images. Moreover, these studies are relevant because of both their relationship to classical lens imaging and their role in holographic interferometry, where the modified fringes are representative of the superimposed mechanical object deformations with optical image deformation. Therefore, a good part of this book is devoted to the analysis of these effects. This aspect is complemented by a discussion of other aspects of holography such as the diffraction process, especially as it applies to volume holograms. This process, also important to other areas, was first studied in 1969 by *Kogelnik* [1.9] who made use of the coupled wave theory. A general overview is also given of cylindrical composite holography in which image deformation is always present.

Holographic interferometry, generally considered a spectacular technique for visualization of the deformation process, is not characterized by a study leading to the qualitative interpretation of interferograms alone. Rather, by choosing a precise geometrical arrangement of the holographic set-up, it is possible to accurately determine quantities such as displacements, strains, and rotations; here fringe properties of order, direction, interspace, and visibility must be considered. Refined analysis of these fringes and their visibilities was achieved by several investigators around 1970, in particular by *Stetson* [1.10, 11], by *Walles* [1.12], and by *Viénot* and collaborators [1.13], who considered the case of opaque bodies, while fringe localization in transparent media, such as gases and liquids, was studied mainly by *Vest* and his co-workers [1.14–16]. In particular, the approach presented here results in an analogy connecting the main quantities encountered in opaque body analysis with those defined for transparent media, where the latter are characterized by terms to be integrated. Further, the recording techniques for moving objects and fluids may exert an influence on the fringes; therefore, time dependent effects are also discussed. An advanced approach to these effects, in particular, has already been presented in the book by *Ostrovsky, Butusov*, and *Ostrovskaya*, published as Vol. 20 in this series [1.17]. In addition to the image, the fringe field may also be influenced by an optical modification at the reconstruction. The practical determination of the desired quantities, e.g., the displacement or the strains, becomes much more flexible when the measurement techniques allow for both control and amplification effects. Automation of the measurements may also be simplified by a modified fringe system. However, fringe control, in general, requires a more complete holographic set-up, e.g., two reference sources or two holographic plates. With respect to their potential industrial applications, publications concerning the heterodyne method of *Dändliker* and his co-workers should be consulted [1.18, 19].

Since holography permits not only the reproduction of images in some plane normal to an optical axis (e.g., on an optical bench), but also allows for reproduction of images of general configurations in space, the description of this research field requires, above all else, adequate algebraic tools such as those found in three-dimensional intrinsic calculus. Furthermore, it is known that an optical phenomenon usually originates from the integrated contribution of an entire bundle of neighboring rays, rather than from a single ray. That is why, in addition to the differential equations for wave propagation and the kinematic equations of the deformation gradient, we shall encounter derivatives of certain quantities which relate to the often encountered collineation centers. Working with several surfaces involves many so-called affine connections between these surfaces, which may include the holographic plane (which may be shifted), the deformed image surface at the reconstruction, interfaces placed between domains possessing different refractive indices, the object surface before and after deformation, the observing plane, and wavefronts along a diffracted ray. The connections are often accompanied by successive normal and oblique projectors which take into account the arbitrary disposition of the surfaces involved. The emphasis placed on linear transformations may, at first, appear to be too specific. However, they systematically lead to relevant, quantitative results from experimental applications, including a non-negligible generality. Furthermore, they permit a more unified treatment of holography from which common elements of both optics and mechanics appear.

# 2. Elements of Analysis, Geometrical Optics, and Kinematics

Before introducing the principal subject of this book, i.e., holography and holographic interferometry, we will briefly review a selection of some elements of analysis, geometrical optics, and kinematics, to which we shall refer throughout.

## 2.1 Review of Some Basic Concepts of Tensor Analysis

We shall first review some basic concepts of analysis which are useful in holography; in particular, tensor calculus in space and on curved surfaces [2.1–5]. This will allow us to apply concise tool of calculus which directly conveys an intrinsic geometrical meaning to the physical situation and is also easy to manipulate in general configurations of holographic set-ups. However, this introduction is shorter than [2.6] and partly complementary.

### 2.1.1 Dyadics, Derivatives, and Projectors

Let us start with elementary calculus in three-dimensional Euclidean space. Here, a vector $u$ is usually referred to a base of three orthogonal unit vectors $i, j, k$, so that

$$u = u_x i + u_y j + u_z k \ , \tag{2.1}$$

as represented by *cartesian* components $u_x, u_y, u_z$. If this vector varies from point to point, these components are functions of the coordinates $x, y, z$. We may also write the vector function in short hand as

$$r \rightarrow u(r) \ ,$$

where $r$ denotes the position vector of the point P in which the vector $u$ is considered. Sometimes, especially on curved surfaces, it is also convenient to vary the base from point to point. Instead of (2.1), we then write

$$u = u^1 g_1 + u^2 g_2 + u^3 g_3 = u^i g_i \ , \tag{2.2}$$

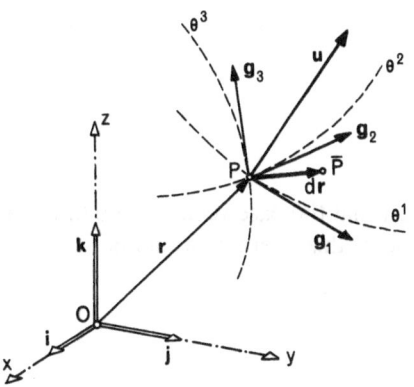

**Fig. 2.1.** Cartesian and curvilinear coordinates with corresponding base vectors

where $g_i = \partial r/\partial \theta^i$ ($i = 1, 2, 3$) are covariant base vectors which lie tangent to the lines of *curvilinear* coordinates $\theta^1$, $\theta^2$, $\theta^3$, and where $u^1$, $u^2$, $u^3$ are contravariant components of $u$ (Fig. 2.1). Note that in (2.2), the repeated index $i$ in $u^i g_i$ suggests the sum from 1 to 3.

In addition to scalars and vectors, we shall often encounter linear transformations or mappings. A special mapping of this type is the following: Let $p$ and $q$ be two given vectors and $u$ any arbitrary vector. Then

$$u \rightarrow w = (q \cdot u)p = Tu \tag{2.3}$$

defines a linear transformation $T$ which maps $u$ into a vector $(q \cdot u)p$ parallel to $p$ (Fig. 2.2). Here $q \cdot u$ denotes the scalar product of $q$ and $u$. Equation (2.3) reads in cartesian components as

$$
\begin{aligned}
w_x &= (q_x u_x + q_y u_y + q_z u_z)p_x \\
w_y &= (q_x u_x + q_y u_y + q_z u_z)p_y \\
w_z &= (q_x u_x + q_y u_y + q_z u_z)p_z \ ,
\end{aligned}
$$

which may also be expressed in matrix form by regrouping the factors:

$$
\begin{Bmatrix} w_x \\ w_y \\ w_z \end{Bmatrix} =
\begin{bmatrix} p_x q_x & p_x q_y & p_x q_z \\ p_y q_x & p_y q_y & p_y q_z \\ p_z q_x & p_z q_y & p_z q_z \end{bmatrix}
\begin{Bmatrix} u_x \\ u_y \\ u_z \end{Bmatrix} . \tag{2.4}
$$

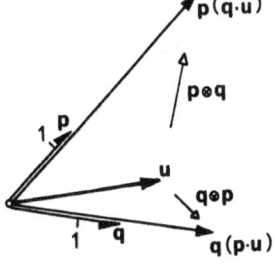

**Fig. 2.2.** A dyadic product $p \otimes q$ and its transpose applied onto a vector $u$

The resultant $3 \times 3$ matrix is the component representation of the operator $T$, sometimes written as a *dyadic product*:

$$T = p \otimes q \triangleq \begin{bmatrix} p_x \\ p_y \\ p_z \end{bmatrix} \otimes [q_x, q_y, q_z] \ .$$

The sign $\triangleq$ should draw attention to the fact that the base vectors are omitted in the last component notation. The transpose of this operator would be

$$T^T = (p \otimes q)^T = q \otimes p \ ,$$

so that we have

$$T^T u = uT = (p \cdot u)q \ . \tag{2.5}$$

In order to conform to the rule of matrix products of the components and for greater flexibility in subsequent treatments, here we alternatively have placed the vector $u$ to the left of the operator $T$. From the stand point of matrix representation, the vector can, therefore, be a row or a line. We shall, however, not use the transposition sign for a vector, as it is sometimes done, because its position relative to the linear operator ($3 \times 3$ matrix) makes its use clear. We may also combine several dyadics. If $T = p \otimes q$ and $U = r \otimes s$ are two dyadics, assuming here $r$ as any vector, the succession of both operators gives:

$$TU = (p \otimes q)(r \otimes s) = (q \cdot r)(p \otimes s) \ . \tag{2.6}$$

General linear transformations or *tensors* are linear combinations of dyadics formed by base vectors. Hence,

$$T = T^{ij} g_i \otimes g_j \tag{2.7}$$

is a representation of such a tensor $T$ by means of contravariant components $T^{ij}$ and covariant base vectors $g_i$, $g_j$ (summation over $i$ and $j$), whereas

$$T = T_{xx}(i \otimes i) + T_{xy}(i \otimes j) + \ldots + T_{zz}(k \otimes k) \quad \text{or}$$

$$T \triangleq \begin{bmatrix} T_{xx} & T_{xy} & T_{xz} \\ T_{yx} & T_{yy} & T_{yz} \\ T_{zx} & T_{zy} & T_{zz} \end{bmatrix} \tag{2.8}$$

would be the corresponding cartesian representation with unit vectors or with a matrix representation of the components.

A special tensor is the identity $I$ with components

$$I \triangleq \begin{bmatrix} 1 & 0 & 0 \\ 0 & 1 & 0 \\ 0 & 0 & 1 \end{bmatrix} ,$$

which may also be written relative to general coordinates in the mixed form

$$I = g^i \otimes g_i .$$

Here, $g^i$ are contravariant base vectors defined by the orthogonality relation

$$g^i \cdot g_j = \delta^i_j = \begin{cases} 1, & \text{if } i = j \\ 0, & \text{if } i \neq j . \end{cases} \tag{2.9}$$

All of these operators may be more specifically termed second-order tensors. This is in contrast to higher-order tensors which can be formed, for instance, as *triadic products*:

$$T = p \otimes q \otimes r = T \otimes r .$$

From here on, we must distinguish between two transpositions; namely, for $T = p \otimes q$ we have

$$(T \otimes r)^T = r \otimes q \otimes p = r \otimes T^T \quad \text{and}$$
$$T \otimes r)^T = p \otimes r \otimes q . \tag{2.10}$$

The latter may be called a *partial transposition*. Here we use the single parenthesis ")$^T$" since "(" cannot be separated from the symbol $T$ when used for the first dyadic or for a general second-order tensor.

Sometimes, though not often, even *quadriadic products* such as

$$T = p \otimes q \otimes r \otimes s$$

may be considered. Here we require a third transposition, so that

$$\begin{aligned} T^T &= (p \otimes q \otimes r \otimes s)^T = s \otimes q \otimes r \otimes p \\ T)^T &= p \otimes q \otimes r \otimes s)^T = p \otimes q \otimes s \otimes r \\ {}^T(T &= {}^T(p \otimes q \otimes r \otimes s = q \otimes p \otimes r \otimes s . \end{aligned} \tag{2.11}$$

The dyadics can be applied twice onto vectors, the triadics three times, and the quadriadics four times. The results are always scalars, but attention must be paid to the order of the factors; with $T = p \otimes q$, we have

$$v \cdot Tu = (v \cdot p)(q \cdot u) , \qquad w : (vTu) = (w \cdot q)(v \cdot p)(r \cdot u) ,$$
$$w \cdot (vTu)z = (w \cdot q)(v \cdot p)(s \cdot u)(r \cdot z) . \tag{2.12}$$

The choice of a symbolic notation such as $Tu$ for a linear transformation seems to be one of personal taste. However, this notation has an advantage over the usual component notation $T^{ij}u_j$, since the base vectors are implicitly included, thereby implying an intrinsic meaning to the tensor. Unless a specific component need be pointed out, no specific coordinates need be introduced explicitly nor any transformation matrices need appear which would indicate the passage from one system to another. So, the geometrical and physical meanings come through without indices and matrices overshadowing them, see e.g. [Ref. 2.7, p. 31]. The calculus is sufficiently flexible in that the tensors simply assimilate the vectors on the side on which they are written. Also, for systematic use, scalars are denoted by ordinary letters, vectors by small bold face letters, second-order tensors by capital bold face letters, third-order tensors by another type of capital bold face letters ($\mathbf{T}$), and fourth-order tensors by heavy capital bold face letters ($\mathbf{T}$).

Finally, we add that the well-known vector product $\boldsymbol{\psi} \times \boldsymbol{u}$ between two vectors $\boldsymbol{\psi}$ and $\boldsymbol{u}$ also constitutes a linear transformation $\boldsymbol{u} \rightarrow \boldsymbol{v} = (\boldsymbol{\psi} \times )\boldsymbol{u}$ with the antimetric operator

$$\boldsymbol{\Psi} = - \boldsymbol{\psi} \times \ldots = \boldsymbol{E}\boldsymbol{\psi} \ , \tag{2.13}$$

where $\boldsymbol{E}$ is the so-called (third-order) permutation tensor with the cartesian components:

$$\boldsymbol{E} \triangleq \begin{bmatrix} 0 & 0 & 0 & | & 0 & 0 & -1 & | & 0 & 1 & 0 \\ 0 & 0 & 1 & | & 0 & 0 & 0 & | & -1 & 0 & 0 \\ 0 & -1 & 0 & | & 1 & 0 & 0 & | & 0 & 0 & 0 \end{bmatrix}. \tag{2.14}$$

Let us now consider the *derivatives*. From a scalar function $\phi(\boldsymbol{r})$, ($\boldsymbol{r}$ is again the position vector), we can, of course, easily find the gradient. In order to define the derivative for a vector function $\boldsymbol{u}(\boldsymbol{r})$, we first form the total differential by means of (2.9):

$$d\boldsymbol{u} = \frac{\partial \boldsymbol{u}}{\partial \theta^i} d\theta^i = d\theta^i \boldsymbol{g}_i \cdot \boldsymbol{g}^k \frac{\partial \boldsymbol{u}}{\partial \theta^k} = d\theta^i \boldsymbol{g}_i \left( \boldsymbol{g}^k \frac{\partial}{\partial \theta^k} \otimes \boldsymbol{u} \right) = d\boldsymbol{r}(\nabla \otimes \boldsymbol{u}) \ . \tag{2.15}$$

Above all, there appears here the derivative operator

$$\nabla = \boldsymbol{g}^k \frac{\partial}{\partial \theta^k} \ , \tag{2.16}$$

which reads in cartesian coordinates as

$$\nabla = \boldsymbol{i} \frac{\partial}{\partial x} + \boldsymbol{j} \frac{\partial}{\partial y} + \boldsymbol{k} \frac{\partial}{\partial z} \ .$$

Then, again in (2.15), the operator $\nabla$ is associated with $\boldsymbol{u}$ by a formal dyadic product (or better, a sum of dyadic products) which acts as a linear transformation onto $d\boldsymbol{r}$ on the left in order to form $d\boldsymbol{u}$ (the symbol "$d$" is always placed on the left). On the other hand, as is usual in analysis, the partial differentiation $\partial/\partial \theta^k$ acts onto the vector function on the right. If we want to write explicit components, we must pay attention to the fact that generally the base varies from point to point. So, in $u = u^l g_l$, both components and base vectors must be differentiated. The result is

$$\nabla \otimes u = u_{l\|k} g^k \otimes g^l , \tag{2.17}$$

where $u_{l\|k} = u_{l,k} - \Gamma^j_{kl} u_j$ is termed the covariant derivative, $u_{l,k} = \partial u_l / \partial \theta^k$ and $\Gamma^j_{kl} = g^j \cdot g_{k,l}$ are Christoffel symbols. In cartesian coordinates, these symbols disappear and we have, of course, simply the matrix representation

$$\nabla \otimes u \triangleq \begin{bmatrix} \dfrac{\partial u_x}{\partial x} & \dfrac{\partial u_y}{\partial x} & \dfrac{\partial u_z}{\partial x} \\[2mm] \dfrac{\partial u_x}{\partial y} & \dfrac{\partial u_y}{\partial y} & \dfrac{\partial u_z}{\partial y} \\[2mm] \dfrac{\partial u_x}{\partial z} & \dfrac{\partial u_y}{\partial z} & \dfrac{\partial u_z}{\partial z} \end{bmatrix} . \tag{2.18}$$

The trace of the tensor $\nabla \otimes u$ is the *divergence* $\partial u_x / \partial x + \partial u_y / \partial y + \partial u_z / \partial z$ and reads in general coordinates by a contraction of the mixed components in the form:

$$\mathrm{tr}\{\nabla \otimes u\} = \nabla \cdot u = u^j{}_{\|k} g^k \cdot g_j = u^j{}_{\|j} . \tag{2.19}$$

Finally, using (2.13) the *curl* becomes here

$$\nabla \times u = - \nabla E u = \varepsilon^{ikl} u_{l\|k} g_i . \tag{2.20}$$

Again considering general calculations, when the derivative operator is applied to products of vectors and tensors, it should be noted that the links and sequences between the factors must be maintained. However, since $\nabla$ is conventionally placed to the left of the function on which it acts as a derivative, tensor transpositions must sometimes be used. Examples are as follows:

$$\begin{aligned} \nabla(u \cdot v) &= (\nabla \otimes u)v + (\nabla \otimes v)u \\ \nabla \cdot (Tu) &= (\nabla T) \cdot u + (\nabla \otimes u) \cdot T^{\mathrm{T}} \\ \nabla \otimes (Tu) &= (\nabla \otimes T)u + (\nabla \otimes u)T^{\mathrm{T}} \\ \nabla(u \otimes v) &= (\nabla \cdot u)v + (\nabla \otimes v)^{\mathrm{T}}u \\ \nabla(TS) &= (\nabla T)S + (\nabla \otimes S)^{\mathrm{T}} \cdot T . \end{aligned} \tag{2.21}$$

Higher derivatives are found from successive applications of the operator $\nabla$. As an example,

$$\nabla \otimes \nabla \phi = \phi_{\|lk} g^k \otimes g^l \tag{2.22}$$

is a second-order tensor with covariant components $\phi_{\|lk}$. Its trace, $\nabla \cdot \nabla \phi = \Delta \phi = \phi|^k_k$, is the well-known *Laplace operator*, applied on $\phi$, also written $\nabla^2 \phi$.

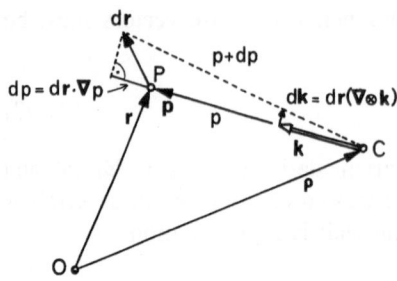

Fig. 2.3. Increments of length $p$ and of unit direction vector $k$ relative to a fixed point C

Some special derivatives will often be considered in later chapters. They concern the successive derivatives of the distance $p(r)$ of a variable point to a fixed center C (Fig. 2.3). The gradient of $p$ is

$$\nabla p = \nabla \sqrt{p \cdot p} = \frac{1}{2p} \nabla(p \cdot p) = \frac{1}{p}(\nabla \otimes p)p = \frac{1}{p}Ip = k \ , \tag{2.23}$$

where $k = p/p$ denotes the unit vector in the direction of $p$. This almost obvious result may also be recognized directly in Fig. 2.3 by means of increments.

Next, we have

$$\nabla \otimes k = \nabla \otimes \left(\frac{p}{p}\right) = \frac{1}{p} \nabla \otimes p - \frac{1}{p^2} \nabla p \otimes p = \frac{1}{p}(I - k \otimes k) = \frac{1}{p}K \ , \tag{2.24}$$

where the symmetric second-order tensor

$$K = I - k \otimes k \tag{2.25}$$

is a *normal projector*. In fact, for any vector $u$, $Ku = u - k(k \cdot u)$ is the projection of $u$ onto the plane normal to $k \triangleq (k_x, k_y, k_z)$ (Fig. 2.4). The cartesian components of $K$ are

$$K \triangleq \begin{bmatrix} 1 - k_x^2 & -k_xk_y & -k_xk_z \\ -k_yk_x & 1 - k_y^2 & -k_yk_z \\ -k_zk_x & -k_zk_y & 1 - k_z^2 \end{bmatrix}$$

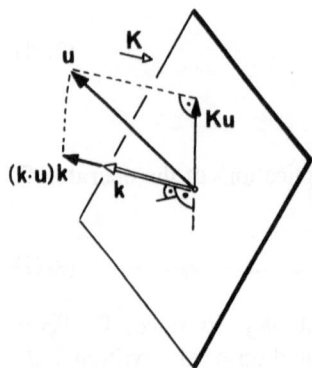

Fig. 2.4. Normal projection $Ku$ of vector $u$ onto the plane perpendicular to the unit vector $k$

Thus, (2.24) indicates that the derivative of a unit direction vector constitutes a second-order tensor: namely, a projector combined with a shortening by the factor $1/p$. This result may also be reconstructed from the increments in Fig. 2.3.

The derivative of (2.24) is less trivial than the preceeding. It leads to

$$\nabla \otimes \left(\frac{1}{p}K\right) = -\frac{1}{p^2}\nabla p \otimes K - \frac{1}{p}(\nabla \otimes k) \otimes k - \frac{1}{p}(\nabla \otimes k) \otimes k]^\mathrm{T}$$

$$= -\frac{1}{p^2}[k \otimes K + K \otimes k + K \otimes k)^\mathrm{T}] = -\frac{1}{p^2}\boldsymbol{K} \ . \tag{2.26}$$

The third-order tensor (its components contain triple products $k_x^3$, $k_x^2 k_y$, ...)

$$\boldsymbol{K} = k \otimes K + K \otimes k + K \otimes k)^\mathrm{T} \tag{2.27}$$

may also be called a *superprojector*, since, for any $v$, $\boldsymbol{K}v = k \otimes Kv + K(k \cdot v) + Kv \otimes k$, contains both the "lateral" projection $Kv$ and the "longitudinal" projection $k \cdot v$. Finally, let us form the fourth-order tensor (its components contain quadruple products $k_x^4$, $k_x^3 k_y$, ...):

$$\nabla \otimes \left(-\frac{1}{p^2}\boldsymbol{K}\right) = -\nabla\left(\frac{1}{p}\right) \otimes \frac{1}{p}\boldsymbol{K} - \frac{1}{p}\nabla \otimes \left[k \otimes \frac{1}{p}K + \frac{1}{p}K \otimes k + \frac{1}{p}K \otimes k)^\mathrm{T}\right]$$

$$= \frac{1}{p^3}[k \otimes \boldsymbol{K} + {}^\mathrm{T}(k \otimes K + K \otimes k)^\mathrm{T} + K \otimes k$$

$$- K \otimes K - K \otimes K)^\mathrm{T} - (K \otimes K)^\mathrm{T}] = \frac{1}{p^3}\boldsymbol{K} \ . \tag{2.28}$$

$\boldsymbol{K}$ might be called a *hyperprojector*. In $K \otimes K)^\mathrm{T}$, note that the second and fourth factors, and in $(K \otimes K)^\mathrm{T}$, the first and fourth factors are transposed.

### 2.1.2 Calculus on Surfaces

A surface is given by a vector valued function of two parameters

$$\theta^1, \theta^2 \to r = r(\theta^1, \theta^2) \ ,$$

where $\theta^1$, $\theta^2$ are curvilinear coordinates on the surface. The covariant base vectors are usually denoted by $a_\alpha (= \partial r/\partial \theta^\alpha)$ [Ref. 2.2, p. 33] instead of $g_\alpha$, $\alpha$ varying from 1 to 2. If $\theta^3$ is the distance from the surface, then $g_3 = n$ becomes the unit normal (Fig. 2.5), as determined by

$$n = \frac{a_1 \times a_2}{|a_1 \times a_2|} \ . \tag{2.29}$$

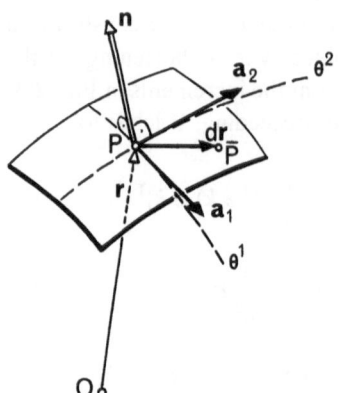

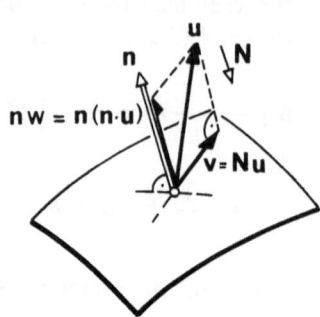

**Fig. 2.5.** Curvilinear coordinates $\theta^1$, $\theta^2$ and corresponding base vectors $a_1$, $a_2$ related to a surface

**Fig. 2.6.** Decomposition of a vector $u$ into its interior part $v$ (the projection $Nu$) and its exterior part $wn$ ($n$: unit normal vector)

Similar to (2.25), we now form the associated normal projector (Fig. 2.6):

$$N = I - n \otimes n = a^\alpha \otimes a_\alpha = a_{\alpha\beta} a^\alpha \otimes a^\beta \ . \tag{2.30}$$

The components $a_{\alpha\beta} = a_\alpha \cdot a_\beta$ of the tensor $N$ are often called the "metric tensor" of the surface, since for any line element $ds$ we have $(ds)^2 = dr \cdot dr = a_{\alpha\beta} d\theta^\alpha d\theta^\beta$ (first fundamental form).

Any vector $u$ in a point $P$ on the surface then allows the decomposition

$$u = Nu + (u \cdot n)n = v + wn \tag{2.31}$$

into an *interior part* $v = Nu$, the *projection* of $u$ onto the tangential plane, and an *exterior part* $wn = (u \cdot n)n$, the projection along the normal to the tangential plane at P. Similarly, any second-order tensor leads to the decomposition

$$T = NTN + (NTn) \otimes n + n \otimes (nTN) + (n \cdot Tn)n \otimes n$$
$$= D + t \otimes n + n \otimes s + o(n \otimes n) \ , \tag{2.32}$$

where $D = NTN$ is its two-dimensional interior part, the full projection, and where $t = NTn$ and $s = nTN$ may be called semi-exterior. As we shall later encounter in holography, we often find the left semi-projection of $T$:

$$NT = D + t \otimes n \ . \tag{2.33}$$

To illustrate this separation, let us choose a local cartesian system at P with $x$, $y$ coordinates in the tangential plane, so that $n \triangleq (0, 0, 1)$. Then, from (2.32), we get a matrix representation with a separation of the third line and the third row:

$$T \triangleq \begin{bmatrix} D_{xx} & D_{xy} & t_x \\ D_{yx} & D_{yy} & t_y \\ S_x & S_y & \sigma \end{bmatrix} = \begin{bmatrix} [D] & \{t\} \\ [s] & \sigma \end{bmatrix} .$$

$NT$ is thus obtained by tracing the last line in the above matrix.

A useful decomposition of a third-order tensor is that of the permutation tensor $E$. Still using the local cartesian coordinates, we easily obtain this decomposition from expression (2.14). In fact, we have

$$En \triangleq \begin{bmatrix} 0 & 1 & 0 \\ -1 & 0 & 0 \\ 0 & 0 & 0 \end{bmatrix} \quad nE \triangleq \begin{bmatrix} 0 & 1 & 0 \\ -1 & 0 & 0 \\ 0 & 0 & 0 \end{bmatrix} \quad E)^\mathrm{T}n \triangleq \begin{bmatrix} 0 & -1 & 0 \\ 1 & 0 & 0 \\ 0 & 0 & 0 \end{bmatrix},$$

so that

$$E = E \otimes n + n \otimes E - E \otimes n)^\mathrm{T} \tag{2.34}$$

where

$$E \triangleq \begin{bmatrix} 0 & 1 & 0 \\ -1 & 0 & 0 \\ 0 & 0 & 0 \end{bmatrix} .$$

Since $v \cdot Ev = 0$ and $|Ev| = v$ (Fig. 2.7), the two-dimensional interior tensor $E$, applied on a vector $v$, i.e., $Ev$, has the effect of a pivot rotation of $v$ by $-\pi/2$ in the tangential plane. The general representation of $E$, again in curvilinear coordinates, is

$$E = E_{\alpha\beta}a^\alpha \otimes a^\beta , \quad \text{with} \quad E_{\alpha\beta} = \sqrt{a} \begin{bmatrix} 0 & 1 \\ -1 & 0 \end{bmatrix}$$

$$\text{and} \quad a = a_{11}a_{22} - (a_{12})^2 . \tag{2.35}$$

The derivative operator $\nabla$ may also be decomposed in the form

$$\nabla = a^\alpha \frac{\partial}{\partial \theta^\alpha} + n \frac{\partial}{\partial \theta^3} = \nabla_\mathrm{n} + n \frac{\partial}{\partial \theta^3} , \quad \text{where} \tag{2.36}$$

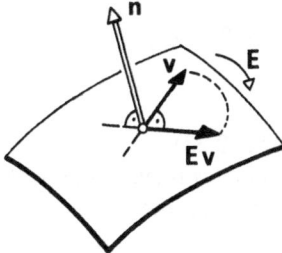

Fig. 2.7. Application of the two-dimensional permutation tensor $E$ onto the interior vector $v$ to yield 90° rotation of $v$

$$\nabla_n = N\nabla \tag{2.37}$$

shows up formally as a normal projection of $\nabla$ onto the tangential plane. As we must later consider several surfaces, the index $n$ of $\nabla_n$ indicates on which surface with unit normal $n$ the operator $\nabla_n$ is acting.

When we express the normal curvature $1/R$ as a function of the tangential unit vector $e$ (second fundamental form), a special symmetrical tensor on the surface is obtained by a derivative of $n$:

$$\frac{1}{R} = \frac{d^2r}{ds^2} \cdot n = \frac{d}{ds}\left(\frac{dr}{ds} \cdot n\right) - \frac{dr}{ds} \cdot \frac{dn}{ds}$$

$$= -\frac{dr}{ds} \cdot (\nabla_n \otimes n)\frac{dr}{ds} = e \cdot Be \tag{2.38}$$

so that

$$B = -\nabla_n \otimes n \ . \tag{2.39}$$

When $e$ varies in the tangential plane at the point P, the minimum $R_1$ and the maximum $R_2$ of $R$ are the principal radii of curvature. Their inverses $1/R_1$, $1/R_2$ are the eigenvalues of $B$. The two invariants of $B$ are the mean curvature $H = (1/R_1 + 1/R_2)/2$ and the Gaussian curvature $K = 1/R_1R_2$. If the section is inclined with respect to the surface normal, we have the following decomposition of the second derivative of $r$

$$\frac{d^2r}{ds^2} = b + \frac{1}{R}n \ , \tag{2.40}$$

where, besides the normal curvature $1/R$, the interior geodesic curvature $|b| = 1/\varrho_g$ appears.

Another derivative of practical use is that of the projector $N$:

$$\nabla_n \otimes N = -(\nabla_n \otimes n) \otimes n - (\nabla_n \otimes n) \otimes n]^T = B \otimes n + B \otimes n]^T \ . \tag{2.41}$$

It helps in directly finding (without components) the decomposition of the derivative of any vector $u = v + wn$ (or of any tensor). In fact, one has the following

$$u = NNu + wn \ ,$$

$$\nabla_n \otimes u = (\nabla_n \otimes N)Nu + (\nabla_n \otimes Nu)N^T + \nabla_n w \otimes n + w\nabla_n \otimes n \ ,$$

$$\nabla_n \otimes u = [(\nabla_n \otimes v)N - Bw] + (Bv + \nabla_n w) \otimes n \ . \tag{2.42}$$

This derivative is, therefore, a semiprojection of type (2.33).

As will be discussed later, in holography we encounter configurations with several surfaces, e.g., the surfaces of an object before and after deformation, an interface between two domains with different indices of refraction, or one or several holograms at the recording and at the reconstruction. In particular, transformations between infinitesimal vicinities on such surfaces, often called *affine connections*, must be studied. So, let us now consider a one-to-one mapping between the points {P} and {P̂} of two surfaces with common convected coordinates $\theta^1$, $\theta^2$ [Ref. 2.8, p. 423]

$$\theta^1, \theta^2 \rightarrow r(\theta^1, \theta^2) \rightarrow \hat{r}(r) = \hat{r}(\theta^1, \theta^2) \ ,$$

and form the differential of $\hat{r}$

$$d\hat{r} = dr(\nabla_n \otimes \hat{r}) \ . \tag{2.43}$$

The tensor $\nabla_n \otimes \hat{r}$, applied onto $dr$ as a linear transformation, see also (2.15), marks the "bridge" between the two sets of neighboring points {P̄}, {P̂} around two given points P, P̂. In the special case where the ensemble of straight lines $\overline{P\hat{P}}$ passes through a common fixed center C (Fig. 2.8) with position vector $\varrho$, we can find $\nabla_n \otimes \hat{r}$ explicitly. For this, we introduce the distances $p$ from C to P and $\hat{p}$ from C to P̂, and the direction unit vector $k$. Differentiating the relation $\hat{r} = \varrho + \hat{p}k$, we find with (2.24 and 37):

$$\nabla_n \otimes \hat{r} = \nabla_n \otimes (\hat{p}k) = \nabla_n \hat{p} \otimes k + \frac{\hat{p}}{p}NK \ . \tag{2.44}$$

In order to determine $\nabla_n \hat{p}$, which is not equal to a simply projected unit vector as in (2.23), we write the condition of normality of $\hat{n}$:

$$0 = d\hat{r} \cdot \hat{n} = dr \cdot (\nabla_n \otimes \hat{r})\hat{n} = dr \cdot \nabla_n \hat{p}(k \cdot \hat{n}) + \frac{\hat{p}}{p}dr \cdot NK\hat{n} \ .$$

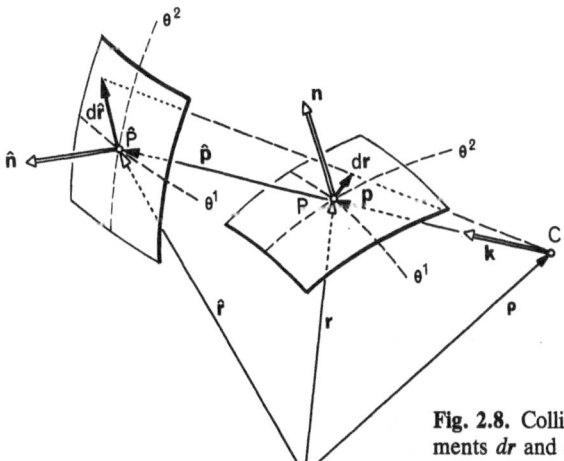

**Fig. 2.8.** Collinear relation between two increments *dr* and *dr̂* defined on two surfaces, also called affine connection

Since $dr$ is arbitrary in the tangential plane, we then get

$$\nabla_n \hat{p} = -\frac{\hat{p}}{p(\hat{n} \cdot k)} N K \hat{n} .$$

Introducing this expression into (2.44), we find

$$\nabla_n \otimes \hat{r} = \frac{\hat{p}}{p} N (I - k \otimes k) \left(I - \frac{\hat{n} \otimes k}{\hat{n} \cdot k}\right) = \frac{\hat{p}}{p} N \left(I - \frac{\hat{n} \otimes k}{\hat{n} \cdot k}\right)$$

or shortened to

$$\nabla_n \otimes \hat{r} = \frac{\hat{p}}{p} N \hat{M} , \quad \text{with} \tag{2.45}$$

$$\hat{M} = I - \frac{1}{\hat{n} \cdot k} (\hat{n} \otimes k) . \tag{2.46}$$

Therefore, in the linear transformation (2.43), we can omit $N$ and write the transpose $\hat{M}^T = I - k \otimes \hat{n}/k \cdot \hat{n}$ in front of $dr$

$$d\hat{r} = \frac{\hat{p}}{p} dr \, N \hat{M} = \frac{\hat{p}}{p} \hat{M}^T dr . \tag{2.47}$$

In the special case of C at infinity, this becomes a proper affine connection

$$d\hat{r} = \hat{M}^T dr . \tag{2.48}$$

Thus, $\hat{M}^T$ represents an oblique projector along $k$ onto the plane perpendicular to $\hat{n}$, whereas $\hat{M}$ is an oblique projector along $\hat{n}$ onto the plane normal to $k$ (Fig. 2.9).

If the roles of the two surfaces are exchanged, i.e., if we consider the inverse function $\hat{r} \rightarrow r = r(\hat{r})$, another oblique projector

$$M = I - \frac{1}{n \cdot k} n \otimes k \tag{2.49}$$

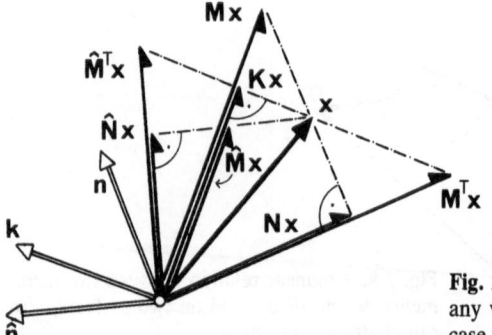

Fig. 2.9. Normal and oblique projections of any vector $x$. This sketch depicts the special case in which all vectors are in the same plane

and its transpose $M^T$ appear. When the derivative of a function $\phi$ must be transferred from one surface to another, a transformation of the derivative operator must be made. As

$$d\phi = dr \cdot \nabla_n\phi = d\hat{r} \cdot \nabla_{\hat{n}}\phi = \frac{\hat{p}}{p}dr \cdot \hat{M}\nabla_{\hat{n}}\phi = \frac{\hat{p}}{p}dr \cdot N\hat{M}\nabla_{\hat{n}}\phi$$

is valid for $\forall\, dr$ in the tangential plane, we obtain

$$\nabla_n = \frac{\hat{p}}{p}N\hat{M}\nabla_{\hat{n}} \quad \text{or also}$$

$$\nabla_{\hat{n}} = \frac{p}{\hat{p}}\hat{N}M\nabla_n \;. \tag{2.50}$$

Some remarks are appropriate here:

a) Contrary to the relation (2.47) for the increments, we cannot omit $N$ and $\hat{N}$ in the relation (2.50) for $\nabla_n$ and $\nabla_{\hat{n}}$.

b) In (2.50), $\nabla_n$ and $\nabla_{\hat{n}}$ are applied at different points P and $\hat{P}$. If we use the formal equation (2.37) in both points, we must distinguish $\nabla$ in P from $\hat{\nabla}$ in $\hat{P}$ and write $\nabla_{\hat{n}} = \hat{N}\hat{\nabla}$.

c) Successive projectors appear in (2.45, 50), as we shall often encounter later. Note that, in general, these combinations cannot be reduced to one single projector. Two exceptions should be mentioned: (i) projectors onto the *same plane* as in

$$MK = K\;, \quad M\hat{M} = \hat{M}\;, \quad \hat{M}M = M\;; \tag{2.51}$$

and (ii) projectors along the *same direction* as in

$$MN = M\;, \quad M^T\hat{M}^T = M^T\;, \quad \hat{M}^TM^T = \hat{M}^T\;. \tag{2.52}$$

d) It is important to note that we have introduced oblique projectors in order to relate the increments $dr$ and $d\hat{r}$. In principle, however, it is not necessary to make an infinitesimal consideration when defining oblique projectors as in (2.46, 49). It should be emphasized that projectors will come into consideration in physical situations only where derivatives are needed.

e) An alternative formula for $M$ is found by means of the Lagrange identity

$$a \times (b \times c) = aE(bEc) = [b \otimes c - (b \otimes c)^T]a$$
$$= (a \cdot c)b - (a \cdot b)c\;, \tag{2.53}$$

which is valid for any three vectors $a$, $b$, $c$. By means of (2.13, 34), this gives for an arbitrary vector $x$

$$Mx = \frac{k \cdot n}{k \cdot n} x - \frac{n}{n \cdot k} k \cdot x = -\frac{1}{n \cdot k} k \times (n \times x)$$

$$= -\frac{1}{n \cdot k} kE(nEx) = -\frac{1}{n \cdot k} E'KENx$$

so that

$$M = -\frac{1}{n \cdot k} E'KEN \ . \tag{2.54}$$

In other words, an oblique projector is composed of two normal projectors, $N, K$, and two $\pi/2$-rotators, $E$ and $E'$, where the latter is in the plane normal to $k$.

A somewhat different kind of connection is the transfer between parallel surfaces, where the related points are on common normals (Fig. 2.10). We shall encounter this situation during the study of propagation of wave fronts. A parallel surface $\bar{r}(\theta^1, \theta^2)$ is related to a given surface $r(\theta^1, \theta^2)$ by

$$\bar{r} = r + zn \ ,$$

where $z = \theta^3$ denotes the distance between the two parallel surfaces, and $n$ is the common unit normal. Referring to (2.39), we first have

$$d\bar{r} = dr + z \, dn = dr(N + z\nabla_n \otimes n) = dr(N - Bz) \ ,$$

and then, since

$$dr \cdot \nabla_n \phi = d\bar{r} \cdot \bar{\nabla}_n \phi$$

is valid for any $\phi$ and any $dr$ in the tangential plane, we find

$$\nabla_n = (N - Bz) \bar{\nabla}_n \ , \tag{2.55}$$

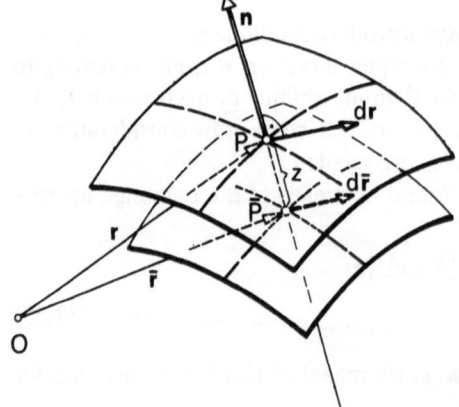

(Z < 0)

Fig. 2.10. Connection between two increments $dr$ and $d\bar{r}$ with the same unit normal $n$ and located on parallel surfaces of relative distance $z$

which corresponds to (2.50). Using the transformation of $\bar{V}_n$ and then applying the operators $V_n$ and $\bar{V}_n$ onto $n$ results in the transformation for the curvature tensors $B = - V_n \otimes n$ and $\bar{B} = - \bar{V}_n \otimes n$

$$B = (N - zB)\bar{B} \ , \tag{2.56}$$

which, in the case of small $z$, can also be easily inverted

$$\bar{B} = B + B^2 z + B^3 z^2 + \dots \ . \tag{2.57}$$

On the other hand if $K \neq 0$, instead of (2.56), we may write a relation for the inverses $B^{-1}$ and $\bar{B}^{-1}$ ("inverse" here means that $B^{-1}B = N$):

$$B^{-1} = \bar{B}^{-1}(N - zB)^{-1} \ ,$$
$$\bar{B}^{-1} = B^{-1} - zN \ . \tag{2.58}$$

In particular, this indicates, that the surfaces have the same centers of curvature for each normal section.

Finally, let us consider the derivatives of functions with several variables and some problems concerning the chain rule in tensor analysis. We first suppose that a scalar function $\phi(\hat{r}, r)$ depends on the momentary independent vector variables $\hat{r}$ and $r$. If $\partial_{\hat{r}} = \hat{N} \ \partial_{\hat{r}}$ and $\partial_r = N \ \partial_r$ are two-dimensional vector operators of the partial derivative with respect to $\hat{r}$ and $r$, the total differential of $\phi$ may be written as

$$d\phi = d\hat{r} \cdot \partial_{\hat{r}}\phi + dr \cdot \partial_r\phi \ . \tag{2.59}$$

On the other hand, a scalar $\phi[u(r)]$ may be a function of a vector $u(r)$, that is itself dependent on the position $r \in \mathbb{R}^3$. The gradient of $\phi$ is then given by

$$\nabla\phi = (\nabla \otimes u)\partial_u\phi \ . \tag{2.60}$$

If a scalar function $\phi[T(r)]$ depends on a tensor variable $T(r)$, being itself a function of the position vector $r \in \mathbb{R}^3$, the gradient of $\phi$ is then expressed by

$$\nabla\phi = (\nabla \otimes T) \cdot \partial_T\phi \ , \tag{2.61}$$

where $\nabla \otimes T$ is a third order tensor and $\partial_T$ a second order tensor [Ref. 2.9, p. 187].

## 2.2 Introduction to Some Elements of Geometrical Optics

Although there exist, of course, many text books that may be consulted for an introduction to classical optics, it seems advisable to recapitulate at least those elements of geometrical optics which later prove useful in holography. We must especially have at our disposal the basic equations in an intrinsic form since, as we have already pointed out, this will ensure the greatest flexibility for general configurations. We treat here those topics which constitute preparation for the aberration theory in the image modification with thin holograms, for holographic interferometry of transparent media with a variable index of refraction, for visibility considerations and finally, for the coupled-wave theory in volume holograms.

### 2.2.1 Wave Equation, Eikonal, and Ray Equation

The theory of geometrical optics emerges from Maxwell's electrodynamic equations for the light waves theory with the assumption of a very small wavelength [2.10–11]. We first recall some basic concepts in a manner similar to that of Born and Wolf [Ref. 2.12, p. 109ff.]. In a non-homogeneous isotropic medium, Maxwell's differential equations are expressed in the MKSA unit system (the four capital bold face letters here denote vectors only used in this paragraph)

$$\nabla \times \boldsymbol{H} - \dot{\boldsymbol{D}} = \boldsymbol{i} , \qquad \nabla \cdot \boldsymbol{B} = 0$$
$$\nabla \times \boldsymbol{E} + \dot{\boldsymbol{B}} = 0 , \qquad \nabla \cdot \boldsymbol{D} = \varrho , \tag{2.62}$$

where $\boldsymbol{E}$ and $\boldsymbol{H}$ are the electric and magnetic vector, respectively, $\boldsymbol{B}$ and $\boldsymbol{D}$ are the magnetic induction and the electric displacement, $\boldsymbol{i}$ is the electric current density, and $\varrho$ the electric charge density. In order to determine each of the four unknown vector fields separately, (2.62) must be supplemented by the constitutive relations

$$\boldsymbol{i} = \sigma \boldsymbol{E} , \qquad \boldsymbol{D} = \varepsilon \varepsilon_0 \boldsymbol{E} , \qquad \boldsymbol{B} = \mu \mu_0 \boldsymbol{H} , \tag{2.63}$$

where $\sigma$ is known as the specific conductivity and $\varepsilon_0$ and $\mu_0$ are the dielectric constant and the magnetic permeability in the vacuum, respectively. We assume here that $\mu = 1$ for the case of non-magnetic substances, although the relative dielectric constant $\varepsilon(\neq 1)$ is, in general, variable in space. By substituting $\boldsymbol{B}$ from the third material equation (2.63c) into (2.62c) and by applying the derivative operator $\nabla \times$ once more, we obtain the wave equation

$$\nabla \times (\nabla \times \boldsymbol{E}) + \mu_0 \sigma \dot{\boldsymbol{E}} + \varepsilon \varepsilon_0 \mu_0 \ddot{\boldsymbol{E}} = 0 . \tag{2.64}$$

If we then consider the harmonic solutions for monochromatic waves

$$E(r, t) = \operatorname{Re}\{e_\omega(r) \exp(-i\omega t)\} , \tag{2.65}$$

the wave equation (2.64) (after cancelling the time factors) becomes "time independent"

$$\nabla \times (\nabla \times e_\omega) - \frac{4\pi^2}{\lambda_0^2} \gamma^2 e_\omega = 0 ,$$

with the new complex coefficient

$$\gamma = \sqrt{\varepsilon + i\frac{\sigma}{\varepsilon_0 \omega}} \qquad (2.66)$$

and the wave length in vacuum $\lambda_0 = 2\pi c/\omega = 2\pi/\omega \sqrt{\varepsilon_0 \mu_0}$, $c = (\varepsilon_0 \mu_0)^{-1/2}$ denoting the velocity of light. Lagrange's identity (2.53) for the operator $\nabla$ and $e_\omega$ [i.e., $\nabla \times (\nabla \times e_\omega) = \nabla(\nabla \cdot e_\omega) - \nabla^2 e_\omega$] leads afterwards to the wave equation in the form

$$\nabla^2 e_\omega - \nabla(\nabla \cdot e_\omega) + \frac{4\pi^2}{\lambda_0^2}\gamma^2 e_\omega = 0 \ . \qquad (2.67)$$

Similarly, we may substitute (2.65) into (2.62d), which gives

$$\nabla \cdot e_\omega + \nabla(\log \varepsilon) \cdot e_\omega = \varrho_\omega \ . \qquad (2.68)$$

We shall often investigate the propagation of waves in non-conducting isotropic transparent substances $(\sigma = 0)$ free from currents $(i = 0)$ and charges $(\varrho = 0)$, each with the index of refraction $n = \sqrt{\varepsilon}$. In particular, for a *homogeneous* medium $(\varepsilon = \text{const})$, we also conclude from (2.68) that the divergence of $e_\omega$ vanishes. Thus, we have a system of two equations, one vectorial and the other scalar:

$$\nabla^2 e_\omega + \frac{4\pi^2}{\lambda^2}e_\omega = 0 \ , \qquad \nabla \cdot e_\omega = 0 \ , \qquad (2.69)$$

where $\lambda = \lambda_0/n$ is the wavelength in the medium under consideration. If we separate $e_\omega$ multiplicatively into a complex scalar function $U(r)$ and a real unit vector function $v(r)$, so that $e_\omega = Uv$, (2.69a) becomes

$$\left(\nabla^2 U + \frac{4\pi^2}{\lambda^2}U\right)v + 2\nabla U(\nabla \otimes v) + U\nabla^2 v = 0 \ .$$

After contraction with $v$ and using the relation $(\nabla \otimes v)v = \nabla(v^2)/2 = 0$, we obtain:

$$\nabla^2 U + \frac{4\pi^2}{\lambda^2}U + U\nabla^2 v \cdot v = 0 \ , \qquad \nabla U \cdot v + U\nabla \cdot v = 0 \ . \qquad (2.70)$$

Special solutions of these scalar equations are those describing *plane waves* (Fig. 2.11) in a direction given by the unit vector $k = k_0 = \text{const}$ [2.13–16]. First, if we keep $v = v_0 = \text{const}$, (2.70a) simplifies into the *Helmholz* equation

$$\nabla^2 U + \frac{4\pi^2}{\lambda^2}U = 0 \ , \qquad (2.71)$$

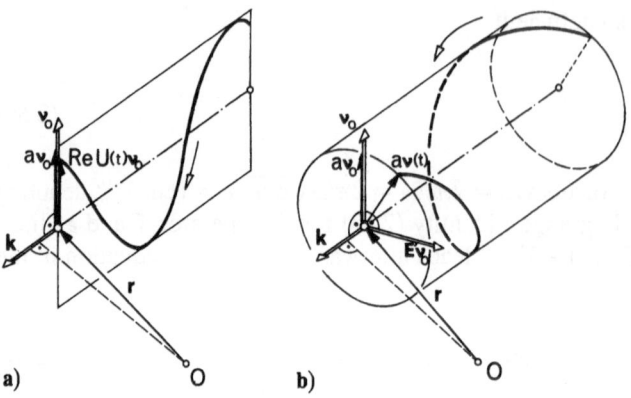

**Fig. 2.11a, b.** Plane waves travelling in the direction of the unit vector $k$: (a) linearly polarized wave ($v_0$: constant unit vector), (b) circularly polarized wave

which is fulfilled by the complex amplitude or wave disturbance: $U = a \exp(2\pi i\, k \cdot r/\lambda)$, since $\nabla U = a \exp(2\pi i\, k \cdot r/\lambda)\, 2\pi i\, \nabla(k \cdot r)/\lambda = 2\pi i\, U(\nabla \otimes r)k/\lambda = 2\pi i\, UIk/\lambda = 2\pi i\, Uk/\lambda$ and $\nabla^2 U = \nabla \cdot \nabla U = -4\pi^2\, Uk \cdot k/\lambda^2 = -4\pi^2\, U/\lambda^2$. Moreover, (2.70b) shows that $k \cdot v = 0$, which confirms the acknowledged property that $k$ is normal to $v$. This result represents a plane *linearly polarized* wave. Alternatively, if we keep $U = a = \text{const}$, (2.70a) now becomes

$$\frac{4\pi^2}{\lambda^2} + \nabla^2 v \cdot v = 0 \ . \tag{2.72}$$

This permits the solution $v = [K \cos(2\pi\, k \cdot r/\lambda) - E' \sin(2\pi k \cdot r/\lambda)]v_0$, where $K = I - k \otimes k$ and $E'^2 = -K$ ($E'$ is the permutation tensor relative to $k$, see also (2.34)). In fact, we have $|v| = 1$ and with (2.21)

$$\nabla^2 v = \nabla(\nabla \otimes v) = \nabla\left\{\frac{2\pi}{\lambda}k \otimes \left[-K \sin\left(\frac{2\pi}{\lambda}k \cdot r\right) - E' \cos\left(\frac{2\pi}{\lambda}k \cdot r\right)\right]v_0\right\}$$

$$= -\frac{2\pi}{\lambda}k\left(\frac{2\pi}{\lambda}k \otimes v\right) = -\frac{4\pi^2}{\lambda^2}v \ .$$

This solution results in the equation describing the case of a plane *circularly polarized* wave (Fig. 2.11).

These examples are two extreme cases, one of steeply varying amplitude and the other of steeply varying direction of $e_\omega$. Of course, the circularly polarized wave may also be obtained by superposition of two orthogonal linearly polarized waves, separated in phase by $\pi/2$. That is why, in the following, the considerations may be often restricted to waves similar to the first type. Later on, we shall have to deal with *"spherical waves"*, which seem to originate at some collineation center C, the "source". Tentatively, we will try with a decreasing, steeply varying complex amplitude, $U(r) = (A/r) \exp(2\pi \mathrm{i}\, r/\lambda)$, along each radius $r$ of the spheres around C and, provisionally, with any unit vector function $\mathbf{v}(r)$, the origin of coordinates now being C. With $\nabla r = \mathbf{k}$, where $\mathbf{k} = \mathbf{r}/r$, see (2.23), we first get

$$\nabla U = \left(\frac{2\pi \mathrm{i}}{\lambda} - \frac{1}{r}\right) U \mathbf{k} \ . \tag{2.73}$$

Thereafter, with $\nabla \cdot \mathbf{k} = \mathrm{tr}\{\nabla \otimes \mathbf{k}\} = \mathrm{tr} K/r = 2/r$, see (2.24), we obtain

$$\nabla \cdot \nabla U = \frac{U}{r^2} \nabla r \cdot \mathbf{k} + \left(\frac{2\pi \mathrm{i}}{\lambda} - \frac{1}{r}\right) \nabla U \cdot \mathbf{k} + \left(\frac{2\pi \mathrm{i}}{\lambda} - \frac{1}{r}\right) U \nabla \cdot \mathbf{k}$$

$$= \frac{U}{r^2} + \left(-\frac{4\pi^2}{\lambda^2} - \frac{4\pi \mathrm{i}}{\lambda r} + \frac{1}{r^2}\right) U + \left(\frac{4\pi \mathrm{i}}{\lambda r} - \frac{2}{r^2}\right) U = -\frac{4\pi^2}{\lambda^2} U \ .$$

Thus, we see that neither of the two equations in (2.70) is fulfilled, since the terms in $\mathbf{v}$ cannot be cancelled at all points. However, as was already mentioned, we assume the wavelength $\lambda$ to be small compared to some characteristic length; for instance, if $r$ is the distance to the source C, then $\lambda \ll r$. Hence, for slowly varying $\mathbf{v}(r)$, the terms $U\nabla^2 \mathbf{v} \cdot \mathbf{v}$ and $U\nabla \cdot \mathbf{v}$ in (2.70) are negligible compared to those containing a factor $1/\lambda$. Since we discuss differential equations, we limit our consideration to a small portion of each sphere of radius $r$ [2.17]. We, therefore, need no longer consider the behavior of the polarization direction $\mathbf{v}$ in detail, but, rather, investigate the scalar part $U$ of the waves. The scalar $U$ satisfies only approximately (2.70a), but rigorously fulfills the Helmholz equation (2.71). This constitutes the so-called scalar wave theory [2.18, 19]. Of course, in fact, an extended spherical wave must be thought of as a superposition of elementary waves travelling in all directions and being determined within the complete electromagnetic theory by the boundary conditions of the "source", which is not a single point. Let us also add that, in the preceeding considerations, one could alternatively work with potentials [2.12] or also with components starting from (2.69) directly [2.12, 18, 19]. However, one should be very careful when applying the operator $\nabla^2$ onto a component of the vector $\mathbf{e}_\omega$ instead of applying it onto a scalar like $U$. For instance, if $\nabla^2$ is expressed by spherical coordinates, additional terms with the other components of $\mathbf{e}_\omega$ and with Christoffel symbols as factors appear, involving supplementary terms like those in (2.70). Since the calculus is rather cumbersome, we prefer the development as described above, which is free of coordinates.

Let us then briefly recall the classical theory for obtaining a general expression for the light disturbance at a point K (which is needed in holographic interferometry) by means of the *Huygens-Fresnel* theory. Consider a closed surface $A_p$, separating K from the light source S, and let $A_\varepsilon$ be a small sphere of radius $\varepsilon$ around K (Fig. 2.12). If $U(\varrho)$ and $U'(\varrho, r)$ are two functions satisfying both of the Helmholz equations in (2.71) with respect to $r$ and $\varrho$, we obtain from Green's theorem, applied to the domain $V$ between $A_p$ and $A_\varepsilon$, [Ref. 2.12, p. 376] and [Ref. 2.18, p. 146]

$$\iiint\limits_{V} (U\nabla^2 U' - U'\nabla^2 U)dV = - \iint\limits_{A_\varepsilon, A_p} \left(U\frac{\partial U'}{\partial n} - U'\frac{\partial U}{\partial n}\right)dA = 0 , \quad (2.74)$$

where $\partial/\partial n = \boldsymbol{n} \cdot \nabla$, and where $\boldsymbol{n}$ is the inward unit normal. We now choose $U' = L^{-1}\exp(2\pi i\, L/\lambda)$ with $L = (\boldsymbol{\varrho} - \boldsymbol{r}) \cdot \boldsymbol{k}$, so that $\partial U'/\partial n = -(2\pi i/\lambda - 1/L)U'(\boldsymbol{k} \cdot \boldsymbol{n})$, where $\boldsymbol{k} = (\boldsymbol{\varrho} - \boldsymbol{r})/L$. For the part on the sphere $A_\varepsilon$ of the surface integral, we then obtain the limit, with $dA = \varepsilon^2 d\Omega$ ($d\Omega$ is the surface element on the unit sphere):

$$\lim_{\varepsilon\to 0}\iint\limits_{A_\varepsilon} U\frac{\partial U'}{\partial n}dA = \lim_{\varepsilon\to 0}\iint\limits_{A_\varepsilon} U\left(\frac{2\pi i}{\lambda} - \frac{1}{\varepsilon}\right)(\boldsymbol{n} \cdot \boldsymbol{n})\frac{1}{\varepsilon}\exp\left(\frac{2\pi i}{\lambda}\varepsilon\right)\varepsilon^2 d\Omega$$

$$= -4\pi U .$$

The other integral on $A_\varepsilon$ tends to zero so that we get from (2.74) the so-called integral theorem of *Helmholz-Kirchhoff* [Ref. 2.12, p. 377]:

$$U(\boldsymbol{\varrho}) = \frac{1}{4\pi}\iint\limits_{A_p}\left\{U(\boldsymbol{r})\frac{\partial}{\partial n}\left[\frac{1}{L}\exp\left(\frac{2\pi i}{\lambda}L\right)\right] - \frac{1}{L}\exp\left(\frac{2\pi i}{\lambda}L\right)\frac{\partial U(\boldsymbol{r})}{\partial n}\right\}dA_p$$

$$\hspace{11cm}(2.75)$$

$$= -\frac{1}{4\pi}\iint\limits_{A_p}\left[U\left(\frac{2\pi i}{\lambda} - \frac{1}{L}\right)(\boldsymbol{n} \cdot \boldsymbol{k}) + \frac{\partial U}{\partial n}\right]\frac{1}{L}\exp\left(\frac{2\pi i}{\lambda}L\right)dA_p .$$

This formula is then used in the Kirchhoff diffraction theory, where the limit $A_p = A \cup B \cup C$ is composed of an opening A, a plane non-transparent screen B, and a half sphere C at infinity (Fig. 2.12). The boundary conditions [Ref. 2.12, p. 379] on A with the spherical wave from S [unit vector $\boldsymbol{h}$ and distance $L_s = \boldsymbol{h} \cdot (\boldsymbol{r} - \boldsymbol{\varrho}_s)$] are given by

$$U = \frac{A_s}{L_s}\exp\left(\frac{2\pi i}{\lambda}L_s\right) , \quad \frac{\partial U}{\partial n} = \frac{A_s}{L_s}\left(\frac{2\pi i}{\lambda} - \frac{1}{L_s}\right)(\boldsymbol{n} \cdot \boldsymbol{h})\exp\left(\frac{2\pi i}{\lambda}L_s\right) ,$$

and on B (no illumination on the side of K), they reduce to

$$U = 0 , \quad \frac{\partial U}{\partial n} = 0 . \hspace{6cm} (2.76)$$

Assuming further that no previous contribution from C (i.e., "in the past") has reached K, with these boundary conditions and with $\lambda \ll L, L_s$, we find from (2.75) the *Fresnel-Kirchhoff diffraction formula* ([Ref. 2.12, p. 382] and [Ref. 2.18, p. 132])

$$U(\boldsymbol{\varrho}) = -\frac{iA_s}{2\lambda}\iint\limits_{A}\frac{1}{L_s L}\boldsymbol{n} \cdot (\boldsymbol{k} + \boldsymbol{h})\exp\left[\frac{2\pi i}{\lambda}(L_s + L)\right]dA . \hspace{2cm} (2.77)$$

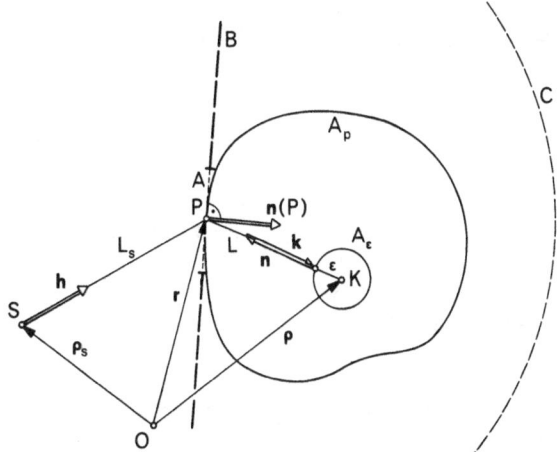

**Fig. 2.12.** Illustration of the Helmholz-Kirchhoff and Fresnel-Kirchhoff diffraction formulas. ($n$, $h$, $k$: unit direction vectors)

Considering now the case of a *nonhomogeneous* but still isotropic medium with a varying index of refraction $n(r)$, we encounter a situation similar to that of the previously described small-wavelength development of spherical waves. We return to (2.67) which, for $\gamma = n$, $\sigma = 0$ and with $\nabla^2 e_\omega = \nabla(\nabla \otimes e_\omega)$, becomes

$$\nabla(\nabla \otimes e_\omega) - \nabla(\nabla \cdot e_\omega) + \frac{4\pi^2 n^2}{\lambda_0^2} e_\omega = 0 \ . \tag{2.78}$$

In the previous case of a homogeneous medium with a constant index of refraction $n$, we described the linear polarized wave by $e_\omega = U v_0 = A \exp [2\pi i \ (k \cdot r)/\lambda] v_0$, which could also be written in the form $a \exp [2\pi i \ n(k \cdot r)/\lambda_0]$, where $n(k \cdot r)$ is the *optical path* along the wave propagation. In the present case of a nonhomogeneous field, we now use the separation in a more generalized form

$$e_\omega = a(r) \ \exp \left[ \frac{2\pi i}{\lambda_0} S(r) \right] , \tag{2.79}$$

where the scalar function $S(r)$ is called the *eikonal*. Thus, on the one hand, (2.68) for $\varrho_\omega \simeq 0$ turns into

$$\nabla \cdot a + \frac{2\pi i}{\lambda_0} \nabla S \cdot a = - \nabla (\log \varepsilon) \cdot a \ . \tag{2.80}$$

On the other hand, with (2.79, 80), and the derivative rules (2.21), the different terms of (2.78) become

$$\nabla[\nabla \otimes e_\omega] = \nabla\left[\exp\left(\frac{2\pi i}{\lambda_0}S\right)\nabla \otimes a + \frac{2\pi i}{\lambda_0}\exp\left(\frac{2\pi i}{\lambda_0}S\right)\nabla S \otimes a\right]$$

$$= \exp\left(\frac{2\pi i}{\lambda_0}S\right)\left[\nabla(\nabla \otimes a) + \frac{2\pi i}{\lambda_0}(2\nabla S(\nabla \otimes a) + \nabla^2 Sa) - \frac{4\pi^2}{\lambda_0^2}|\nabla S|^2 a\right]$$

$$-\nabla(\nabla \cdot e_\omega) = -\nabla\left[\exp\left(\frac{2\pi i}{\lambda_0}S\right)\nabla \cdot a + \frac{2\pi i}{\lambda_0}\exp\left(\frac{2\pi i}{\lambda_0}S\right)\nabla S \cdot a\right]$$

$$= \nabla\left\{\exp\left(\frac{2\pi i}{\lambda_0}S\right)[\nabla(\log\varepsilon) \cdot a]\right\}$$

$$= \exp\left(\frac{2\pi i}{\lambda_0}S\right)\left\{\frac{2\pi i}{\lambda_{ii}}\nabla S[\nabla(\log\varepsilon) \cdot a] + [\nabla \otimes \nabla(\log\varepsilon)]a\right.$$

$$\left. + (\nabla \otimes a)\nabla(\log\varepsilon)\right\}$$

$$= \exp\left(\frac{2\pi i}{\lambda_0}S\right)\left\{\left[\frac{2\pi i}{\lambda_0}\nabla S \otimes \nabla(\log\varepsilon) + \nabla \otimes \nabla(\log\varepsilon)\right]a\right.$$

$$\left. + (\nabla \otimes a)\nabla(\log\varepsilon)\right\} .$$

Therefore, the time independent equation (2.78), together with (2.68), then reads after separating the terms of order $1/\lambda_0^2$, $1/\lambda_0$ and 1:

$$\frac{4\pi^2}{\lambda_0^2}(|\nabla S|^2 - n^2)a - \frac{2\pi i}{\lambda_0}\{2\nabla S(\nabla \otimes a) + \nabla^2 Sa + [\nabla S \otimes \nabla(\log\varepsilon)]a\}$$

$$- \{\nabla(\nabla \otimes a) + [\nabla \otimes \nabla(\log\varepsilon)]a + (\nabla \otimes a)\nabla(\log\varepsilon)\} = 0 . \tag{2.81}$$

We now assume the existence of a simple solution for $S(r)$, when $\lambda_0$ tends to zero. Provided that $\nabla \otimes a$ is not of the order $1/\lambda_0$ ($a$ is supposed to be a slowly varying vector) and that $\nabla S$ is not of the order $\lambda_0$, we conclude that the second and third bracket-terms of (2.81) may be neglected with respect to the first. Thus, we have the necessary condition

$$|\nabla S|^2 - n^2 = 0 , \tag{2.82}$$

called the *eikonal equation*. The term with the factor $2\pi i/\lambda_0$ leads to the so-called equation of transport [Ref. 2.12, p. 118]. Equation (2.82) describes a family of surfaces $S = $ const, called *wave fronts* with unit normals $l = \nabla S/n$ tangent to the *curved light rays*.

Let us then form the derivative of

$$\nabla S = nl , \tag{2.83}$$

which gives

$$\nabla \otimes \nabla S = \nabla n \otimes l + n \nabla \otimes l ,$$

and from which we take the antimetric part ($\nabla \otimes \nabla S$ is symmetric)

$$0 = \nabla n \otimes l - l \otimes \nabla n + n \nabla \otimes l - n(\nabla \otimes l)^{\mathrm{T}} .$$

Applying this expression onto $l$ and observing that $l^2 = 1$ and $l(\nabla \otimes l)^{\mathrm{T}} = \nabla(l^2)/2 = 0$, we get:

$$(l \cdot \nabla n)l - \nabla n + nl(\nabla \otimes l) = 0 . \tag{2.84}$$

As in (2.37), we now introduce the projection $\nabla_1 = L\nabla = (I - l \otimes l)\nabla$ of the derivative operator $\nabla$ in the plane normal to $l$. We also have $l = dr/ds$, where $ds$ is the arc element on the ray so that (2.84) becomes

$$- \nabla_1 n + n\frac{dr}{ds} (\nabla \otimes l) = 0 .$$

Here the increment $dr(\nabla \otimes l) = dl$ of the unit vector $l$ appears. Thus, according to Frenet's formula, we obtain

$$\nabla_1 (\log n) = \frac{dl}{ds} = \frac{m_0}{\varrho_s} . \tag{2.85}$$

This is the *ray equation*. It shows that the lateral gradient of the logarithm of the index of refraction is related to the principal unit normal $m_0$ (in the osculating plane) and to the radius of curvature $\varrho_s$ (Fig. 2.13).

It should be noted that (2.85) does not contain the complete gradient of the logarithm of the index of refraction, $\nabla(\log n)$. The longitudinal component

$$l \cdot \nabla \log n = \frac{d(\log n)}{ds} = -\frac{d(\log v)}{ds}$$

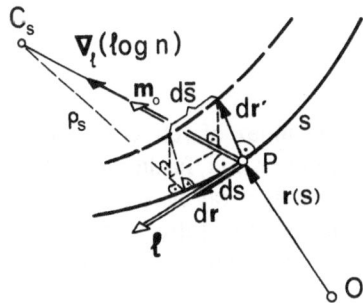

Fig. 2.13. Neighboring curved light rays through a non homogeneous medium with a spatially variable index of refraction $n(r)$. ($ds$, $d\bar{s}$: arc elements; $\varrho_s$: radius of curvature; $m_0$: principal unit normal vector)

gives the change in the velocity $v = c/n$ along the ray. Furthermore, the longitudinal component of (2.83) is $dS/ds = n$, so that

$$S = \int n \, ds + S_0 \tag{2.86}$$

is, in fact, a measure of the optical path, and (2.83) is the generalization of the derivation of a distance, see (2.23), giving a unit vector. In some ways, (2.85), which determines the ray $r(s)$, is partly the geometrical interpretation of (2.82) for the eikonal. With the variations (Fig. 2.13) given by

$$d\bar{s} = ds(1 - dr' \cdot m_0/\varrho_s) \, , \quad \bar{n} = n[1 + dr' \cdot \nabla_1(\log n)] \, , \tag{2.87}$$

eq. (2.86) leads to Fermat's principle of the stationary behavior of the optical path between two given points $P_1$, $P_2$:

$$\delta \int_{\widehat{P_1 P_2}} n \, ds = 0 \, . \tag{2.88}$$

The separation into lateral and longitudinal parts is usually not given explicitly, but instead, a combined equation is deduced. We may obtain it here by adding $l \, dn/ds$ on both sides of (2.85), after multiplication by $n$,

$$\nabla n = l\frac{dn}{ds} + \nabla_1 n = n\frac{dl}{ds} + l\frac{dn}{ds} = \frac{d}{ds}(nl) = \frac{d}{ds}\left(n\frac{dr}{ds}\right) \, . \tag{2.89}$$

Thus, (2.85) is also $d/ds(n \, dr/ds) = \nabla n$ [Ref. 2.12, p. 122].

If the distribution of the refractive index is rotationally symmetric, then $\nabla n = kr$, so that

$$\frac{d}{ds}(r \times nl) = l \times nl + r \times \frac{d}{ds}(nl) = 0 \, ,$$

yields the integal

$$r \times nl = c \, , \tag{2.90}$$

which is Bouguer's formula [Ref. 2.12, p. 123]. The light path is then planar and one has a sort of "law of area" similar to that of the momentum equation in the case of a central force.

## 2.2.2 Refraction at Interfaces

In the preceding subsection, we assumed that the index of refraction varied continuously throughout the medium under consideration and, furthermore, that this function was differentiable along each ray. If a light ray passes from one medium to another of different density, then the index of refraction changes suddenly by a finite step. Of course, this involves the elementary problem of refraction described by Snell's law

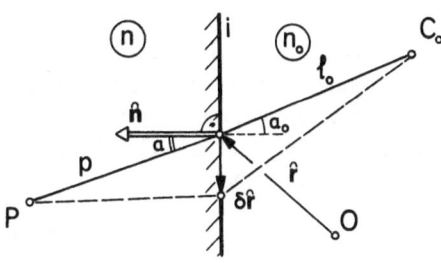

**Fig. 2.14.** Refraction at a plane interface i separating two domains with different indices of refraction $n_0$, $n$. ($\hat{n}$: unit normal)

$$\frac{\sin \alpha}{\sin \alpha_0} = \frac{n_0}{n} \ , \tag{2.91}$$

where $n_0$, $n$ are the indices on both sides of the interface and $\alpha_0$, $\alpha$ are the angles of the rays with respect to the unit normal $\hat{n}$ (Fig. 2.14). In order to have an easier approach in holography later on (while leaving out polarization effects here), we intend to present this topic from a point of view which differs from the usual treatment. If we then consider this case as a limit of the previous one, where, in the vicinity of the interface i, the index $n$ changes rapidly but continuously from $n_0$ to $n$, Fermat's principle, (2.88), may be applied to the optical path between some point source $C_0$ and any point P in the other medium. We then have

$$\delta(n_0 l_0 + np) = 0 \ , \tag{2.92}$$

where $l_0$, $p$ are the variable distances from the fixed points $C_0$, P to a point J on the interface i. By means of the increment $\delta\hat{r}$ of the position vector $\hat{r}$ of J, this equation becomes

$$\delta\hat{r} \cdot (n_0 \nabla_{\hat{n}} l_0 + n \nabla_{\hat{n}} p) = 0 \ , \tag{2.93}$$

where $\nabla_{\hat{n}} = \hat{N}\nabla$, see (2.37), is the two-dimensional derivative operator on the interface at J. We provisionally assume homogeneity in both media outside the interface. For the derivatives of the two lengths, we have $\nabla l_0 = k_0$, $\nabla p = -k$, see (2.23), where $k_0$, $k$ are unit direction vectors of the ray. Therefore, since (2.93) is valid for any $\delta\hat{r}$ on the interface, we obtain by means of the projector $\hat{N} = I - \hat{n} \otimes \hat{n}$ that

$$n_0 \hat{N} k_0 = n \hat{N} k \ , \tag{2.94}$$

which is precisely the law of refraction (2.91) determining $k$ when $|k| = 1$ is used. In the more general case where, for example, $n$ varies on the left side of i, we simply replace $np$ by $-S$ in (2.92). Using (2.83), we obtain $\nabla S_J = nl_{|J} = nk$, so that (2.94) is also valid in the non homogeneous case. $k$ then denotes the tangential unit vector to the curved rays at J (Fig. 2.15). Of course, note that a reflection also appears but we do not consider it here.

Since generally more than one ray arrives from $C_0$, thus forming a bundle called (in the homogeneous case) a rectilinear congruence, the question arises as to how this congruence is transformed in the other non-homogeneous medium. This is a classical problem. We are especially interested in the behavior of transformed neighboring rays, created by the *narrow pencil* around $C_0 J$. Thus, we write (2.94) for a point $\bar{J}$, separated from J by the increment $d\hat{r}$:

$$n_0\bar{\hat{N}}\bar{k}_0 = \bar{n}\bar{\hat{N}}\bar{k} \ . \tag{2.95}$$

The bar marks the quantities of the neighboring ray. Subtracting (2.94) from (2.95), we get

$$n_0(d\hat{N}k_0 + \hat{N}\ dk_0) = n(d\hat{N}k + \hat{N}d\hat{k}) + dn\ \hat{N}k \ . \tag{2.96}$$

For these differentials, analogously to (2.24 and 15), we first have ($K_0 = I - k_0 \otimes k_0$ is a projector like $\hat{N}$)

$$dk_0 = d\hat{r}(\nabla_{\hat{n}} \otimes k_0) = d\hat{r}\,\frac{1}{l_0}\hat{N}K_0 \ ; \tag{2.97}$$

and, secondly, from (2.41), we also have

$$\begin{aligned} d\hat{N} &= -\,d\hat{r}[(\nabla_{\hat{n}} \otimes \hat{n}) \otimes \hat{n} + (\nabla_{\hat{n}} \otimes \hat{n}) \otimes \hat{n})^{\mathrm{T}}] \\ &= d\hat{r}[\hat{B} \otimes \hat{n} + \hat{B} \otimes \hat{n})^{\mathrm{T}}] \ , \end{aligned} \tag{2.98}$$

where $\hat{B} = -\,\nabla_{\hat{n}} \otimes \hat{n}$ is now the tensor of curvature of the interface i. Contrary to the vector $dk_0$, the increment $d\hat{k}$ cannot be referred to a collineation center for which the bar is a reminder. In fact, the straight virtual rays corresponding to $k$ and $\bar{k}$ are *skew* (Fig. 2.15). However, we may resolve (2.96) and, using (2.97, 98), we find

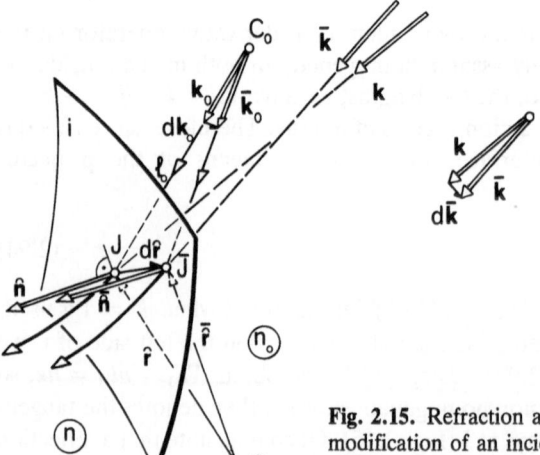

**Fig. 2.15.** Refraction at a curved interface i, showing the modification of an incident rectilinear congruence. ($k_0$, $k$: unit direction vectors)

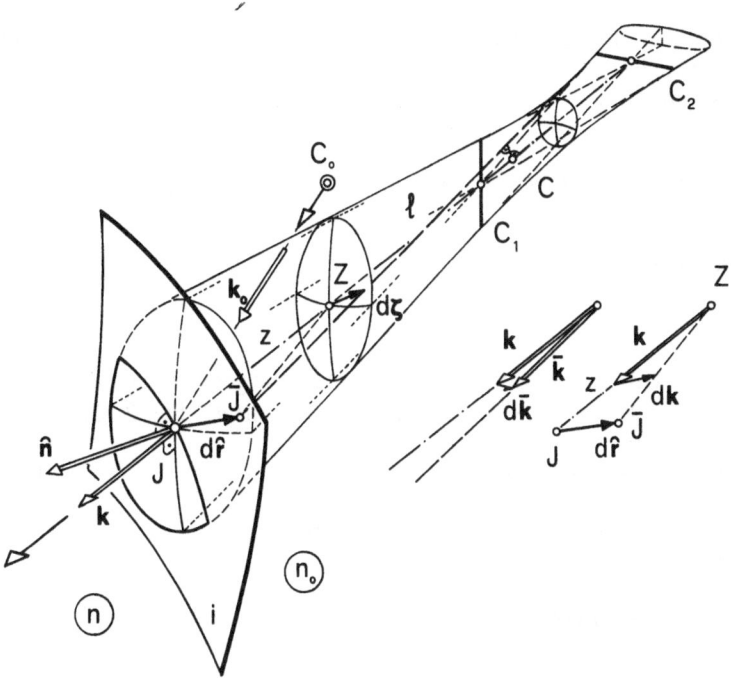

**Fig. 2.16.** Refracted astigmatic pencil and corresponding wavefronts. ($C_0$: center of the incident spherical wave. Focal lines through $C_1$, $C_2$; $d\zeta$: increment on the wavefront through point Z; $d\hat{r}$: increment on the oblique interface i at point J. $l$: distance JC; $z$: distance JZ)

$$\hat{N} \, d\bar{k} = \frac{1}{n} \, d\hat{r}\hat{N} \left\{ \frac{n_0}{l_0} K_0 + \hat{B} \left[ (k_0 n_0 - kn) \cdot \hat{n} \right] - \nabla n \otimes k \right\} \hat{N} \; . \tag{2.99}$$

Hereafter, let us choose a point Z at the distance $z$ from J on the virtual reference ray and make a connection between the interface i and the unit sphere around Z:

$$d\hat{r} = z\hat{M}^T dk \; . \tag{2.100}$$

The increment $dk$, which should not be confused with $d\bar{k}$, represents the increment of $k$, when the line ZJ changes into the line Z$\bar{J}$. Both increments are, of course, normal to $k$ (Fig. 2.16). Further, $\hat{M} = I - \hat{n} \otimes k/\hat{n} \cdot k$ is an oblique projector as previously described, see (2.49).

Using the transposed equation (2.100) and operating from the right with the projector $\hat{M}^T = I - k \otimes \hat{n}/k \cdot \hat{n}$, (2.99) with $\hat{M}\hat{N} = \hat{M}$, see (2.52), and with $\hat{M} \, d\bar{k} = d\bar{k}$, becomes:

$$d\bar{k} = \frac{z}{n} \, dk\hat{M} \left\{ \frac{n_0}{l_0} K_0 + \hat{B} \left[ (k_0 n_0 - kn) \cdot \hat{n} \right] - \nabla n \otimes k \right\} \hat{M}^T \; . \tag{2.101}$$

Let us then introduce an abbreviation for the resultant characteristic tensor which is symmetric because $- \nabla n \otimes k = n(\nabla \otimes k) - \nabla \otimes \nabla S$, (2.83), and which lies in the plane normal to $k$:

$$W = \frac{1}{n} \hat{M} \left\{ \frac{n_0}{l_0} K_0 + \hat{B}[(k_0 n_0 - kn) \cdot \hat{n}] - \nabla n \otimes k \right\} \hat{M}^{\mathrm{T}} . \tag{2.102}$$

In abbreviated form, (2.101) is now

$$d\bar{k} = zW \, dk . \tag{2.103}$$

If we consider the virtual wavefront through point Z, the geometrical meaning of the tensor $W$ becomes clear. Using the normal increment $d\zeta$ (Fig. 2.16),

$$d\zeta = K \, d\hat{r} - z \, d\bar{k} = z \, dk(K - zW) \tag{2.104}$$

on this wavefront [$\hat{M}K = K$, see (2.51); $d\bar{k}$ is a particular increment of $k$ but not of $\bar{k}$] we may write

$$d\bar{k} = d\zeta(\nabla_k^* \otimes k) = z \, dk(K - zW)(\nabla_k^* \otimes k) . \tag{2.105}$$

Here $\nabla_k^*$ is the two-dimensional derivative operator, $k$ and $*$ reminding us that $\nabla_k^*$ operates in a surface perpendicular to the unit vector $k$ at point Z. Therefore, $- \nabla_k^* \otimes k = - W^*$ is the tensor of curvature of the wavefront at Z, see (2.39). On the other hand, comparing (2.103 and 105), we find the relation

$$W = (K - zW)(\nabla_k^* \otimes k) = (K - zW)W^* ; \tag{2.106}$$

in particular, when $z \to 0$, we obtain

$$W = \nabla_k \otimes k , \tag{2.107}$$

where $\nabla_k$ now denotes the derivative operator on the surface through J, which is parallel to that through Z. Hence, $- W$ is the *tensor of curvature of the wavefront* through J and (2.106) may be compared to (2.56), where $K$, $W$, $W^*$, $k$ take the roles of $N$, $-B$, $-\bar{B}$, $n$ respectively. In these considerations, we have once again chosen the intrinsic method rather than the calculus with components, in order to keep a greater flexibility later on in holography. This is particularly advantageous, since the tensors of curvature $\hat{B}$ and $- W$ have, in general, different principal directions. Further, let us look at a point C situated at a distance $z = l$ from J, for which $dk \cdot d\zeta = 0$. This point is located very close to the shortest distance between the slightly skewed rays passing through J and $\bar{J}$. The above formulated condition to determine point C may also be written with (2.104)

$$dk \cdot \left( \frac{1}{l} K - W \right) dk = 0 , \tag{2.108}$$

or in the particular form

$$\frac{1}{l} = m \cdot Wm \ , \tag{2.109}$$

when $m = dk/|dk|$ is a unit vector in the direction $dk$. Comparing (2.109) with (2.38), we see that $l$ is the radius of curvature of the normal section of the wavefront in the direction $m$. The maximum $l_2$ and the minimum $l_1$ of the whole set $\{l\}$ (generated by the variation of direction of the vector $m$) are the principal radii of curvature. Their inverses $1/l_1$, $1/l_2$ are the eigenvalues of $W$. Roughly stated, the image of $C_0$ is thus an astigmatic interval $\{C\}$. In Fig. 2.16, we recognize the so-called "focal lines" passing through the end points $C_1$ and $C_2$, where all of the skewed rays around the central ray intersect.

In the special case of a homogeneous medium and of a plane interface ($\hat{B} \equiv 0$), we simply have

$$W = \frac{n_0}{nl_0}\hat{M}K_0\hat{M}^T = \frac{n_0}{nl_0}\left(I - \frac{\hat{n} \otimes k}{\hat{n} \cdot k}\right)(I - k_0 \otimes k_0)\left(I - \frac{k \otimes \hat{n}}{k \cdot \hat{n}}\right) . \tag{2.110}$$

Using the symmetry of this special case, we conclude that the eigenvectors $m_1$, $m_2$ are both normal and parallel to the plane which contains $k_0$, $k$ and $\hat{n}$ (Fig. 2.17). The corresponding eigenvalues (or principal curvatures) are then calculated from (2.109) and, after some computations with the above dyadics, the following values are obtained:

$$\frac{1}{l_1} = \frac{n_0}{nl_0} \ , \qquad \frac{1}{l_2} = \frac{n_0}{nl_0}\left(\frac{\hat{n} \cdot k_0}{\hat{n} \cdot k}\right)^2 . \tag{2.111}$$

Again, in the more general case of curved interfaces but with the simplifying assumption of homogeneity in each medium, let us consider what happens with the rectilinear congruence, i.e., the astigmatic pencil, when it passes through the second interface i' and ends up in a third medium of refractive index $n'$. Of course, we must first determine the intersection point J' of the central ray with this second interface. The point J' is situated at the distance $e$ from J. Applying (2.95) to the new interface, we obtain the analogous relation ($\hat{N}' = I - \hat{n}' \otimes \hat{n}'$, $\hat{n}'$ is the unit normal to i'):

$$n\hat{N}'k = n'\hat{N}'k' \ , \tag{2.112}$$

Fig. 2.17. Refraction at a plane interface i; representation of eigenvectors $m_1$, $m_2$ of the astigmatic interval $C_1C_2$

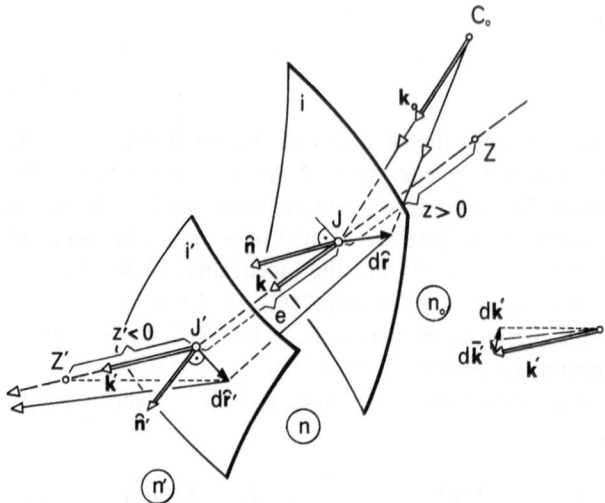

**Fig. 2.18.** Refraction at a second interface $i'$ with unit normal $\hat{n}'$. ($n_0$, $n$, $n'$: indices of refraction; $k_0$, $k$, $k'$: unit direction vectors)

which, together with $|k'| = 1$, determines the unit direction vector $k'$ (Fig. 2.18).

In addition to this elementary consideration, (2.99) must be replaced by

$$\hat{N}'d\bar{k}' = \frac{n}{n'}\hat{N}'d\bar{k} + \frac{1}{n'}d\hat{r}'\hat{B}'[(kn - k'n') \cdot \hat{n}'] \; , \tag{2.113}$$

since the exit *and* entering pencils are both astigmatic. With a second point $Z'$ on the virtual ray $k'$, remembering that $Z$ is on the virtual ray $k$, we simultaneously have

$$d\bar{k} = (z + e)\overline{W} \, dk \; , \qquad d\bar{k}' = z'W'dk' \; , \tag{2.114}$$

and the affine connections

$$d\hat{r}' = z'\hat{M}'^{\mathrm{T}}dk' \; , \qquad dk = \frac{1}{z + e}K \, d\hat{r}' = \frac{z'}{z + e}K\hat{M}'^{\mathrm{T}}dk' \; , \tag{2.115}$$

where $\hat{M}' = I - \hat{n}' \otimes k'/\hat{n}' \cdot k'$.

Note that $-\overline{W}$ is the curvature tensor at $J'$ of the wavefront relative to $k$, whereas $-W'$ is the curvature tensor at this same point $J'$, but for the new wavefront which is relative to $k'$. Introducing (2.114, 115) into (2.113), in analogy to (2.101) and with oblique projectors $\hat{M}'$, $\hat{M}'^{\mathrm{T}}$ on both sides, we get

$$z' dk' W' = \frac{z'}{n'} dk' \hat{M}' \{n\overline{W} + \hat{B}'[(kn - k'n') \cdot \hat{n}']\} \hat{M}'^{\mathrm{T}} \; .$$

Since $dk'$ is arbitrary in the plane normal to $k'$, $W'$ may be expressed similarly to (2.102) as

$$W' = \frac{1}{n'}\hat{M}'\{n\overline{W} + \hat{B}'[(kn - k'n') \cdot \hat{n}']\}\hat{M}'^{\mathrm{T}} \ . \tag{2.116}$$

As a final step, we need to express $\overline{W}$ by $W$. To this end, we apply (2.106) for $z = -e$, i.e.,

$$W = (K + eW)\overline{W} \ . \tag{2.117}$$

Similarly to what was done in (2.57), eq. (2.117) may be approximately inverted in the case where $e$ is sufficiently small (e.g., in the theory of a moderately thin lens). This inversion reads:

$$\overline{W} = W - eW^2 + e^2 W^3 + \ldots \ .$$

Using (2.102) in the above expression for $\overline{W}$ and than substituting into (2.116) we obtain

$$W' = \frac{1}{n'}\hat{M}'\left\{\hat{M}\left[\frac{n_0}{l_0}K_0 + \hat{B}((k_0 n_0 - kn) \cdot \hat{n})\right]\hat{M}^{\mathrm{T}}\right.$$
$$\left. + \hat{B}'[(kn - k'n') \cdot \hat{n}'] - neW^2\right\}\hat{M}'^{\mathrm{T}} \ . \tag{2.118}$$

A trivial example is seen in the case of a very thin circular lens, where $C_0$ lies on the optical axis. Here,

$$\hat{n} = \hat{n}' = k_0 = k = k' \ , \quad n' = n_0 \ , \quad e \to 0 \ ,$$

$$\hat{B} = \frac{1}{R}\hat{N} \ , \quad \hat{B}' = \frac{1}{R'}\hat{N} \ ,$$

so that

$$W' = \frac{1}{n_0}\left(\frac{n_0}{l_0} + \frac{n_0 - n}{R} - \frac{n_0 - n}{R'}\right)\hat{N}' \ .$$

The two eigenvalues of $W'$ are identical, $1/l'_1 = 1/l'_2 = 1/l'$, the image $C'$ of $C_0$ is stigmatic, and we get the well-known relation for the so-called Gaussian image [Ref. 2.12, p. 163]

$$-\frac{1}{l'} + \frac{1}{l_0} = \frac{n - n_0}{n_0}\left(\frac{1}{R} - \frac{1}{R'}\right) \quad (\text{with} \quad R' < 0) \ . \tag{2.119}$$

In this field of study, we must also consider an affine connection between two surfaces, as was already emphasized in Sect. 2.1.2, but with the additional diffi-

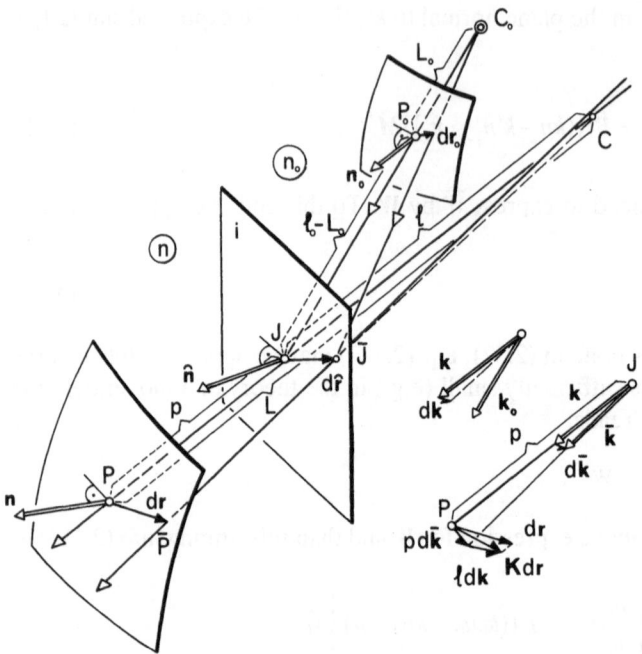

**Fig. 2.19.** Affine connection between increments $dr_0$, $dr$ situated on two surfaces separated by a refracting interface i. ($n_0$, $\hat{n}$, $n$: unit normals)

culty incurred by the presence of a refracting interface running between them (Fig. 2.19). In order to find the relation between the two increments $dr_0$ and $dr$ in the corresponding points $P_0$ and P of the two surfaces in question, we first write the equation between $dr_0$ and $d\hat{r}$:

$$d\hat{r} = \frac{l_0}{L_0} M_0^T dr_0 \ . \tag{2.120}$$

Here $l_0$ represents the distance from $C_0$ to J and the oblique projector $M_0 = I - \hat{n} \otimes k_0 / \hat{n} \cdot k_0$ differs from the other oblique projector $\hat{M}$ as defined earlier by $\hat{M} = I - \hat{n} \otimes k / \hat{n} \cdot k$. Secondly, for the skewed rays through P and $\overline{P}$ and the intersecting rays at C, we have

$$K \, dr = p \, d\bar{k} + l \, dk = l(pW + K)dk = (pW + K)K \, d\hat{r} \ . \tag{2.121}$$

By using the following properties of projectors, $\hat{M}^T K = \hat{M}^T$, $\hat{M}^T M_0^T = M_0^T$, $M^T K \, dr = M^T dr = dr$, and by combining (2.120) with (2.121), we obtain

$$dr = \frac{l_0}{L_0} M^T (pW + K) M_0^T dr_0 \ , \tag{2.122}$$

where $M = I - n \otimes k/n \cdot k$ is the oblique projector related to the object surface, see also (2.49) and where $W$ is given by (2.102). In the special case of a plane interface and $\nabla n = 0$, the above relation, always with $K_0 = I - k_0 \otimes k_0$, becomes

$$dr = \frac{l_0}{L_0} \, M^T \hat{M} \left( \frac{n_0 p}{n l_0} K_0 + K \right) M_0^T dr_0 \; . \tag{2.123}$$

The above affine connection represents a one-one mapping $dr \rightleftarrows dr_0$, only if the unit normal vectors $n$, $n_0$, $\hat{n}$ are known in advance.


## 2.3 Kinematics of Deformation

In every case discussed so far, we have considered a simple configuration in space, although several surfaces were involved. However, later we must simultaneously investigate several configurations in space. In discussing holography in Chap. 3, we first encounter the set-up of the holographic plate, ready to record the object, and later the hologram in its new position which reconstructs the image of the object. Secondly, in holographic interferometry, we still study the object or its image in both undeformed and deformed states. Therefore, we want to recall briefly some basic concepts of the kinematics of deformation in a manner which corresponds to its use in continuum mechanics (see particularly [Ref. 2.7, Chap. 6], [Ref. 2.8, Chap. 3] and [2.9, 20]).


### 2.3.1 General Equations for the Deformation of a Body

In Fig. 2.20 we show two configurations of a body i.e., the undeformed state {P} of the ensemble of material points at time $t_0 = 0$ and the deformed state {P'} of the same material points in new positions at some time $t > 0$. If $r$, $r'$ denote the two position vectors of P, P', the transformation $r \rightarrow r'(r, t)$ is called the *deformation* of the body in a Lagrangean representation ($r$ is the independent variable, $t$ is the parameter). Conversely, $r' \rightarrow r(r', t)$ would be the Eulerian representation of this deformation with $r'$ as the independent variable. If no dislocations are present, we may write infinitesimal transformations

$$dr' = dr(\nabla \otimes r') = F \, dr \quad \text{or} \quad dr = F^{-1} dr' \; , \tag{2.124}$$

where the invertible tensor

$$F = (\nabla \otimes r')^T = I + (\nabla \otimes u)^T \tag{2.125}$$

is expressed here by the derivative of the *displacement vector* $u = r' - r$ of P. That is why $F$ is called the *deformation gradient*. In cartesian components $u$, $v$, $w$ of $u$, this Lagrangian representation becomes:

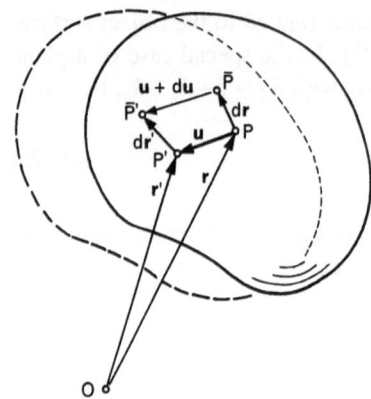

**Fig. 2.20.** Deformation of a body. P,P′: the same material point in the undeformed and deformed configuration. ($\overline{P}$,$\overline{P}'$: an arbitrary material point in the vicinity of P,P′; ; $r$, $r'$: position vectors, $u$: displacement vector)

$$F \triangleq \begin{bmatrix} \left(1 + \dfrac{\partial u}{\partial x}\right) & \dfrac{\partial u}{\partial y} & \dfrac{\partial u}{\partial z} \\[2ex] \dfrac{\partial v}{\partial x} & \left(1 + \dfrac{\partial v}{\partial y}\right) & \dfrac{\partial v}{\partial z} \\[2ex] \dfrac{\partial w}{\partial x} & \dfrac{\partial w}{\partial y} & \left(1 + \dfrac{\partial w}{\partial z}\right) \end{bmatrix} . \tag{2.126}$$

An alternative representation of $F$ is the following: In terms of two sets of general covariant base vectors $g_i$, $g'_j$, but with only one set of convected coordinates $\theta^k$ (the "address" of P), we have $dr = g_i d\theta^i$, $dr' = g'_j d\theta^j$. With the associated base vectors $g_i$, $g^k$ and $g'_j$, $g'^l$, and noting that $dr$ (or $dr'$) are arbitrary, (2.124) yields

$$F = g'_j \otimes g^j \quad \text{and} \quad F^{-1} = g_i \otimes g'^i . \tag{2.127}$$

This clearly shows that $F$ marks the "bridge" between the two configurations. Moreover, in order to emphasize the one-to-one mapping $dr \rightleftarrows dr'$ in both directions, we may also write $(F^T)^{-1} = \nabla' \otimes r = I - \nabla' \otimes u$, where $\nabla' = g'^i \partial/\partial\theta^j$ is the derivative operator in the deformed configuration. Since for any scalar function $\phi(\theta^1, \theta^2, \theta^3)$, we have $dr \cdot \nabla\phi = dr' \cdot \nabla'\phi$, valid for any $dr$, we arrive at the transformation rule (similar to (2.50) in the case of surfaces):

$$\nabla' = (F^T)^{-1}\nabla , \qquad \nabla = F^T \nabla' . \tag{2.128}$$

In many cases, $F$ differs little from the identity, so that the transposed inverse of $F$ becomes, as in (2.57):

$$(F^T)^{-1} \cong I - \nabla \otimes u + (\nabla \otimes u)(\nabla \otimes u) ; \tag{2.129}$$

thereby, $F^T(F^T)^{-1} = I$ is verified up to the third-order terms. In (2.128a), $\nabla'$ can thus be expressed approximately in a Lagrangean way:

$$\nabla' \cong [I - \nabla \otimes u + (\nabla \otimes u)(\nabla \otimes u)] \nabla . \tag{2.130}$$

Usually the polar decomposition separates $F$ into an orthogonal tensor $Q$ and a symmetric tensor $U$. The deformation gradient then becomes:

$$F = QU = U'Q , \tag{2.131}$$

and consists of a stretch tensor $U$ (or the symmetric tensor $U'$), which are related to the dilatation and a rotation tensor $Q$.

The strain tensor is defined by

$$\tilde{E} = \tfrac{1}{2}(F^{\mathrm{T}}F - I) = \tfrac{1}{2}(U^2 - I) , \tag{2.132}$$

and can be expressed with the decomposition (2.125) in the form

$$\tilde{E} = \tfrac{1}{2}[\nabla \otimes u + (\nabla \otimes u)^{\mathrm{T}} + (\nabla \otimes u)(\nabla \otimes u)^{\mathrm{T}}] . \tag{2.133}$$

In the case of "small" deformations, $\varepsilon \ll 1$, the linear dilatation or strain of an elemental distance $ds$, in the direction $e$, is

$$\varepsilon \simeq \frac{(ds')^2 - (ds)^2}{2(ds)^2} = \frac{1}{2} e \cdot (F^{\mathrm{T}}F - I)e = e \cdot \tilde{E}e ,$$

and the angular dilatation or shearing strain considering two perpendicular directions, $e_1$ and $e_2$, reads

$$\gamma_{12} \simeq \frac{dr_1' \cdot dr_2'}{ds_1' ds_2'} \simeq e_1 \cdot F^{\mathrm{T}}Fe_2 = 2e_1 \cdot \tilde{E}e_2 .$$

In a cartesian system with the unit base vectors $i, j, k$, the components of the tensor $\tilde{E}$ are identical to the usual strains and shearing strains relative to the axes

$$\tilde{E} \triangleq \begin{bmatrix} \tilde{\varepsilon}_x & \tfrac{1}{2}\tilde{\gamma}_{xy} & \tfrac{1}{2}\tilde{\gamma}_{xz} \\ \tfrac{1}{2}\tilde{\gamma}_{yx} & \tilde{\varepsilon}_y & \tfrac{1}{2}\tilde{\gamma}_{yz} \\ \tfrac{1}{2}\tilde{\gamma}_{zx} & \tfrac{1}{2}\tilde{\gamma}_{zy} & \tilde{\varepsilon}_z \end{bmatrix} .$$

In the special case in which the strains and rotations are small and of the same order of magnitude, we introduce the expansion for the strain tensor $\tilde{E}$ or for the tensor $U$:

$$\tilde{E} = \eta E_1 + \eta^2(E_2 + \tfrac{1}{2}E_1^2) + \dots , \qquad U = I + \eta E_1 + \eta^2 E_2 + \dots . \tag{2.134}$$

We see that $U$ differs little from the identity when $\eta = \sigma_{max}/E \ll 1$. The value $\sigma_{max}$ represents the largest stress component appearing in the considered body and $E$ is the modulus of elasticity. In the same way, we develop $Q$ with another small parameter $\chi$, whose physical meaning appears later,

$$Q = I + \chi \boldsymbol{\Omega}_1 + \chi^2 \boldsymbol{\Omega}_2 + \dots , \tag{2.135}$$

and where $\boldsymbol{\Omega}$ is skew symmetric. The polar decomposition of the deformation gradient becomes

$$F = QU = I + (\eta \boldsymbol{\mathcal{E}}_1 + \chi \boldsymbol{\Omega}_1) + \dots$$
$$= I + (\nabla \otimes u)^{\mathrm{T}} .$$

If the parameters $\eta$ and $\chi$ are both of the same order of magnitude and if the terms of higher order are neglected, we can decompose $(\nabla \otimes u)^{\mathrm{T}}$ into a symmetric part and a skew-symmetric part and write the linear kinematic relations as follow

$$\eta \boldsymbol{\mathcal{E}}_1 \simeq \tfrac{1}{2}[\nabla \otimes u + (\nabla \otimes u)^{\mathrm{T}}] = \boldsymbol{\mathcal{E}} \simeq \tilde{\boldsymbol{\mathcal{E}}}$$
$$- \chi \boldsymbol{\Omega}_1 \simeq \tfrac{1}{2}[\nabla \otimes u - (\nabla \otimes u)^{\mathrm{T}}] = \boldsymbol{\Omega} , \tag{2.136}$$

which approximate the strain $\tilde{\boldsymbol{\mathcal{E}}}$ and the rotation.

Thus, we also have

$$U \simeq I + \boldsymbol{\mathcal{E}} , \qquad Q \simeq I - \boldsymbol{\Omega}$$
$$\nabla \otimes u = \boldsymbol{\mathcal{E}} + \boldsymbol{\Omega} , \qquad F = I + \boldsymbol{\mathcal{E}} - \boldsymbol{\Omega} . \tag{2.137}$$

Indeed, in the neighborhood of a fixed point at a clamped edge of a body, where $Q = I$, the maximum value of the components of $\nabla \otimes u$ could be of order $\eta$. For any other point of the body, for which the maximum value of these components is also of order $\eta$, we conclude from the polar decomposition

$$Q \simeq I - \chi \boldsymbol{\Omega}_1 \simeq FU^{-1} \simeq [I + (\nabla \otimes u)^{\mathrm{T}}](I - \eta \boldsymbol{\mathcal{E}}_1) ,$$

that $\chi$ is of the order $\eta$ anywhere.

### 2.3.2 Deformation of a Surface and a Thin Body

Since we are interested in the deformation of the surface of an opaque body, we shall conveniently transform the strain and the rotation tensor, describing the deformation of the volume element underneath it. Using some basic concepts of the shell theory [2.21–31], the main relations concerning the surface of a body will be obtained by projections. The case in which the parameters $\eta$ and $\chi$ are small and of the same order of magnitude is considered in [Ref. 2.6, Sect. 2.2.2] and will, therefore, be mentioned only briefly. Indeed, the projections separate the significant part from the secondary parts of tensors. For increments $dr$ and $dr'$, situated in the plane $T$ (tangential to the undeformed body) and in the plane $T'$ (tangential to the deformed body), respectively (Fig. 2.21), we have

$$dr' = N' dr' = F\, dr = FN\, dr = N'FN\, dr . \tag{2.138}$$

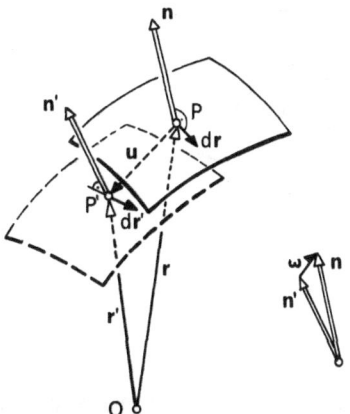

**Fig. 2.21.** Deformation of a surface. ($n$, $n'$: unit normals at the same material point in both undeformed and deformed configurations)

The linear transformation shows how the "surface-deformation gradient", $N'FN = FN$ is formed. It is simultaneously a mixed projection and a semiprojection of the three-dimensional deformation gradient. The right semiprojection may be separated by the polar decomposition in the following manner:

$$FN = QUN = Q(NUN + n \otimes nUN) \ . \tag{2.139}$$

If $n$ is here a principal direction of $U$, then $n \otimes nUN = 0$, and if the rotation of the volume element is resolved into an in-plane (a pivot rotation $Q_p$ around $n$) and an out-of-plane rotation (an inclination $Q_i$), we have $FN = Q_iQ_pV$, where $V = NUN$ is a two-dimensional symmetric tensor.

The strain tensor of a surface element is given by the two-dimensional tensor

$$\tilde{\gamma} = \tfrac{1}{2}(NF^TFN - N) = N\tilde{\mathcal{E}}N \ . \tag{2.140}$$

In particular, with (2.136), in the case of small rotations, the symmetric tensor of small surface strain becomes:

$$\gamma = N\mathcal{E}N = V - N = \tfrac{1}{2}N[\nabla \otimes u + (\nabla \otimes u)^T]N \ . \tag{2.141}$$

Using the decomposition of the displacement, $u = v + wn$, and the decomposition (2.42) of its derivative, we obtain the following linear kinematic relation at the surface

$$\gamma = \tfrac{1}{2}N[\nabla \otimes v + (\nabla \otimes v)^T]N - Bw \ . \tag{2.142}$$

Let us also recall the additive decomposition of the surface deformation gradient

$$\nabla_n \otimes u = N\mathcal{E}N + N\Omega N + (N\mathcal{E}n + N\Omega n) \otimes n$$
$$= \gamma + \Omega E + \omega \otimes n \ , \tag{2.143}$$

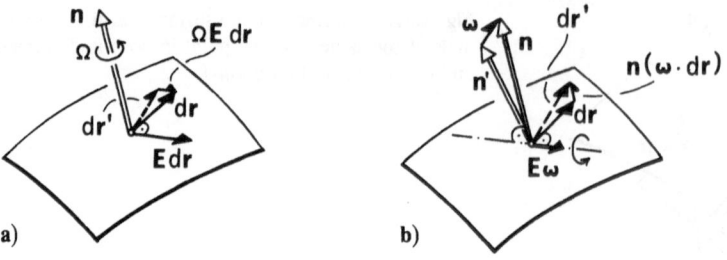

**Fig. 2.22.** (a) Pivot motion, (b) inclination of a surface element

where the scalar $\Omega$ quantifies the pivot motion around the unit normal vector $n$ and where the vector $\omega$ is the inclination of the surface element (Fig. 2.22). With $E \cdot E = -2$, the kinematic relations for the rotation are

$$\Omega = \tfrac{1}{4}\{N[\nabla \otimes v - (\nabla \otimes v)^T]N\} \cdot E$$

$$\omega = Bv + \nabla_n w \ . \tag{2.144}$$

However, the parameters $\eta$ and $\chi$ are not always of the same order of magnitude. In the case of plates and shells, when large displacements take place, we have to consider parameters with other orders of magnitude such as $\eta = O(\chi^2)$. This case is representative of the deformation theory obtained by restricting strains to be small everywhere, but by considering moderate rotations. We shall see later that this is only possible in a thin body, when we exclude the rigid motion ($\eta = 0$) of an entirely free body without edge clamping. The expansion (2.134) for the strain tensor $\tilde{E}$ or for the tensor $U$ is maintained with the same meaning for $\eta$ ($= \sigma_{max}/E \ll 1$).

We also develop the rotation tensor $Q$ with the parameter $\chi$ as in (2.135). From $Q^T Q = I$, it is easy to see that $\Omega_1$ is a skew symmetric and $\Omega_2$ a symmetric tensor. This relation corresponds to the expansion of trigonometric functions of angles of rotation into Taylor series for the rotation tensor describing the exact theory of finite rotations. According to the polar decomposition, we have:

$$F = I + (\nabla \otimes u)^T = QU = I + (\chi\Omega_1 + \eta\mathcal{E}_1) + \eta^2\mathcal{E}_2 + \eta\chi\mathcal{E}_1\Omega_1 + \chi^2\Omega_2 + \dots \ ,$$

and while $\eta = O(\chi^2)$,

$$\nabla \otimes u = O(\eta)\mathcal{E}_1 - \chi\Omega_1 + \chi^2\Omega_2^T + O(\chi^3)\dots \ .$$

The symmetric and skew-symmetric parts of $\nabla \otimes u$ are now compared with (2.136):

$$\mathcal{E} = \tfrac{1}{2}[\nabla \otimes u + (\nabla \otimes u)^T] \simeq \eta\mathcal{E}_1 + \tfrac{1}{2}\chi^2(\Omega_2 + \Omega_2^T)$$

$$\Omega = \tfrac{1}{2}[\nabla \otimes u - (\nabla \otimes u)^T] \simeq -\chi\Omega_1 - \tfrac{1}{2}\chi^2(\Omega_2 - \Omega_2^T) = -\chi\Omega_1 \ . \tag{2.145}$$

The strain tensor defined by (2.132) then becomes

$$\tilde{\pmb{E}} \simeq \pmb{E} - \tfrac{1}{2}\pmb{\Omega}^2 \ . \tag{2.146}$$

As opposed to the linear case, this equation shows that the linear part $\pmb{E}$ also contains terms of rotation, whereas the nonlinear part depends on the rotation alone. On the other hand, we may express $\nabla \otimes \pmb{u}$ with (2.145, 146) as

$$\nabla \otimes \pmb{u} = \tilde{\pmb{E}} + \pmb{\Omega} + \tfrac{1}{2}\pmb{\Omega}^2 \ . \tag{2.147}$$

The skew-symmetric tensor $\pmb{\Omega}$ may still be expressed by a three-dimensional rotation vector $\pmb{\omega}_r = \pmb{E}\pmb{\omega}_i + \pmb{\Omega}\pmb{n}$ (Fig. 2.23):

$$\pmb{\Omega} = \pmb{E}\pmb{\omega}_r = \pmb{E}\pmb{E}\pmb{\omega}_i + \pmb{\Omega}\pmb{E}\pmb{n} = \pmb{\omega}_i \otimes \pmb{n} - \pmb{n} \otimes \pmb{\omega}_i + \pmb{\Omega}\pmb{E} \ . \tag{2.148}$$

It must be noted here that $\pmb{\omega}$ specifically means the inclination $\pmb{\omega}_i$ of a surface element only in the case where the normality hypothesis requiring vanishing shear strain along the normal $\pmb{n}$ near the stress-free surface for an infinitesimal volume underneath is fulfilled.

The strain tensor of an element of the body surface is again the full projection

$$\tilde{\pmb{\gamma}} = \pmb{N}\tilde{\pmb{E}}\pmb{N} = \pmb{N}\pmb{E}\pmb{N} - \tfrac{1}{2}\pmb{N}\pmb{\Omega}^2\pmb{N} \ . \tag{2.149}$$

On inserting (2.145, 148) into (2.149), we get the nonlinear kinematic relations

$$\tilde{\pmb{\gamma}} = \pmb{\gamma} + \tfrac{1}{2}(\pmb{\omega} \otimes \pmb{\omega} + \pmb{\Omega}^2\pmb{N}) \ , \tag{2.150}$$

where (2.141 or 142) still hold for $\pmb{\gamma}$.

We may now calculate the projection $\pmb{N}'$ of the deformed free surface and relate it to the undeformed surface:

$$\pmb{N}' = \pmb{a}'^\alpha \otimes \pmb{a}'_\alpha = (\pmb{F}^{\mathrm{T}})^{-1}\pmb{a}^\alpha \otimes \pmb{a}_\alpha\pmb{F}^{\mathrm{T}} = (\pmb{F}^{\mathrm{T}})^{-1}\pmb{N}\pmb{F}^{\mathrm{T}} \ . \tag{2.151}$$

Using the decompositions (2.125, 147–150) and the fact that $\pmb{N}\tilde{\pmb{E}}\pmb{n} = 0$, on the one hand, we get

$$\pmb{N}\pmb{F}^{\mathrm{T}} = \pmb{N} + \pmb{N}\tilde{\pmb{E}} + \pmb{N}\pmb{\Omega} + \tfrac{1}{2}\pmb{N}\pmb{\Omega}^2 = \pmb{N} + \pmb{\gamma} + \pmb{\Omega}\pmb{E} + \pmb{\omega} \otimes \pmb{n} + \tfrac{1}{2}\pmb{\Omega}\pmb{E}\pmb{\omega} \otimes \pmb{n} \ ; \tag{2.152}$$

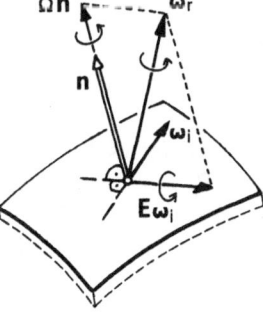

Fig. 2.23. Decomposition of the three-dimensional rotation vector $\pmb{\omega}_r$, describing a surface element of a body, into the inclination $\pmb{\omega}_i$ and the pivot motion $\pmb{\Omega}\pmb{n}$

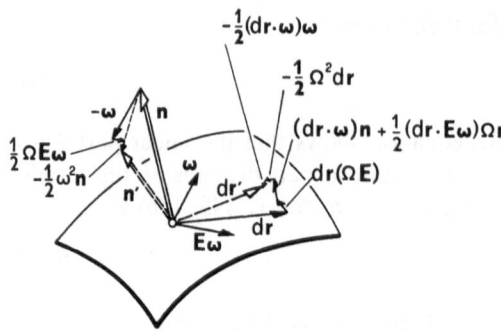

**Fig. 2.24.** Modification of the unit normal $n$ and the increment $dr$ in the case of a moderate rotation of the surface. This drawing illustrates the non-linear terms

and, on the other hand, by using (2.129) and again (2.147), we get

$$(F^T)^{-1}N = N - \gamma - \Omega E - \omega \otimes \omega - \Omega^2 N + n \otimes \omega + \tfrac{1}{2}n \otimes \Omega E \omega \ . \tag{2.153}$$

Thus, by always neglecting the third-order terms, $N'$ becomes

$$N' = N + n \otimes \omega + \omega \otimes n - \omega \otimes \omega - \tfrac{1}{2}[(n \otimes \Omega E \omega) + (\Omega E \omega \otimes n)]$$
$$+ \omega^2(n \otimes n) \ . \tag{2.154}$$

As this expression must also be equal to the normal projection in the deformed configuration $I - n' \otimes n'$, a comparison shows that the unit normal of the deformed surface is

$$n' = n - \omega + \tfrac{1}{2}\Omega E \omega - \tfrac{1}{2}\omega^2 n \ , \tag{2.155}$$

which can also be found by simple geometrical consideration in Fig. 2.24.

This last relation for $n'$ also enables us to define the curvature tensor $B' = - \nabla'_n \otimes n'$ of the deformed surface, as well as the curvature change of the outer surface of an arbitrary body (or of the middle surface of a shell). Similarly to the difference of the squares $ds'^2 - ds^2$, which constitutes the difference $I' - I$ of the first fundamental forms of two surfaces, we now consider the difference of the normal curvatures, i.e., the reciprocals of the curvature radii $R$ and $R'$ in an arbitrary direction with the unit vectors $e$ and its corresponding direction $e'$ parallel to $dr$ and $dr'$, respectively:

$$\frac{1}{R'} - \frac{1}{R} = e' \cdot B'e' - e \cdot Be = \frac{dr' \cdot B'dr'}{ds'^2} - \frac{dr \cdot B \, dr}{ds^2} = \frac{II'}{I'} - \frac{II}{I} \ . \tag{2.156}$$

From this difference arise the first and second fundamental forms I and II, respectively. In order to relate the curvature tensor $B' = - \nabla'_n \otimes n'$ to the undeformed state, we must refer to the preceeding transformations $\nabla = F^T \nabla'$ and $NF^T N' = NF^T$ and derive the expression (2.155) of the unit vector $n'$ by using the derivative rules. Thus,

$$- \nabla_n \otimes n' = - \nabla_n \otimes (n - \omega + \tfrac{1}{2}\Omega E\omega - \tfrac{1}{2}\omega^2 n)$$

$$= B - \kappa + B\omega \otimes n - \tfrac{1}{2}(\nabla_n\Omega \otimes E\omega + \Omega B E\omega \otimes n$$

$$+ \Omega\kappa E + 2\kappa\omega \otimes n + \omega^2 B) \ ,$$

where we have used the abbreviation $- (\nabla_n \otimes \omega)N = \kappa$. The operator $\nabla_n'$ is transformed consistently with (2.128) and with a mixed projection:

$$\nabla_n' = N'(F^{-1})^T \nabla = N'(F^{-1})^T N\nabla$$

$$= (N - \gamma - \Omega E - \omega \otimes \omega - \Omega^2 N + n \otimes \omega + \tfrac{1}{2}n \otimes \Omega E\omega) \nabla_n \ .$$

Finally, by referring to the simultaneous symmetry of $B'$ and $B$, we get the expression for the curvature tensor $B'$:

$$B' = B - \tfrac{1}{2}[\kappa + \kappa^T - \Omega(BE - EB) + (B\gamma + \gamma B) + n \otimes B\omega + B\omega \otimes n]$$

$$- \tfrac{1}{2}\{2\Omega^2 B + \omega^2 B + \omega \otimes B\omega + B\omega \otimes \omega - \Omega(E\kappa - \kappa^T E)$$

$$+ \tfrac{1}{2}[\Omega(\kappa E - E\kappa^T) + (\nabla_n\Omega \otimes E\omega + E\omega \otimes \nabla_n\Omega)] + (\kappa + \kappa^T)\omega \otimes n$$

$$+ n \otimes \omega(\kappa + \kappa^T) + \Omega E B\omega \otimes n + n \otimes \Omega E B\omega\} + (\omega \cdot B\omega)n \otimes n \ . \qquad (2.157)$$

We need, however, use only the part $NF^T B'FN$ in the second fundamental form in (2.156). Once again using (2.152) and its transpose, we find

$$NF^T B'FN = B - \tfrac{1}{2}\{(\kappa + \kappa^T) + \Omega(BE - EB) - (B\gamma + \gamma B) - \omega \otimes B\omega$$

$$- B\omega \otimes \omega + \omega^2 B + \tfrac{1}{2}[\nabla_n\Omega \otimes E\omega + E\omega \otimes \nabla_n\Omega - \Omega(\kappa E - E\kappa^T)]\} \ . \qquad (2.158)$$

On the other hand, the first fundamental form in the denominator of (2.156) reads:

$$(ds')^2 = dr' \cdot dr' = ds\, e \cdot NF^T FN e\, ds \simeq (1 + 2e \cdot \tilde{\gamma}e)ds^2 \ . \qquad (2.159)$$

Therefore, the difference of curvature, described in (2.156) with (2.150, 158 and 159), becomes

$$\frac{1}{R'} - \frac{1}{R} = \left(\frac{ds}{ds'}\right)^2 e \cdot NF^T B'FN e - e \cdot Be$$

$$= -\tfrac{1}{2}e \cdot \{\kappa + \kappa^T + \Omega(BE - EB) - (B\tilde{\gamma} + \tilde{\gamma}B)$$

$$- \tfrac{1}{2}(\omega \otimes B\omega + B\omega \otimes \omega) + \omega^2 B + \Omega^2 B$$

$$+ \tfrac{1}{2}[\nabla_n\Omega \otimes E\omega + E\omega \otimes \nabla_n\Omega - \Omega(\kappa E - E\kappa^T)]\}e$$

$$- 2(e \cdot \tilde{\gamma}e)(e \cdot Be) \ , \qquad (2.160)$$

or in abbreviated form

$$\frac{1}{R'} - \frac{1}{R} = -e \cdot \tilde{\kappa}_c e - 2(e \cdot \tilde{\gamma}e)(e \cdot Be) . \tag{2.161}$$

We are now also able to define a symmetric tensor or so-called "reduced curvature", which was introduced in the case of linear shell theory by *Sanders* [2.28] and *Koiter* [2.29], and generalized as follows [2.32]:

$$\tilde{\kappa} = \tfrac{1}{2}\{\kappa + \kappa^{\mathrm{T}} + \Omega(BE - EB) - \tfrac{1}{2}(\omega \otimes B\omega + B\omega \otimes \omega) + \Omega^2 B$$
$$+ \omega^2 B + \tfrac{1}{2}[\nabla_n\Omega \otimes E\omega + E\omega \otimes \nabla_n\Omega - \Omega(\kappa E - E\kappa^{\mathrm{T}})]\} . \tag{2.162}$$

As this summarizes all terms in (2.160), except those in the strain $\tilde{\gamma}$, it describes the exact curvature change independently of the rotation, even though it is expressed by $\omega$, $\Omega$, and its derivatives alone. The tensors $\tilde{\kappa}$ and $\tilde{\kappa}_c$ vanish in the case in which the surface is neither strained nor bent.

Let us look further at a parallel surface near the free surface of a body where the normal $n$ is a principal direction of $\tilde{E}$. For this parallel surface, the displacement vector of a point $\bar{P}$ is $\bar{u} = \bar{r}' - \bar{r} = u + z(n' - n)$, where $n'$ is given by (2.155) (Fig. 2.25). During the deformation, parallel surfaces remain roughly parallel, and the distance between them does not change much. Consequently we may refer to transformations (2.55, 57) in order to define the operator $\bar{\nabla}_n$ on the parallel surface

$$\bar{\nabla}_n = (N + Bz + B^2z^2 + \ldots)\nabla_n . \tag{2.163}$$

The surface strain tensor in $\bar{P}$ then reads with (2.133) and $\bar{N} \equiv N$ as

$$\tilde{\bar{\gamma}} = \tfrac{1}{2}N[\bar{\nabla} \otimes \bar{u} + (\bar{\nabla} \otimes \bar{u})^{\mathrm{T}} + (\bar{\nabla} \otimes \bar{u})(\bar{\nabla} \otimes \bar{u})^{\mathrm{T}}]N . \tag{2.164}$$

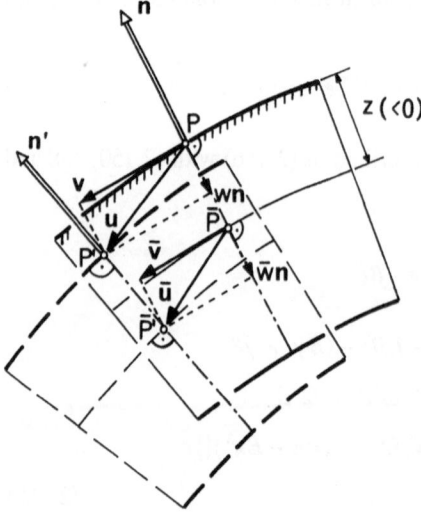

**Fig. 2.25.** Deformation of a thin shell element when the normality hypothesis is fulfilled

Neglecting the terms in $z^2$, the displacement gradient $\overline{\nabla}_n \otimes \bar{u}$ on the parallel surface becomes

$$\overline{\nabla}_n \otimes \bar{u} = \gamma + \Omega E + \omega \otimes n + \tfrac{1}{2}\Omega E\omega \otimes n + z[B\gamma + \Omega BE + \kappa$$
$$+ \kappa\omega \otimes n + \Omega BE\omega \otimes n + \tfrac{1}{2}(\nabla_n\Omega \otimes E\omega + \Omega\kappa E + \omega^2 B)] \ .$$

Thus, $\bar{\bar{\gamma}}$ may be written with the tensors $\bar{\kappa}$ and $\tilde{\gamma}$ defined by (2.162 and 150), respectively, [2.33] on the form

$$\bar{\bar{\gamma}} = \tilde{\gamma} + z\bar{\kappa} + \tfrac{1}{2}z(B\tilde{\gamma} + \tilde{\gamma}B) \ . \tag{2.165}$$

However, the order of magnitude of the strain $\bar{\bar{\gamma}}$ is still determined by the factor $\eta$, whereas the rotations $\omega$ and $\Omega$ were assumed to be of order $\chi = O(\sqrt{\eta})$. Therefore, the physical meaning of $\chi$ will be determined by the fact that all the terms are of order $\eta$ in (2.165). In particular, the linear part $z[\kappa + \kappa^T + \Omega(BE - EB)]/2$ of (2.165) must also be of order $\eta$. The terms $z\kappa$ and $z\Omega BE$ are both of order $h\chi/L = O(\eta)$ and $h\chi/R = O(\eta)$, respectively, where $L$ is the smallest wavelength of deformation patterns, $R$ is the smallest radius of curvature, and $h$ is the largest possible distance from the free surface. Since $\eta = O(\chi^2)$, we conclude that the small parameter $\chi$ describing the rotation can be defined by either $h/L$ or $h/R$.

Thus the body must be limited to not far away from the free surface $z = 0$. For instance, if $z = h$ is another (parallel) free surface, we have a thin plate or shell of thickness $h = L O(\chi)$ or $h = R O(\chi)$. In fact, we join the geometrically nonlinear theory of thin plates and shells, for which there exists a classification on the order of magnitude for the rotations, with a restriction of small strains everywhere in the shell [2.34]. We assume here that the so-called normality hypothesis is still valid. In a refined treatment, this condition should be replaced by an asymptotic analysis [2.35–37].

# 3. Elements of Holography and Image Modification

As was already pointed out in the introduction, holography has become such an extended field of optics that it is impossible to thoroughly treat the subject within such a limited space. For a complete theoretical treatment and information about equipment and material, other books such as [3.1–40] should be consulted.

Here, we first look briefly at the basic principles, the formation of the image of a point source, its aberration caused by a change of the optical set-up at the reconstruction, and some aspects of coupled-wave theory. Secondly, we investigate the corresponding deformation of a whole image in space, followed by a consideration of small optical modifications.

## 3.1 Formation and Aberration of the Image at the Reconstruction. Coupled-Wave Theory

The basic principles of holography concern the recording of an object by an interference pattern and reconstruction of the image in the relatively simple case in which the optical set-up does not change. This may be called standard holography and will be presented in the first section. Thereafter, we consider the more complicated problem of aberrations, but, in particular, those created from a point source. We will consider cases of optical modifications at the reconstruction for thin holograms. In the last section, we consider the coupled-wave theory, describing the diffraction in volume holograms. The basic three-dimensional equations are deduced in convenient form by a small wavelength approach.

### 3.1.1 Image Formation in Standard Holography

Beginning with the main subject of this book, we consider the optical arrangement illustrated in Fig. 3.1a. A beam of light emitted by a *laser* L is divided into two parts by a beam splitter BS. Each narrow ray bundle is then expanded by a lens of very short focal distance and passes through a pinhole which approximates a point source. One of the resulting spherical waves is directly guided onto a photographic plate and is called the *reference wave*. Its complex amplitude, considered at any point H of the plate with coordinate vector $\vec{r}$ and expressed relative to the source Q, reads

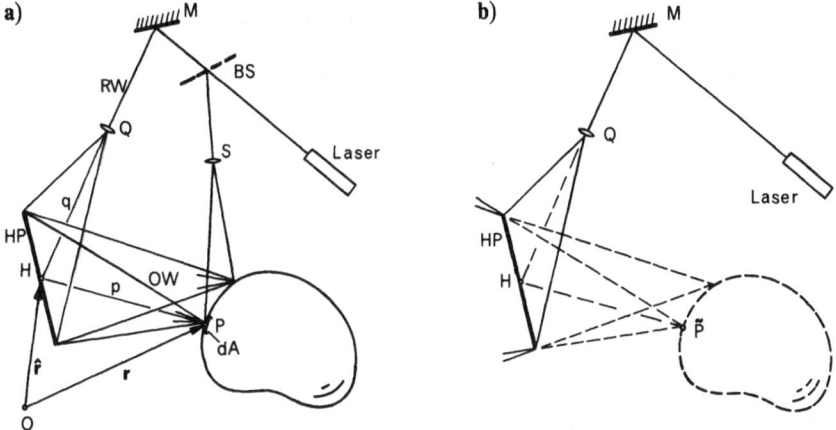

**Fig. 3.1. (a)** Recording of an object. (BS: beam splitter; HP: hologram plate; RW: reference wave with point source Q; OW: object wave with point source S; P: object point with surface element $dA$; H: hologram point with position vector $\hat{r}$; $q$, $p$: distances from Q, P to H, respectively). **(b)** Reconstruction of an ideal virtual image of the object by a wave identical to the reference wave (standard holography)

$$V(\hat{r}) = \frac{A_Q}{q} \exp\left(\frac{2\pi i}{\lambda} q\right) , \qquad (3.1)$$

where $A_Q = |A_Q| \exp(i\psi)$ is a complex constant containing the phase $\psi$ of the reference source Q, and where $q(\hat{r})$ denotes the variable distance from Q to H. The other wave is reflected by the surface of an opaque object O and forms the so-called *object wave*. Each surface point P with the position vector $r$ acts then as a secondary point source of a spherical wave. The total complex amplitude of the object wave at H may be written in the form

$$U(\hat{r}) = \iint_A \frac{\tilde{a}_P}{p} \exp\left(\frac{2\pi i}{\lambda} p\right) dA , \qquad (3.2)$$

where $\tilde{a}_P(r, \ldots)$ is a specific complex factor, which is dependent on $r$ and on the inclination of the rays at P, as in (2.77). Further, $p$ denotes the distance from P to H, $A$ is the illuminated part of the object surface, and $dA = |a_1 \times a_2| d\theta^1 d\theta^2$ is the representation of the surface element by means of the curvilinear coordinates $\theta^1$, $\theta^2$ and tangential base vectors $a_1$, $a_2$ (Sect. 2.1). The intensity produced by superposition of these two fields is

$$I(\hat{r}) = \tfrac{1}{2}(U + V)(U + V)^* = \tfrac{1}{2}(|U|^2 + |V|^2 + UV^* + U^*V) , \qquad (3.3)$$

where the asterisk denotes the complex conjugate. If the photographic plate is exposed during the time interval $\tau$, the energy received is $E = \tau I$. After development, the plate has the *amplitude transmittance* $T = f(E)$, where $f$ represents some characteristic function (Fig. 3.2). In addition, the arrangement should be placed so as to keep $E$ within a small interval. We may then linearize $f(E)$ around a mean value, preferably in the vicinity of the point of inflection on the graph in Fig. 3.2. This can be made over the whole plate, if

$$|V| \sim \text{const} \quad \text{and} \quad |U| \ll |V| \;,$$

so that, with

$$\beta = - dT/dE|_{E = E_0} \;, \qquad T_0 = \frac{\tau}{2} |V|^2_{E = E_0} \;,$$

we may write

$$T \cong T_0 - \frac{\beta\tau}{2} (UV^* + U^*V) \;. \tag{3.4}$$

In the special case where the "object" is a single point source $P$, we have $U = (|A_P|/p) \exp[2\pi i p/\lambda + i\phi]$; together with (3.1), here we explicitly obtain:

$$T = T_0 - \frac{\beta\tau}{2} \left\{ \frac{A_Q^* A_P}{qp} \exp\left[ \frac{2\pi i}{\lambda} (p - q) \right] + \frac{A_Q A_P^*}{qp} \exp\left[ \frac{2\pi i}{\lambda} (q - p) \right] \right\}$$

$$= T_0 - \beta\tau \frac{|A_Q| |A_P|}{qp} \cos\left[ \frac{2\pi}{\lambda} (p - q) + (\phi - \psi) \right] \;. \tag{3.5}$$

In this simple case, we see that the variation of the relative phase from point to point on the photographic plate is tranformed into an interference pattern retaining the complete information of the object wave. The plate is therefore called a *hologram* and the entire process is the *recording*.

In order to produce a visible image of the object, which constitutes the *reconstruction* (Fig. 3.1b), let us illuminate the hologram by a wave $\tilde{V}$. The complex amplitude of the transmitted wave just behind the hologram is then described by

$$\tilde{V}T = \tilde{V}T_0 - \frac{\beta\tau}{2} (UV^*\tilde{V} + U^*V\tilde{V}) \;. \tag{3.6}$$

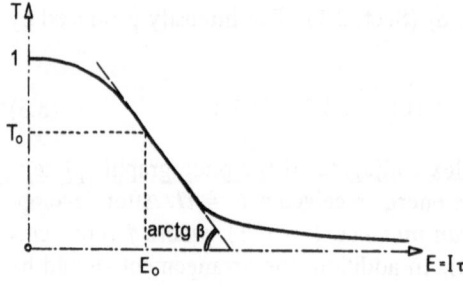

**Fig. 3.2.** Amplitude transmittance $T$ of the photographic plate as a function of the received energy $E$

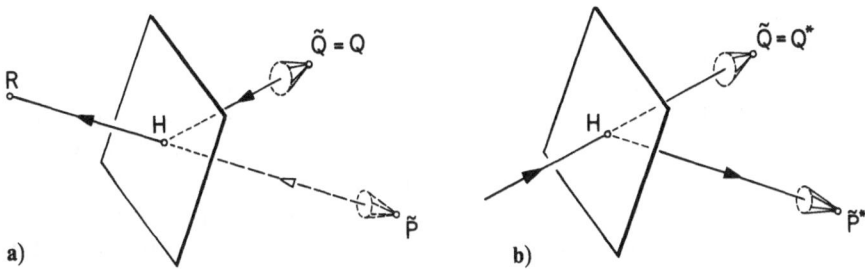

**Fig. 3.3. (a)** Reconstruction of an ideal virtual image point $\check{P}$. R: center of the observing instrument. **(b)** Reconstruction of an ideal real image point $\check{P}^*$

The term $\tilde{V}T_0$ in (3.6) is the part of the light which passes directly through the hologram without any direction changes of the rays. The remaining two terms represent the diffracted parts; $-(\beta\tau/2)UV^*\tilde{V}$ is the *primary-image* wave field, whereas $-(\beta\tau/2)U^*V\tilde{V}$ is the *conjugate-image* wave field.

Two special cases are of interest:

a) The reconstruction wave $\tilde{V}$ is identical to the reference wave $V$: $\tilde{V} \equiv V$. The primary wave field is $-(\beta\tau/2)|V|^2 U$, i.e., except for a constant real factor it is equal to the object wave $U$. The observer, therefore, sees a *virtual image* of the object as if the hologram was a window (Fig. 3.3a).

b) The reconstructing wave $\tilde{V}$ is identical to the conjugate of the reference wave $V$: $\tilde{V} \equiv V^*$. The conjugate wave field here is $-(\beta\tau/2)|V|^2 U^*$. Since, as far as the corresponding time dependent waves are concerned, we have

$$\mathrm{Re}\{U^* \exp(-i\omega t)\} = \mathrm{Re}\{U\exp(i\omega t)\}^* \ ,$$

we conclude that this wave field with reverse travelling light reconstructs a *real image* of the object in the same position (Fig. 3.3b).

The above description illustrates only one among many possible recording arrangements. If the two interfering wave fields are travelling in almost the same direction, one has a Gabor or in-line hologram [3.41–43]. Contrary to this, Fig. 3.3 represents the case of a Leith-Upatnieks or off-axis hologram [3.44–46]. The two interfering waves may also come from different sides, resulting in a Lippmann or reflection hologram [3.47–49]. Moreover, one can distinguish between Fresnel and Fraunhofer holograms depending on the near or far position of the object. If a lens is used to produce a far-field pattern at the recording plate, with the lens positioned halfway between, one gets a so-called Fourier hologram. We must then also discriminate between thin (plane) holograms and thick (volume) holograms. Finally, one should differentiate between amplitude holograms, as in the above case, and phase holograms, where a phase modulation is imposed on the wave. For this classification, the reader is referred to [3.30, 33, 36, 50]. The curve in Fig. 3.2 depends, in fact, on the type of recording materials used, which are described in detail in [3.50] and also in [3.51–67]. Nonlinearity and polarization effects may be found in [3.68–83].

There are only the above two cases (a), (b) in which the image configuration is identical to the object configuration: (1) when a virtual image is produced with the primary wave field ($\tilde{V} \equiv V$), and (2) when a real image is produced with the conjugate wave field ($\tilde{V} \equiv V^*$). In any other case, the reconstructed wave field differs from the recording wave field, causing a distortion (or deformation) and aberrations of the image. Such *modifications* will take up a major part in this book; they also play a certain role in holographic interferometry. We shall mainly assume the use of *thin Fresnel holograms*, since their recording geometry best permits a three-dimensional investigation of the image and the possibility of modifications. This hologram type therefore offers the greatest flexibility in making precise measurements. In contrast to this "exterior" holography, we shall also mention the theory describing the wave formation in the "interior" of the hologram. Section 3.1.3 will be devoted to the diffraction processes in thick holograms considered in the coupled-wave theory.

### 3.1.2 Image Formation of a Point Source for a Modification at the Reconstruction

We now intend to study the formation of the image of a point source P, when the optical arrangement at the reconstruction differs from that at the recording. The first and simplest example describes the use of different wavelengths: $\lambda$ at the recording and $\tilde{\lambda}$ at the reconstruction (by use of another laser at the recording) [3.41–43, 84–94]. In another case, one could envision a geometrical modification, e.g., when the position of the source $\tilde{Q}$ differs from that of Q at the recording [3.84, 87–89, 95–97]. At this stage, we assume that the hologram remains fixed and is developed without any distortion. The actual modified, non-spherical wavefront is then described by the primary part of (3.6) (Fig. 3.4a):

$$\tilde{U} = -\frac{\beta\tau}{2}\tilde{V}V^*U = -\frac{\beta\tau}{2}\frac{A_{\tilde{Q}}A_Q^*A_P}{\tilde{q}qp} \ \exp\left[2\pi i\left(\frac{\tilde{q}}{\tilde{\lambda}} + \frac{p}{\lambda} - \frac{q}{\lambda}\right)\right] . \qquad (3.7)$$

If we provisionally assume the existence of a virtual image point $\tilde{P}$ of P, we must then compare this wavefront with a virtual spherical wavefront

$$\tilde{U}_{\tilde{p}} = \frac{A_{\tilde{P}}}{\tilde{p}} \ \exp\left(2\pi i\frac{\tilde{p}}{\tilde{\lambda}}\right) , \qquad (3.8)$$

where $A_{\tilde{P}} = |A_{\tilde{P}}| \exp(i\tilde{\phi})$ is the complex amplitude with the phase $\tilde{\phi}$ at the expected point $\tilde{P}$, and where $\tilde{p}$ denotes the distance $\tilde{P}H$ to the hologram point H [3.84, 98–101]. With this comparison, we reason similarly as with the calculus of the geometrical aberration theory of rotationally symmetrical optical systems [3.102, 103] and [Ref. 3.104, p. 102]. We return to this problem later when

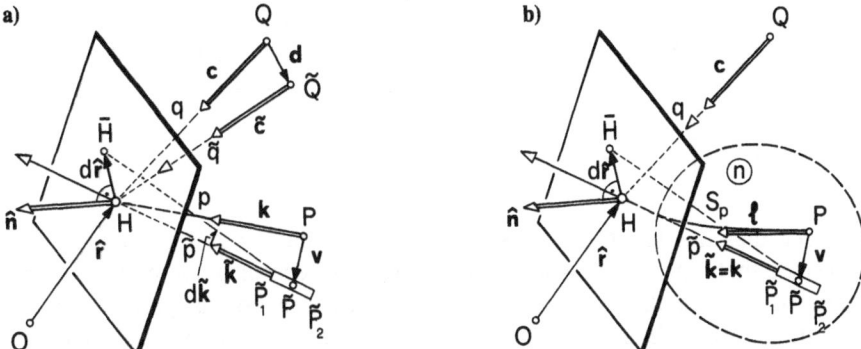

**Fig. 3.4.** (a) Set of images {P̌} of an object point source P when the reconstruction source Q̌ differs from the reference source Q (optical modification). (p̌, p, q̌, q: corresponding distances of P̌, P, Q̌ and Q to H; ǩ, k, č, c: associated unit vectors; n̂: unit normal of the hologram surface). (b) Set of images {P̌} of an object point source P which, at the recording, was immersed in a nonhomogeneous medium with spatially varying index of refraction. (l: unit vector tangential to the curved ray through P)

dealing with higher-order aberration terms. In particular, the phase terms of (3.7, 8) should be related at the point H; for example, with (3.5), we write

$$2\pi \left[ \frac{1}{\tilde{\lambda}} \, (\tilde{p} - \tilde{q}) - \frac{1}{\lambda} \, (p - q) \right] = (\phi - \psi) - (\tilde{\phi} - \tilde{\psi}) \, . \tag{3.9}$$

This equation relates the diffraction at H during the reconstruction with the interference at this same point at the recording. In other words, we may say that the "interference" produced by Q̌ and P̌ is the same as that produced by Q and P. This important equation will, therefore, be named the *interference identity*, defining the phase φ̃.

For a given hologram surface, of course, it is not possible for (3.9) to hold for any point H in the hologram for a fixed point P̌ and at constant values of φ, ψ, φ̃, ψ̃, since the phase difference

$$\theta(\hat{r}) = 2\pi \left[ \frac{1}{\tilde{\lambda}} \, (\tilde{p} - \tilde{q}) - \frac{1}{\lambda} \, (p - q) \right] \tag{3.10}$$

is, in general, a function of the position vector r̂ of H [3.84, 105–107]. In case of a modification, this means that the different zones of the hologram do not reconstruct the same single image point P̌. Let us add that the same situation occurs if, at the recording, the point P is immersed in a non-homogeneous medium with a variable index of refraction (Fig. 3.4b). We then must replace (3.7) by

$$\tilde{U} = - \frac{\beta \tau}{2} \, \frac{A_{\tilde{Q}} A_Q^*}{\tilde{q}q} a_p \, \exp \left[ 2\pi i \left( \frac{\tilde{q}}{\tilde{\lambda}} + \frac{S_p}{\lambda_0} - \frac{q}{\lambda} \right) \right] \, , \tag{3.11}$$

with the eikonal $S_p$ considered (at H) and with $\lambda_0$ being the wavelength in vacuum. The complex $a_p$ could be found from the transport equation, see the term with factor $2\pi i/\lambda_0$ in (2.81). Similarly, (3.10) is replaced by

$$\theta(\hat{r}) = 2\pi \left[ \frac{1}{\tilde{\lambda}} (\tilde{p} - \tilde{q}) - \frac{S_p}{\lambda_0} + \frac{q}{\lambda} \right] . \tag{3.12}$$

We see that this function varies even if $\tilde{q} = q$ and $\tilde{\lambda} = \lambda$. Consequently, we conclude that a non-homogeneous medium at the recording also creates aberrations in the standard case, because the reconstruction is performed in a homogeneous medium [3.84, 108–115]. Although no single point $\tilde{P}$ exists, we may now wonder if at least a small region $\{\overline{H}\}$ around a *given* point H roughly reproduces such an image "point" (Fig. 3.4a).

The observer or the observation system is always provided with a small diaphragm which lets a narrow bundle of rays pass. Thus, we develop $\theta(\hat{r})$ into a Taylor series around H. With $d^2\hat{r} = 0$, $d^3\hat{r} = 0,\dots$ ($\hat{r}$ is the independent variable), we have

$$\bar{\theta} = \theta + d\theta + \frac{1}{2!}d^2\theta + \frac{1}{3!}d^3\theta + \frac{1}{4!}d^4\theta + \dots$$

$$= \theta + d\hat{r} \cdot \nabla_{\!\hat{r}}\theta + \frac{1}{2}d\hat{r} \cdot (\nabla_{\!\hat{r}} \otimes \nabla_{\!\hat{r}}\theta)d\hat{r} + \dots . \tag{3.13}$$

Now, a condition needed to satisfy (3.9) approximately in $\overline{H}$ is given by the stationary behavior of $\bar{\theta}$, which primarily requires the disappearance of the first order term $d\hat{r} \cdot \nabla_{\!\hat{r}}\theta$ for any increment $d\hat{r}$. This leads to

$$\nabla_{\!\hat{r}}\theta = \hat{N}\nabla\theta = 0 . \tag{3.14}$$

The derivatives of the distances $\tilde{p}, \tilde{q}, p, q$ from H to the fixed points $\tilde{P}, \tilde{Q}, P, Q$ all lead to unit vectors: $\nabla\tilde{p} = \tilde{k}$, $\nabla\tilde{q} = \tilde{c}$, $\nabla p = k$, $\nabla q = c$, see (2.23). However, we also have $\nabla S_p = nk$ at point H, see (2.83) and Fig. 3.4b. Therefore, in both homogeneous and non-homogeneous media, (3.14) becomes

$$\hat{N}\left[ \frac{1}{\tilde{\lambda}} (\tilde{k} - \tilde{c}) - \frac{1}{\lambda} (k-c) \right] = 0 . \tag{3.15}$$

For any given vectors $c$, $k$ and $\tilde{c}$, (3.15), together with the auxiliary condition

$$|\tilde{k}| = 1 ,$$

determines the unknown direction $\tilde{k}$ to the expected image $\tilde{P}$. The supplementary condition $|\hat{N}\tilde{k}| < 1$ determines the existence of this image $\tilde{P}$. If this condition is not fulfilled, the reconstructed wave is an evanescent one which propagates along the hologram [3.100, 101, 118, 126–128]. When the unit vectors $k$, $c$, $\tilde{c}$, and $\hat{n}$ lie

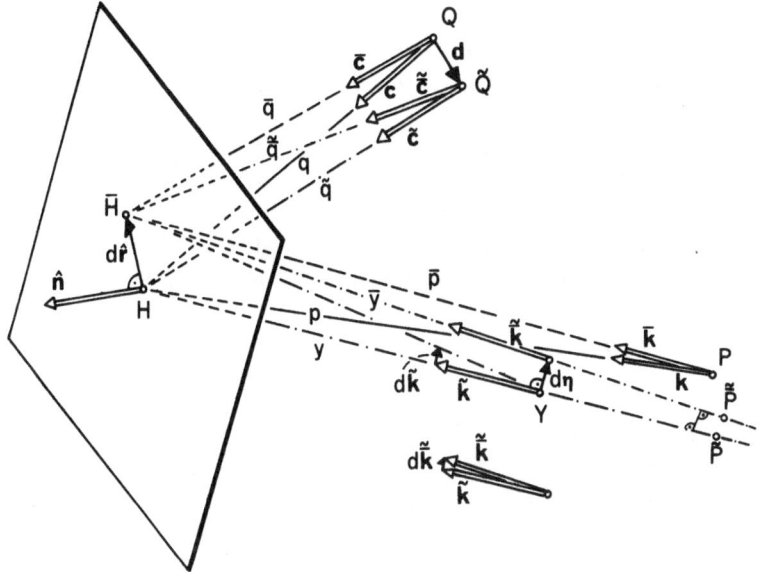

**Fig. 3.5.** Transverse ray aberration $d\eta$ between skewed image forming rays which pass through neighboring points H, $\tilde{H}$

in the same plane, (3.15) is only one-dimensional and can be advantageously expressed by a sinus relation as in ([Ref. 3.105, Eq. (11)], [Ref. 3.116, Eq. (13)], [Ref. 3.117, Eq. (12)]). In the general case, however, (3.15) [see also (3.31, 107)] is two-dimensional and has already been given alternatively by two trigonometric relations ([Ref. 3.100, Eq. (1.13)], [Ref. 3.101, Eq. (16)], [Ref. 3.106, Eq. (5.2)], [Ref. 3.118, Eqs. (9) and (10)], [Ref. 3.119, Eqs. (2) and (3)], [3.120–122]) or by the vector product ([Ref. 3.123, Eq. (2)], [3.124, 125]).

Note also that (3.15) looks like (2.94). Therefore, the following reasoning will be on a similar line of thought as in Sect. 2.2.2. However, here we shall give a somewhat more detailed version. In fact, in order to find information about the still unknown distance $\tilde{p}$ from the image $\tilde{P}$ to H, two possibilities arise: first, we may reinforce the condition of stationary behavior of $\bar{\theta}$ to the second order term in (3.13) and write

$$d^2\theta = d\hat{r} \cdot (\nabla_{\hat{n}} \otimes \nabla_{\hat{n}}\theta)d\hat{r} = 0 \ , \tag{3.16}$$

which will keep us closer to the concepts of *wave aberration* concerning the phase difference function $\theta$ for a fixed point $\tilde{P}$ (Fig. 3.5); secondly, we can alternatively apply (3.14 or 15) to a neighboring point $\tilde{H}$ [as in (2.95)] and write $\nabla_{\hat{n}}\tilde{\theta} = 0$, i.e.,

$$\hat{N}\left[\frac{1}{\tilde{\lambda}}(\tilde{\tilde{k}} - \tilde{c}) - \frac{1}{\lambda}(\bar{k}-\bar{c})\right] = 0 \ . \tag{3.17}$$

This leads us to the concept of *transverse ray aberration* [3.106, 129, 130], because now a new skewed ray $\tilde{P}\overline{H}$ of direction $\bar{\bar{k}}$ is involved ($\tilde{P}$ differs from $\bar{P}$), coming from another image $\tilde{P}$ (Fig. 3.5). That is why $\bar{\bar{\theta}}$, relative to $\tilde{P}\overline{H}$, must be distinguished from $\bar{\theta}$, relative to $\bar{P}\overline{H}$.

Before going into any detailed calculus, we can first view both concepts with a generality which shows that both concepts, in fact, describe the same phenomenon. Let us choose a point Y on the reference ray at a distance $y$ from H. For the second-order tensor, resulting from the second derivative of (3.10), but now referred to the arbitrary point Y, we have: ($p/\lambda$ could also be $S_p/\lambda_0$)

$$\nabla_{\hat{n}} \otimes \nabla_{\hat{n}}\theta_Y = \frac{2\pi}{\tilde{\lambda} y} \hat{N}\tilde{K}\hat{N} - 2\pi \nabla_{\hat{n}} \otimes \nabla_{\hat{n}}\left[\frac{1}{\tilde{\lambda}}\tilde{q} + \frac{1}{\lambda}(p-q)\right] , \tag{3.18}$$

since $\hat{N}$ is a constant projector. Here, $\nabla \otimes \bar{k} = \tilde{K}/y$ is a normal projector relative to $\bar{k}$, see (2.24). On the other hand, for the direction change of the skewed rays through H and $\overline{H}$ with $d\bar{k} = \bar{\bar{k}} - \bar{k}$ and (3.18), we obtain by combining (3.15 and 17) that

$$\hat{N} d\bar{k} = \tilde{\lambda} \, d\hat{r} \, \nabla_{\hat{n}} \otimes \nabla_{\hat{n}}\left[\frac{1}{\tilde{\lambda}}\tilde{q} + \frac{1}{\lambda}(p-q)\right]$$

$$= \frac{1}{y} d\hat{r} \, \tilde{K}\hat{N} - \frac{\tilde{\lambda}}{2\pi} d\hat{r}(\nabla_{\hat{n}} \otimes \nabla_{\hat{n}}\theta_Y) . \tag{3.19}$$

Further, let us introduce the first-order transverse ray aberration $d\eta$ perpendicular to $\bar{k}$ at Y, which is (Fig. 3.5):

$$d\eta = \tilde{K} \, d\hat{r} - y \, d\bar{k} . \tag{3.20}$$

Contracting then (3.19 and 20) with $d\hat{r}$ and eliminating the term $d\hat{r} \cdot \tilde{K} d\hat{r}$, we find with (3.9, 10) the illustrative relation

$$d^2\{\theta_Y - [(\phi - \psi) - (\tilde{\phi}_Y - \tilde{\psi})]\} = d^2\theta_Y = \frac{2\pi}{\tilde{\lambda} y} d\hat{r} \cdot d\eta \tag{3.21}$$

connecting the second-order wave aberration $d^2\theta_Y$ and the first-order transverse ray aberration $d\eta$. In particular, if $y = \tilde{p}$ ($Y \equiv \tilde{P}$), we have, by definition (3.16): $d^2\theta_{\tilde{p}} = 0$, or

$$d\hat{r} \cdot d\eta_{\tilde{p}} = d\eta_{\tilde{p}} \cdot \tilde{K} \, d\hat{r} = 0 . \tag{3.22}$$

If $|d\eta_{\tilde{p}}| \ll |\tilde{K} \, d\hat{r}|$, (i.e., if the two rays are only slightly skew), the approximation $d\eta_{\tilde{p}} \cdot \tilde{K} \, d\hat{r} \cong d\eta_{\tilde{p}} \cdot \tilde{K}(d\hat{r} - d\eta_{\tilde{p}}) \cong 0$ shows that $\tilde{P}$ is very close to the *shortest distance* that separates them (Fig. 3.5).

Conversely, we may calculate $d\eta$ in Y by means of the affine connection

$$d\hat{r} = y\tilde{M}^T d\tilde{k} , \tag{3.23}$$

where $\tilde{M} = I - \hat{n} \otimes \tilde{k}/\hat{n} \cdot \tilde{k}$ is an oblique projector, but where $d\tilde{k} = \bar{\tilde{k}} - \tilde{k}$ should not be confused with $d\tilde{k} = \tilde{\bar{k}} - \tilde{k}$. Also, (3.19) with $\tilde{M}\tilde{K} = \tilde{K}$, $\tilde{M}\tilde{N} = \tilde{M}$ and $d\bar{\tilde{k}}\tilde{M}^{\mathrm{T}} = d\bar{\tilde{k}}$, becomes

$$d\bar{\tilde{k}} = \tilde{\lambda} y \, d\tilde{k}\tilde{M}\left[\nabla \otimes \nabla\left(\frac{\tilde{q}}{\tilde{\lambda}} + \frac{p}{\lambda} - \frac{q}{\lambda}\right)\right]\tilde{M}^{\mathrm{T}} . \tag{3.24}$$

The characteristic tensor

$$\tilde{T} = \tilde{\lambda}\tilde{M}\left[\nabla \otimes \nabla\left(\frac{\tilde{q}}{\tilde{\lambda}} + \frac{p}{\lambda} - \frac{q}{\lambda}\right)\right]\tilde{M}^{\mathrm{T}} \tag{3.25}$$

is symmetric, lies in the plane normal to $\tilde{k}$, and should be considered at H, since it does not depend on $y$. With this definition of the tensor $\tilde{T}$, (3.24) reads

$$d\bar{\tilde{k}} = y \, d\tilde{k}\tilde{T} , \tag{3.26}$$

so that (3.20), again with (3.23), becomes

$$d\eta = y \, d\tilde{k}(\tilde{K} - y\tilde{T}) = \frac{\tilde{\lambda} y^2}{2\pi} d\tilde{k} \, \tilde{M}(\nabla \otimes \nabla\theta_Y)\tilde{M}^{\mathrm{T}} . \tag{3.27}$$

Furthermore, regarding $d\eta$ as an increment of the wavefront, such as $d\zeta$ in (2.104, 105) (Fig. 2.16), we obtain

$$d\bar{\tilde{k}} = d\eta(\nabla_{\tilde{k}}^* \otimes \tilde{k}) = y \, d\tilde{k}(\tilde{K} - y\tilde{T})(\nabla_{\tilde{k}}^* \otimes \tilde{k}) , \tag{3.28}$$

where $\nabla_{\tilde{k}}^*$ is the two-dimensional derivative operator on this surface, and $-\nabla_{\tilde{k}}^* \otimes \tilde{k} = -\tilde{T}_Y$ is the tensor of curvature. A comparison of (3.26 and 28) allows us to write

$$\tilde{T} = (\tilde{K} - y\tilde{T})\tilde{T}_Y , \qquad \tilde{T} = \tilde{T}_Y|_{y=0} . \tag{3.29}$$

This result is similar to (2.106, 107) and provides the physical meaning of the tensor $\tilde{T}$. Indeed, $-\tilde{T}$ represents the tensor of curvature of the wavefront at H. The ensemble of points $\{\tilde{P}\}$ are the centers of curvature for each normal section of direction $\tilde{m} = d\tilde{k}/|d\tilde{k}|$. For the distance $\tilde{p}$, we have

$$\frac{1}{\tilde{p}} = \tilde{m} \cdot \tilde{T}\tilde{m} . \tag{3.30}$$

Both, the maximum and minimum values $1/\tilde{p}_1$ and $1/\tilde{p}_2$ are the eigenvalues of $\tilde{T}$, see [3.31, 107]. When the reference source Q, the reconstruction source $\tilde{Q}$, the hologram point H and the object point P lie in a plane including the hologram unit normal $\hat{n}$, (3.30) can then be alternatively expressed with cosines as already done in references ([Ref. 3.105, Eq. (12)], [Ref. 3.116, Eq. (13)]).

In the homogeneous case, the tensor $\tilde{T}$ may be given explicitely. As the second derivatives of the three lengths $\tilde{q}$, $p$, and $q$ are reduced projectors $\nabla \otimes \nabla \tilde{q} = \tilde{C}/\tilde{q}$,   $\nabla \otimes \nabla p = K/p$,   $\nabla \otimes \nabla q = C/q$,   respectively,   and   where $\tilde{C} = I - \tilde{c} \otimes \tilde{c}$, see (2.24), we simply get

$$\tilde{T} = \tilde{\lambda}\tilde{M}\left(\frac{1}{\tilde{\lambda}\tilde{q}}\tilde{C} + \frac{1}{\lambda p}K - \frac{1}{\lambda q}C\right)\tilde{M}^{\mathrm{T}} . \tag{3.31}$$

If all the direction unit vectors coincide, e.g., $\tilde{c} = c = \tilde{k} = k$, (3.15) is fulfilled and (3.30) gives a unique value for a stigmatic image point $\tilde{P}$

$$\frac{1}{\tilde{p}} = \tilde{\lambda}\left(\frac{1}{\tilde{\lambda}\tilde{q}} + \frac{1}{\lambda p} - \frac{1}{\lambda q}\right) ; \tag{3.32}$$

that is to say, $\tilde{P}$ would be the *Gaussian image*. Of course, this case is not realistic, since the direct reconstruction light would overlap the image. However, in analogy to geometric optics of rotational symmetric systems, we may keep this definition for a Gaussian image $\tilde{P}_0$, for which the distance

$$\frac{1}{\tilde{p}_0} = \tilde{\lambda}\left(\frac{1}{\tilde{\lambda}\tilde{q}} + \frac{1}{\lambda p} - \frac{1}{\lambda q}\right) \tag{3.33}$$

gives a reference point on the ray of direction $\tilde{k} = \tilde{k}_0$ in the general case where $\tilde{c} \neq c \neq k \neq \tilde{k}$ [3.117, 118, 131–135]. Relative to $\tilde{P}_0$, we can decompose the tensor $\tilde{T}$ in the form:

$$\tilde{T} = \frac{1}{\tilde{p}_0}\tilde{K} + \tilde{S} = \frac{1}{\tilde{p}_0}\tilde{K} + \left[\frac{1}{2}\left(\frac{1}{\tilde{p}_1} + \frac{1}{\tilde{p}_2}\right) - \frac{1}{\tilde{p}_0}\right]\tilde{K} + \frac{1}{2}\left(\frac{1}{\tilde{p}_1} - \frac{1}{\tilde{p}_2}\right)\tilde{D} , \tag{3.34}$$

where, based on (3.31 and 33), $\tilde{S}$ may be alternatively expressed as

$$\tilde{S} = \tilde{\lambda}\tilde{M}\left[\frac{1}{\tilde{\lambda}\tilde{p}_0}(\tilde{k} \otimes \tilde{k}) - \frac{1}{\tilde{\lambda}\tilde{q}}(\tilde{c} \otimes \tilde{c}) - \frac{1}{\lambda p}(k \otimes k) + \frac{1}{\lambda q}(c \otimes c)\right]\tilde{M}^{\mathrm{T}} . \tag{3.35}$$

The two-dimensional tensor $\tilde{D}$ is a deviator for which tr $\{\tilde{D}\} = 0$ by definition. The components of $\tilde{K}$ and $\tilde{D}$ in the principal directions are

$$\tilde{K} \triangleq \begin{bmatrix} 1 & 0 \\ 0 & 1 \end{bmatrix} , \quad \tilde{D} \triangleq \begin{bmatrix} 1 & 0 \\ 0 & -1 \end{bmatrix} . \tag{3.36}$$

For an arbitrary direction (angle $\beta$ with respect to the principal direction), $\tilde{D}$ would have the form

$$\tilde{D} \triangleq \begin{bmatrix} \cos 2\beta & \sin 2\beta \\ \sin 2\beta & -\cos 2\beta \end{bmatrix} .$$

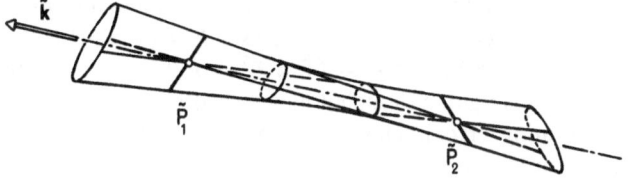

**Fig. 3.6.** Astigmated pencil of rays at the reconstruction. ($\tilde{P}_1$, $\tilde{P}_2$: endpoints of the astigmatic interval through which the focal lines pass)

The part $\frac{1}{2}(1/\tilde{p}_1 - 1/\tilde{p}_2)\tilde{D}$ of $\tilde{S}$ characterizes the *astigmatism* [3.136–138], whereas the isotropic part $[\frac{1}{2}(1/\tilde{p}_1 + 1/\tilde{p}_2) - 1/\tilde{p}_0]\tilde{K}$ of $\tilde{S}$ contains the mean curvature, sometimes called *field curvature*. The eigenvalues of $\tilde{S}$ are the curvature changes $1/\tilde{p}_1 - 1/\tilde{p}_0$, $1/\tilde{p}_2 - 1/\tilde{p}_0$.

On the other hand, for the eigenvectors $\tilde{m}_1$, $\tilde{m}_2$ normal to $\tilde{k}$, we have the linear system

$$\left(\tilde{T} - \frac{1}{\tilde{p}_\alpha}\tilde{K}\right)\tilde{m}_\alpha = 0 \qquad \alpha = 1, 2 \, , \tag{3.37}$$

so that, from (3.27), we find

$$d\boldsymbol{\eta}_{\tilde{P}_\alpha} \cdot \tilde{m}_\alpha = 0$$

for any $d\tilde{k}$. Within this approximation, all rays intersect the so-called focal lines near $\tilde{P}_1$, $\tilde{P}_2$ (Fig. 3.6).

Let us add here that one must pay attention when seeking the eigenvalues of the involved tensors by means of $3 \times 3$ matrices in an arbitrary cartesian system $x$, $y$, $z$, whose axes $x$, $y$ are not necessarily in the plane normal to $\tilde{k}$. Then $\tilde{T} + \tilde{k} \otimes \tilde{k}$ has the three eigenvalues $1/\tilde{p}_1$, $1/\tilde{p}_2$, and 1, so that we conveniently write

$$\left|\tilde{T} + \tilde{k} \otimes \tilde{k} - \frac{1}{\tilde{p}}I\right| \left(\frac{1}{\tilde{p}} - 1\right)^{-1} = 0 \, , \tag{3.38}$$

where $|\ldots|$ denotes the $3 \times 3$ determinant [3.139].

We shall now investigate a more general situation in which, at the reconstruction, other elements of the optical set-up differ from those at the recording. For instance, it could happen, in addition to $\tilde{Q} \neq Q$, $\tilde{\lambda} \neq \lambda$, that the hologram (which could also be curved) is deformed at the reconstruction (Fig. 3.7) [3.140–142]. Here, we still disregard the additional extended source effects [3.99, 143, 144]. In analogy to (2.124, 125), the deformation of the hologram is described by the affine connection

$$d\hat{r} = \hat{F}\, d\tilde{r} \quad \text{or} \quad d\tilde{r} = \hat{\tilde{F}}\, d\hat{r} \, , \tag{3.39}$$

where $\hat{r}$, $\hat{r}$ are the position vectors of the same hologram point, denoted now by $\hat{H}$ at the recording and by $\hat{H}$ at the reconstruction. The tensor $\hat{F} = (\hat{\nabla} \otimes \hat{r})^T$ is the deformation gradient of the hologram and $\hat{F} = (\hat{\nabla} \otimes \hat{r})^T = \hat{F}^{-1}$ its inverse. In most cases, the hologram remains rigid, although it may be shifted and rotated at the reconstruction (repositioning error). Here the tensor $\hat{F} = (\hat{F}^T)^{-1} = \hat{Q}$ is simply an orthogonal operator. Since we still consider only surface holograms, the vectors $d\hat{r}$ and $d\hat{r}$ remain in their respective tangent planes, so that only the mixed projection of the deformation gradients, see (2.138), is relevant:

$$\hat{N}\hat{F}\hat{N} = \hat{F}\hat{N} , \quad \hat{N}\hat{F}\hat{N} = \hat{F}\hat{N} . \tag{3.40}$$

We write, for example, (3.39b) in the form

$$d\hat{r} = \hat{F}\hat{N} \; d\hat{r} . \tag{3.41}$$

The condition of the interference identity (3.9) now concerns the related points $\hat{H}$ and $\hat{H}$. Similar to (3.10), we define a phase difference function

$$\theta[\hat{r}, \; \hat{r}(\hat{r})] \;=\; \frac{2\pi}{\tilde{\lambda}} \, (\tilde{p} - \tilde{q}) - \frac{2\pi}{\lambda} \, (p-q) , \tag{3.42}$$

for which the position vector $\hat{r}$ of $\hat{H}$ is considered here as the independent variable. The stationary behavior of $\theta$ then gives the necessary condition

$$d\theta \;=\; \frac{2\pi}{\tilde{\lambda}} d\hat{r} \; \cdot \; \nabla_{\hat{n}}(\tilde{p} - \tilde{q}) - \frac{2\pi}{\lambda} d\hat{r} \cdot \nabla_{n}(p-q) = 0 \tag{3.43}$$

or

$$\frac{1}{\tilde{\lambda}}\hat{N}(\tilde{k} - \tilde{c}) - \frac{1}{\lambda}\hat{N}\hat{F}^T(k-c) = 0 , \tag{3.44}$$

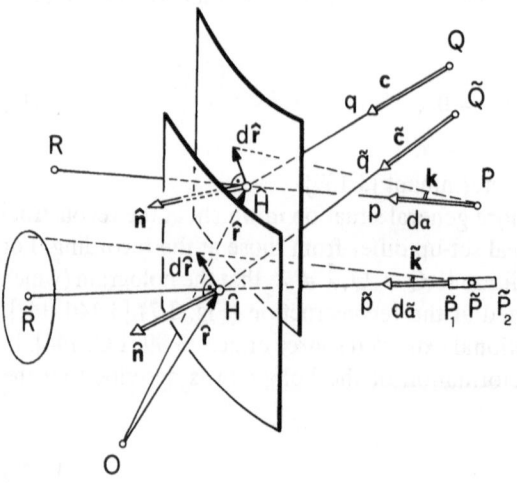

Fig. 3.7. Optical modification in the case of a general deformation of a curved hologram. ($\hat{r}$, $\hat{r}$: position vectors; $\hat{n}$, $\hat{n}$: unit normals at the same material hologram point in the two configurations at both the recording and the reconstruction)

which is the generalization of (3.15). As before, together with the auxiliary condition $|\tilde{k}| = 1$, (3.44) determines the direction $\tilde{k}$ to the sought image point as a function of the transposed inverse deformation gradient, $\hat{F}^{\mathrm{T}} = (\hat{V} \otimes \hat{r}) = (\hat{V} \otimes \hat{r})^{-1}$, and of the direction unit vectors $k$, $c$, and $\tilde{c}$.

In order to determine the distance $\tilde{p}$, we may next calculate, as previously, the second differential of the phase difference $\theta$, starting from (3.43)

$$d^2\theta_{\tilde{p}} = \frac{2\pi}{\tilde{\lambda}} d\hat{r} \cdot \nabla_{\hat{n}} \otimes \nabla_{\hat{n}}(\tilde{p} - \tilde{q})d\hat{r} - \frac{2\pi}{\lambda} d\hat{r} \cdot \nabla_{\hat{n}} \otimes \nabla_{\hat{n}}(p - q)d\hat{r}$$

$$+ \frac{2\pi}{\tilde{\lambda}} d^2\hat{r} \cdot \nabla_{\hat{n}}(\tilde{p} - \tilde{q}) - \frac{2\pi}{\lambda} d^2\hat{r} \cdot \nabla_{\hat{n}}(p - q) \ . \tag{3.45}$$

Such a second differential should be performed along a materialized curve in each position on the hologram through the points $\hat{H}$ and $\hat{H}$. However, we may choose a part of a geodesic curve through $\hat{H}$ for which its osculating plane at $\hat{H}$ is normal to the hologram surface. The vector $d^2\hat{r}$ is then parallel to $\hat{n}$, which implies that $d^2\hat{r} \cdot \nabla_{\hat{n}}(\tilde{p} - \tilde{q}) = 0$. In the second line of (3.45), the only remaining term is

$$-\frac{2\pi}{\lambda} |d\hat{r}|^2 \hat{b} \cdot \nabla_{\hat{n}}(p - q) \ ,$$

where $\hat{b}$ is the in-plane vector of geodesic curvature or, in other words, a direction change of the tangent, see (2.40). However, if the hologram deforms little in its plane, $\hat{b}$ is very small, so that this last term becomes negligible. In other words, we assume that, in practice, a hologram may be bent without altering its Gaussian curvature. For instance, a plane hologram could be deformed into a cylindrical or any other developable surface. Analogous problems might be encountered with curved gratings in thick holograms, which are treated in the following paragraph.

Returning to the first line in (3.45), using (2.21 and 41) for curved holograms, we have

$$(\nabla_{\hat{n}} \otimes \nabla_{\hat{n}}\tilde{p})\hat{N} = \hat{N}[(\hat{V} \otimes \hat{N}) \hat{V}\tilde{p}]\hat{N} + \hat{N}[\hat{V} \otimes \hat{V}\tilde{p}]\hat{N} = \hat{B}(\hat{n} \cdot \tilde{k}) + 1/\tilde{p} \ \hat{N}\tilde{K}\hat{N} \ ,$$

and similar expressions for the three other lengths. Therefore, for (3.45), we find

$$d^2\theta_{\tilde{p}} = \frac{2\pi}{\tilde{\lambda}} d\hat{r} \cdot \hat{N}\left[\frac{1}{\tilde{p}}\tilde{K} - \frac{1}{\tilde{q}}\tilde{C} + \hat{B}(\hat{n} \cdot \tilde{k} - \hat{n} \cdot \tilde{c})\right]\hat{N} \ d\hat{r}$$

$$- \frac{2\pi}{\lambda} d\hat{r} \cdot \hat{N}\left[\frac{1}{p}K - \frac{1}{q}C + \hat{B}(\hat{n} \cdot k - \hat{n} \cdot c)\right]\hat{N} \ d\hat{r} \ , \tag{3.46}$$

where $\hat{B} = -\nabla_{\hat{n}} \otimes \hat{n}$ and $\hat{B} = -\nabla_{\hat{n}} \otimes \hat{n}$ are the two curvature tensors of the hologram surface at the reconstruction and at the recording, respectively. In practice, when the hologram is subjected to moderately small deformations, the tensor $\hat{B}$ may be expressed as a function of $\hat{B}$ and other kinematic terms, as calculated in (2.157) [3.145, 146]. Next, we replace $d\hat{r}$ and $d\hat{r}$ by an increment, $d\tilde{k} = \tilde{m}\, d\tilde{a}$, with a unit vector $\tilde{m}$ normal to $\tilde{k}$, using the following infinitesimal linear transformations

$$d\hat{r} = \tilde{p}\hat{M}^{\mathrm{T}}d\tilde{k} = \tilde{p}\hat{M}^{\mathrm{T}}\tilde{m}\, d\tilde{a} \;,$$

$$d\hat{r} = \tilde{p}\hat{F}\hat{M}^{\mathrm{T}}d\tilde{k} = \tilde{p}\hat{F}\hat{M}^{\mathrm{T}}\tilde{m}\, d\tilde{a} \;. \tag{3.47}$$

Here, $\hat{M} = I - \hat{n} \otimes \tilde{k}/\hat{n}\cdot\tilde{k}$ is an oblique projector (referring to $\tilde{k}$ and not to $k$, as in $\hat{M}$ appearing in (2.101)). Equation (3.46) may be written

$$d^2\theta_{\tilde{P}} = \tilde{p}^2\frac{2\pi}{\tilde{\lambda}}\tilde{m} \cdot \left(\frac{1}{\tilde{p}}\tilde{K} - \tilde{T}\right)\tilde{m}\, d\tilde{a}^2$$

with the characteristic tensor [Ref. 3.147, Eq. (14)]

$$\tilde{T} = \tilde{\lambda}\hat{M}\left\{\frac{1}{\tilde{\lambda}}\left[\frac{1}{\tilde{q}}\tilde{C} - \hat{B}(\hat{n}\cdot\tilde{k} - \hat{n}\cdot\tilde{c})\right]\right.$$

$$\left. + \frac{1}{\lambda}\,\hat{F}^{\mathrm{T}}\left[\frac{1}{p}K - \frac{1}{q}C + \hat{B}(\hat{n}\cdot k - \hat{n}\cdot c)\right]\hat{F}\right\}\hat{M}^{\mathrm{T}}\;. \tag{3.48}$$

Note that we have used the mixed projection of $\hat{F}$, (3.40) or, rather, the relation of an oblique projected deformation

$$\hat{M}\hat{F}^{\mathrm{T}}\hat{N} = \hat{M}\hat{N}\hat{F}^{\mathrm{T}}\hat{N} = \hat{M}\hat{N}\hat{F}^{\mathrm{T}} = \hat{M}\hat{F}^{\mathrm{T}}\;,$$

$$\hat{N}\hat{F}\hat{M}^{\mathrm{T}} = \hat{N}\hat{F}\hat{N}\hat{M}^{\mathrm{T}} = \hat{F}\hat{N}\hat{M}^{\mathrm{T}} = \hat{F}\hat{M}^{\mathrm{T}}\;. \tag{3.49}$$

Expression (3.48) is the generalization of (3.31); $-\tilde{T}$ maintains the meaning of "tensor of curvature" of the wavefront, now at $\hat{H}$. The condition $d^2\theta_{\tilde{P}} = 0$ associates a point $\tilde{P}$, of distance $\tilde{p}$ from $\hat{H}$, with each vector $\tilde{m}$. With the projection propriety $\hat{M}\tilde{K}\hat{M}^{\mathrm{T}} = \tilde{K}$ we find, similar to (3.30), the distance

$$\frac{1}{\tilde{p}} = \tilde{m}\cdot\tilde{T}\tilde{m}\;. \tag{3.50}$$

The extreme values, or eigenvalues $1/\tilde{p}_1$ and $1/\tilde{p}_2$, are the inverses of the distances $\tilde{p}_1$ and $\tilde{p}_2$ to the focal lines.

A particular case of (3.48) which has already been studied in detail by other investigators [3.97, 148–163], consists of a reconstruction with a curved, undeformed hologram ($\hat{B} \equiv \hat{B}$, $\hat{M} \equiv \tilde{M}$, $\hat{F} = I$, $\hat{n} = \hat{n}$). Only the wavelength and/or position of the reference source is changed. Then, (3.48) simply becomes

$$\tilde{T} = \tilde{\lambda}\tilde{M}\left(\frac{1}{\tilde{\lambda}}\frac{1}{\tilde{q}}\tilde{C} + \frac{1}{\lambda p}K - \frac{1}{\lambda q}C\right)\tilde{M}^{\mathrm{T}}$$

$$- \tilde{\lambda}\tilde{M}\hat{B}\tilde{M}^{\mathrm{T}}\left\{\hat{n} \cdot \left[\frac{1}{\tilde{\lambda}}(\tilde{k} - \tilde{c}) - \frac{1}{\lambda}(k - c)\right]\right\} .$$

The first term is analogous to (3.31) and could be decomposed as in (3.34). The second term contains a scalar, $\{\hat{n} \cdot [(\tilde{k} - \tilde{c})/\tilde{\lambda} - (k - c)/\lambda]\}$, similar to one we shall encounter in the Bragg condition (Sect. 3.1.3) and a product of tensors containing the curvature of the hologram $\hat{B}$ and the oblique projector $\tilde{M}$ resulting from the oblicity of the rays with regard to the tangential plane at the considered hologram point. The second term could partially compensate for the first one. Thus, by careful choice of a curved surface on which the hologram is placed, it is possible to reduce, or even eliminate, both the field curvature and astigmatism. In particular, the astigmatism that a plane hologram would produce may be reduced by the deviator part of the tensor $\tilde{\lambda}\{\hat{n} \cdot [(\tilde{k} - \tilde{c})/\tilde{\lambda} - (k - c)/\lambda]\}\tilde{M}\hat{B}\tilde{M}^{\mathrm{T}}$. For a sphere, the two principal curvature radii are the same, $(R_1 = R_2 = R)$ and the tensor $\hat{B}$ becomes simply $\hat{B} = \hat{N}/R$. In this special case, the astigmatism could be reduced by the presence of the deviator of $\tilde{\lambda}\{\hat{n} \cdot [(\tilde{k} - \tilde{c})/\tilde{\lambda} - (k - c)/\lambda]\}$ $\tilde{M}\tilde{M}^{\mathrm{T}}/R$, i.e. by the oblicity only. On the other hand, if $\tilde{k} = \hat{n}$ (pseudo-optical axes), we have $\tilde{M} = \hat{N}$ and, thus, $\tilde{M}\hat{B}\tilde{M}^{\mathrm{T}} = \hat{B}$. In this case, the same astigmatism may be reduced by the presence of the deviator part of $\hat{B}$, which differs from zero for a general surface, $R_1 \neq R_2$ [3.148]. Of course, it is possible to develop the tensors $\tilde{M}$ and $\hat{B}$ near the optical axis of a rotational symmetric system. Such techniques might be applied in order to reduce aberrations in holographic microscopy.

### 3.1.3 Bragg Condition and Coupled-Wave Theory for Volume Holograms

Up to this point, we have discussed the modification at the reconstruction in the case of thin holograms. At the beginning of Sect. 3.1.2, where the modification consists of a shift of the reference source Q and a wavelength change, the basic equation (3.15) determines the unit direction vector $\tilde{k}$ of the primary image wave. This equation expresses a two-dimensional stationary behavior of the phase difference function $\theta$ in the small region covered by the rays through the aperture of the observation system. However, in the case of a *thick* hologram (also called a volume hologram), first introduced by *Denisyuk* [3.47] and *van Heerden* [3.48], it must be noted that, for each ray, the diffraction occurs along a whole segment $H_1H_2$ (Fig. 3.8), and not only at a single point. Each elementary, non-spherical wave that contributes to the formation of the image point $\tilde{P}$ is diffracted at an interior point X with position vector $x$ between $H_1$ and $H_2$ and has a form similar to (3.7), namely,

$$d\tilde{U} = - B(x, \ldots)\exp\left[\frac{2\pi\mathrm{i}}{\tilde{\lambda}_0}S_{\tilde{q}} + \frac{2\pi\mathrm{i}}{\lambda_0}(S_p - S_q) + \mathrm{i}(\tilde{\psi} + \phi - \psi)\right]ds_{\tilde{p}} , \quad (3.51)$$

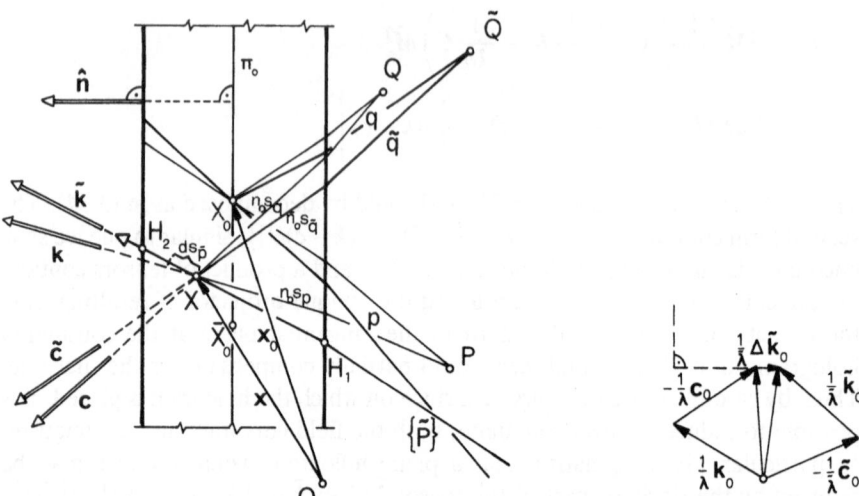

**Fig. 3.8.** Optical modification when using a volume hologram, illustrating the Bragg condition. $x$: position vector of any interior point X in the interval $\langle H_1 H_2 \rangle$ where diffracted rays accumulate. [$x_0$: position vector of a point $X_0$ in the reference plane $\Pi_0$; $c_0, \tilde{c}_0, k_0, \tilde{k}_0$: corresponding unit direction vectors; $(1/\tilde\lambda)\,\Delta\tilde{k}_0$: mismatch]

where $B$ is generally some locally variable factor, and where $S_{\tilde{q}} = \tilde{q} + \tilde{n}_0 s_{\tilde{q}}$, $S_q = q + n_0 s_q$ and $S_p = p + n_0 s_p$, denote the optical paths or eikonals relative to the points $\tilde{Q}$, Q, and P, respectively. The increment $ds_{\tilde{p}}$ at X must be taken along the ray $H_1 H_2$. As before, if we expect an image point $\tilde{P}$, we must compare the complex amplitude of this wave with that of a virtual wave, similar to that expressed in (3.8), given by

$$d\tilde{U}_{\tilde{p}} = B_{\tilde{p}} \exp\left(\frac{2\pi i}{\tilde{\lambda}_0} S_{\tilde{p}} + i\tilde{\phi}\right) ds_{\tilde{p}} \ . \tag{3.52}$$

The condition of interference identity, also called phase matching, is then

$$\frac{2\pi}{\tilde{\lambda}_0}(S_{\tilde{p}} - S_{\tilde{q}}) - \frac{2\pi}{\lambda_0}(S_p - S_q) = (\phi - \psi) - (\tilde{\phi} - \tilde{\psi}) \ . \tag{3.53}$$

In order that the amplitudes, roughly stated, "accumulate properly", we should now write tentatively the condition for a *three-dimensional* stationary behavior around some point $X_0$ of the phase-difference function $\theta = 2\pi(S_{\tilde{p}} - S_{\tilde{q}})/\tilde{\lambda}_0 - 2\pi(S_p - S_q)/\lambda_0$, which reads

$$\nabla\theta = 0 \ . \tag{3.54}$$

Assuming here that the index of refraction does not vary much, even after development, this three-dimensional condition leads to the so-called *Bragg condition* [3.100, 164–177]

$$\frac{1}{\tilde{\lambda}} (\tilde{k} - \tilde{c}) - \frac{1}{\lambda} (k - c) = 0 \ , \tag{3.55}$$

where the unit direction vectors $\tilde{k}$, $\tilde{c}$, $k$, $c$, and the wavelengths $\tilde{\lambda}$, $\lambda$ have to be taken for the paths *inside* the hologram (at the surface there is a refraction so that the corresponding unit vectors outside differ from those considered here).

Since condition (3.55) is more restrictive than that formulated in (3.15), it is generally not possible to find a unit direction vector $\tilde{k}$ for the independently given quantities $k$, $c$, $\tilde{c}$, $\lambda$, $\tilde{\lambda}$. Therefore, we must generally tolerate a deviation from the Bragg condition or a mismatch [3.178–184]:

$$\frac{1}{\tilde{\lambda}} \Delta \tilde{k} = \frac{1}{\tilde{\lambda}} (\tilde{k} - \tilde{c}) - \frac{1}{\lambda} (k - c) \ . \tag{3.56}$$

If the radius $\mathring{r}$ of the aperture is slightly larger than the thickness of the hologram ($\mathring{r} > h$), we should keep the validity of the two-dimensional expression (3.15) at one point $X_0$ or (3.17) at any point $\overline{X}_0$ of a reference plane $\Pi_0$, $\Delta \tilde{k}_0$ must then be a vector parallel to the unit normal $\hat{n}$ (Fig. 3.8):

$$\Delta \tilde{k}_0 = \Delta k \hat{n} \ . \tag{3.57}$$

In order to understand the meaning of amplitudes that "accumulate properly", it is necessary to refer to the so-called *coupled-wave theory*, introduced by *Kogelnik* [3.185]. This theory explains, by means of the wave equation, how and which part of the incoming wave is diffracted at the reconstruction by the interference grating formed at the recording. Within the scope of this book, it is not our intention to exhaustively treat this theory, but rather consider its basic aspects. The reader may, for example, find an extended treatment of this field in the book by *Solymar* and *Cooke* [3.38] and for general gratings in [3.186]. Information about recording media may be found in [3.38, 50, 187–204].

As a short introduction, it is appropriate to present here the so-called *one-dimensional* case [3.38, 185, 205–230]. The sources Q, P, $\tilde{Q}$, and the image point $\tilde{P}$ lie on the same plane (plane of incidence which is perpendicular to the hologram plane) and are situated far away from the hologram ($p$, $q \gg H_1 H_2$) (Fig. 3.9). Therefore, the light waves are nearly plane waves and the pattern produced inside the homogeneous medium of the hologram during the recording is uniform, straight, and characterized by a constant *grating vector*:

$$\frac{2\pi}{\lambda} h = \frac{2\pi}{\lambda} (k - c) \ . \tag{3.58}$$

The complex index of refraction $\gamma$, see (2.66), may now be represented by

$$\gamma = \gamma_0 + \gamma_1 \cos \left( \frac{2\pi}{\lambda} h \cdot x \right) \ , \tag{3.59}$$

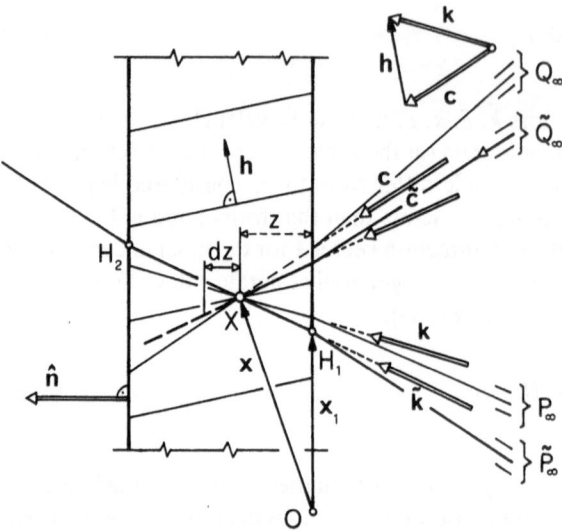

**Fig. 3.9.** Grating formed at the recording by plane waves in a homogeneous sheet (one-dimensional case for the coupled wave theory). ($2\pi h/\lambda$: grating vector)

where $\gamma_0$ and $\gamma_1$ are both complex constants. We make the assumption $|\gamma_1| \ll |\gamma_0|$, characteristic of a weak grating. In expression (3.59), only the *scalar* variable $h \cdot x$ appears and not the complete vector $x$. Hence, $h \cdot x$ could be replaced by $z(h \cdot \tilde{k}/\hat{n} \cdot \tilde{k}) + h \cdot x_1$, where $z$ is the variable distance of X from the entering face (Fig. 3.9).

The wave equation (2.67) is now applied to this hologram medium and, in the scope of this one-dimensional case, we consider a sum of plane waves as a solution for $e_\omega$. Furthermore, we set the polarization vector of these waves perpendicular to the incident plane (TE polarization), so that (2.67) is reduced to a scalar equation as in (2.71). For TM polarization (electric vector in the plane of incidence), we have just to consider the magnetic vector instead of the electric vector [3.231–237]. With $\gamma^2 = \gamma_0^2 + 2\gamma_0\gamma_1 \cos(2\pi h \cdot x/\lambda)$, this scalar equation then reduces to:

$$\nabla^2 \tilde{U} + \frac{4\pi^2}{\tilde{\lambda}_0^2} \gamma_0^2 \tilde{U} = -\frac{8\pi^2}{\tilde{\lambda}_0^2} \gamma_1\gamma_0 \cos\left(\frac{2\pi}{\lambda} h \cdot x\right) \tilde{U} . \qquad (3.60)$$

With the separation of the complex $\gamma_0$ into a real part $\tilde{n}_0$, the actual index of refraction of the medium, and into an imaginary part, $-i\chi_0\tilde{n}_0\tilde{\lambda}/2\pi$, $\gamma_0$ now equals $\tilde{n}_0(1 - i\chi_0\tilde{\lambda}/2\pi)$. Another complex constant is defined as $\pi\gamma_1\gamma_0\tilde{\lambda}/\tilde{\lambda}_0^2 \sim \pi\gamma_1/\tilde{\lambda}_0 = \kappa_1$, so that (3.60) then becomes ($\tilde{n}_0\tilde{\lambda} = \tilde{\lambda}_0$):

$$\nabla^2 \tilde{U} + \left( \frac{4\pi^2}{\tilde{\lambda}^2} - \frac{4\pi i}{\tilde{\lambda}} \chi_0 \right) \tilde{U}$$

$$= - \frac{4\pi}{\tilde{\lambda}} \kappa_1 \left[ \exp\left( \frac{2\pi i}{\lambda} h \cdot x \right) + \exp\left( -\frac{2\pi i}{\lambda} h \cdot x \right) \right] \tilde{U} . \tag{3.61}$$

The function $\tilde{U} = U_{\tilde{p}} + U_{\tilde{q}}$ is a superposition of the incoming and outgoing waves. We try a plane wave solution for (3.61) of the form ($a_{\tilde{p}}$, $a_{\tilde{q}}$ are complex with a different constant phase factor)

$$U_{\tilde{p}} = a_{\tilde{p}}(z) \exp\left( \frac{2\pi i}{\tilde{\lambda}} \tilde{k} \cdot x \right) , \quad U_{\tilde{q}} = a_{\tilde{q}}(z) \exp\left( \frac{2\pi i}{\tilde{\lambda}} \tilde{c} \cdot x \right) . \tag{3.62}$$

Applying the usual derivative rules for $\nabla U_{\tilde{p}}$ and $\nabla^2 U_{\tilde{p}}$, we obtain

$$\nabla U_{\tilde{p}} = \nabla a_{\tilde{p}} \exp\left( \frac{2\pi i}{\tilde{\lambda}} \tilde{k} \cdot x \right) + a_{\tilde{p}} \frac{2\pi i}{\tilde{\lambda}} \nabla(\tilde{k} \cdot x) \exp\left( \frac{2\pi i}{\tilde{\lambda}} \tilde{k} \cdot x \right)$$

$$= \left( \hat{n} \frac{da_{\tilde{p}}}{dz} + \frac{2\pi i}{\tilde{\lambda}} \tilde{k} a_{\tilde{p}} \right) \exp\left( \frac{2\pi i}{\tilde{\lambda}} \tilde{k} \cdot x \right) ,$$

$$\nabla^2 U_{\tilde{p}} = \nabla \cdot \nabla U_{\tilde{p}} = \left[ \frac{d^2 a_{\tilde{p}}}{dz^2} + \frac{4\pi i}{\tilde{\lambda}} (\hat{n} \cdot \tilde{k}) \frac{da_{\tilde{p}}}{dz} - \frac{4\pi^2}{\tilde{\lambda}^2} a_{\tilde{p}} \right] \exp\left( \frac{2\pi i}{\tilde{\lambda}} \tilde{k} \cdot x \right) .$$

Similarly, for $U_{\tilde{q}}$ we have

$$\nabla^2 U_{\tilde{q}} = \left[ \frac{d^2 a_{\tilde{q}}}{dz^2} + \frac{4\pi i}{\tilde{\lambda}} (\hat{n} \cdot \tilde{c}) \frac{da_{\tilde{q}}}{dz} - \frac{4\pi^2}{\tilde{\lambda}^2} a_{\tilde{q}} \right] \exp\left( \frac{2\pi i}{\tilde{\lambda}} \tilde{c} \cdot x \right) .$$

By substituting these expressions into (3.61), we obtain, by multiplying by $\exp(-2\pi i \tilde{k} \cdot x / \tilde{\lambda})$, recalling that $h = k - c$,

$$\left[ \frac{d^2 a_{\tilde{p}}}{dz^2} + \frac{4\pi i}{\tilde{\lambda}} (\hat{n} \cdot \tilde{k}) \frac{da_{\tilde{p}}}{dz} - \frac{4\pi i}{\tilde{\lambda}} \chi_0 a_{\tilde{p}} \right]$$

$$+ \left[ \frac{d^2 a_{\tilde{q}}}{dz^2} + \frac{4\pi i}{\tilde{\lambda}} (\hat{n} \cdot \tilde{c}) \frac{da_{\tilde{q}}}{dz} - \frac{4\pi i}{\tilde{\lambda}} \chi_0 a_{\tilde{q}} \right] \exp\left[ \frac{2\pi i}{\tilde{\lambda}} (\tilde{c} - \tilde{k}) \cdot x \right]$$

$$= - \frac{4\pi}{\tilde{\lambda}} \kappa_1 a_{\tilde{p}} \left\{ \exp\left[ \frac{2\pi i}{\lambda} (k - c) \cdot x \right] + \exp\left[ \frac{2\pi i}{\lambda} (c - k) \cdot x \right] \right\}$$

$$- \frac{4\pi}{\tilde{\lambda}} \kappa_1 a_{\tilde{q}} \left( \exp\left\{ 2\pi i \left[ \frac{1}{\lambda} (k - c) - \frac{1}{\tilde{\lambda}} (\tilde{k} - \tilde{c}) \right] \cdot x \right\} \right.$$

$$\left. + \exp\left\{ 2\pi i \left[ \frac{1}{\lambda} (c - k) - \frac{1}{\tilde{\lambda}} (\tilde{k} - \tilde{c}) \right] \cdot x \right\} \right) . \tag{3.63}$$

If the vector $\tilde{k}$ nearly fulfills the Bragg condition (3.55), we have in (3.63) the two following groups of expressions: (1) slowly varying ones, such as the first bracket on the left side and the third exponential on the right side; and, (2) steeply varying ones, such as all other terms. Moreover, if $a_{\tilde{p}}(z)$ is a slowly varying function, we have

$$\left| \frac{d^2 a_{\tilde{p}}}{dz^2} \right| \ll \left| \frac{d a_{\tilde{p}}}{dz} \right| / \tilde{\lambda} \ . \tag{3.64}$$

This last consideration allows us to write with (3.57) the following first-order coupled differential equations, where the second equation is derived in the same way except that the slowly varying terms of (3.63) are multiplied by the factor $\exp\left[2\pi i\,(\tilde{k} - \tilde{c}) \cdot x / \tilde{\lambda}\right]$:

$$\frac{d a_{\tilde{p}}}{dz} - \frac{\chi_0}{\hat{n} \cdot \tilde{k}} a_{\tilde{p}} = \frac{i\kappa_1}{\hat{n} \cdot \tilde{k}} a_{\tilde{q}} \exp\left( -\frac{2\pi i}{\tilde{\lambda}} \varDelta kz \right) ,$$

$$\frac{d a_{\tilde{q}}}{dz} - \frac{\chi_0}{\hat{n} \cdot \tilde{c}} a_{\tilde{q}} = \frac{i\kappa_1}{\hat{n} \cdot \tilde{c}} a_{\tilde{p}} \exp\left( +\frac{2\pi i}{\tilde{\lambda}} \varDelta kz \right) . \tag{3.65}$$

These first-order differential equations, with a real coefficient of absorption $\chi_0$ and a complex number $\kappa_1$ as a coupling parameter, involve the sought after interaction between $U_{\tilde{p}}$ and $U_{\tilde{q}}$.

The solution is generally discussed for various values of the parameters $\chi_0$, $\kappa_1$, $\hat{n} \cdot \tilde{k}$, $\hat{n} \cdot \tilde{c}$, and for different boundary conditions. Since the system (3.65) of two first-order differential equations is equivalent to one of second-order, this discussion is very similar to those of oscillations in mechanics. For example, in the special case of a loss-free phase hologram ($\chi_0 = 0$), where, in addition, the Bragg condition is fulfilled ($\varDelta k = 0$), $a_{\tilde{p}}$ and $a_{\tilde{q}}$ are harmonic in $z$ with a relative phase shift. However, in all other cases, attenuation effects appear. A more general case is the two-dimensional one in which the amplitudes depend on two variables, but a scalar wave theory is still used [3.38, 238–248].

Let us go directly to the so-called *three-dimensional* case, where the amplitudes depend on the complete vector $x$ [3.249, 250]. This is caused by a non-uniform amplitude distribution of the beams (e.g., Gaussian distribution) or by a non-uniform grating. Here, the *polarization* direction should also be taken into account ([Ref. 3.38, p. 229], [3.251–254] and for polarization in general [3.175, 255–262]). The arrangement of terms in the preceeding equations, with respect to the order 1, $1/\tilde{\lambda}$, $1/\tilde{\lambda}^2$, and especially the neglect of $d^2 a_{\tilde{p}}/dz^2$, see (3.64), indicates that we have performed a "small wavelength approach", as in Sect. 2.2. We could therefore start directly from (2.81), but in order to avoid the derivative $\nabla(\log \tilde{\varepsilon})$, which is not small, it is advisable to consider the general wave equation (2.67). We assume the following general expression to be a solution of the wave equation (2.67): (we can set here $\tilde{\phi} = \tilde{\psi} = \phi = \psi = 0$ without loss of generality)

$$e_\omega = a_{\bar{p}}(x) \exp\left[\frac{2\pi i}{\tilde{\lambda}_0} S_{\bar{p}}(x)\right] + a_{\bar{q}}(x) \exp\left[\frac{2\pi i}{\tilde{\lambda}_0} S_{\bar{q}}(x)\right] , \qquad (3.66)$$

with eikonals $S_{\bar{p}}$, $S_{\bar{q}}$ and slowly varying complex amplitudes $a_{\bar{p}}$, $a_{\bar{q}}$ of the incoming and outgoing waves at the reconstruction. For simplicity, we restrict our consideration to *lossless phase* holograms, where $\gamma_0$ is now a real index of refraction. If $n = \sqrt{\varepsilon}$ is the index of refraction at the recording, and if, with real $a_p$, $a_q$,

$$e_p = a_p \exp\left(\frac{2\pi i}{\lambda_0} S_p\right) , \qquad e_q = a_q \exp\left(\frac{2\pi i}{\lambda_0} S_q\right)$$

are the two interfering waves also at the recording, we may write by introducing some factor $K$ [Ref. 3.38, p. 231] that

$$\tilde{\varepsilon} = \varepsilon + \Delta\varepsilon = \varepsilon + K|e_p + e_q|^2$$

$$= \tilde{\varepsilon}_0 + 2a_p \cdot a_q K \cos\left[\frac{2\pi}{\lambda_0} (S_p - S_q)\right] , \qquad (3.67)$$

where $\tilde{\varepsilon}_0 = \varepsilon + K(|a_p|^2 + |a_q|^2)$.

By comparing this equation to the square of an equation of type (3.59), $\tilde{\varepsilon} = \gamma^2 \cong \gamma_0^2 + 2\gamma_0\gamma_1 \cos\left[2\pi(S_p - S_q)/\lambda_0\right]$, we have

$$\gamma_0^2 = \tilde{n}_0^2 = \tilde{\varepsilon}_0 , \qquad \gamma_1 = K\frac{a_p \cdot a_q}{\tilde{n}_0} .$$

Introducing, as before, the variable coefficient $\kappa_1 = \pi\gamma_1/\tilde{\lambda}_0$, we obtain the auxiliary relation

$$\qquad (3.68)$$

$$\tilde{\varepsilon} = \tilde{n}^2 = \tilde{n}_0^2 + \frac{\tilde{n}_0\tilde{\lambda}_0}{\pi} \kappa_1\left\{\exp\left[\frac{2\pi i}{\lambda_0} (S_p - S_q)\right] + \exp\left[-\frac{2\pi i}{\lambda_0} (S_p - S_q)\right]\right\} .$$

Substituting, in a preliminary step, the trial solution $e_\omega$ (expressed by (3.66)) into (2.68), and after multiplication with $\exp(-2\pi i S_{\bar{p}}/\tilde{\lambda}_0)$, we obtain

$$\frac{2\pi i}{\tilde{\lambda}_0}\left[\nabla S_{\bar{p}} \cdot u_{\bar{p}} + \nabla S_{\bar{q}} \cdot a_{\bar{q}} \exp\frac{2\pi i}{\tilde{\lambda}_0} (S_{\bar{q}} - S_{\bar{p}})\right]$$

$$= -[\nabla \cdot a_{\bar{p}} + \nabla(\log\tilde{\varepsilon}) \cdot a_{\bar{p}}] - [\nabla \cdot a_{\bar{q}} + \nabla(\log\tilde{\varepsilon}) \cdot a_{\bar{q}}]\exp\left[\frac{2\pi i}{\tilde{\lambda}_0} (S_{\bar{q}} - S_{\bar{p}})\right]$$

$$+ \varrho_\omega \exp\left(-\frac{2\pi i}{\tilde{\lambda}_0} S_{\bar{p}}\right) . \qquad (3.69)$$

The left side of (3.69) has $1/\tilde{\lambda}_0$ as a factor, which does not appear in the terms on the right side. Therefore, to a first approximation, the term in brackets on the left

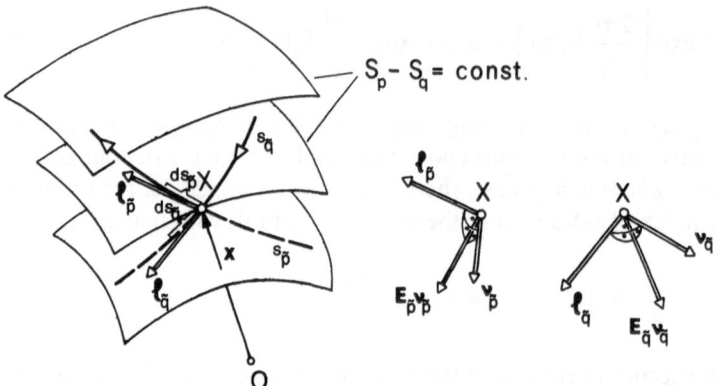

**Fig. 3.10.** Grating formed at the recording in an inhomogeneous medium (three-dimensional case for the coupled wave theory). ($s_q$, $s_p$: arcs; $l_q$, $l_p$: unit direction vectors; $v_q$, $v_p$: complex unit polarization vectors at the reconstruction)

side must disappear. Furthermore, the presence of a steeply varying exponential implies the necessary auxiliary condition expressed by the first of the two following equations

$$\nabla S_{\tilde{p}} \cdot a_{\tilde{p}} = 0 \ , \qquad \nabla S_{\tilde{q}} \cdot a_{\tilde{q}} = 0 \ , \tag{3.70}$$

the second being deduced in a similar manner. The result is that, although the two waves are simultaneously present, their individual polarization directions must be *normal* to their propagation direction, $l_{\tilde{p}}[\nabla S_{\tilde{p}} = |\nabla S_{\tilde{p}}| l_{\tilde{p}}$ (Fig. 3.10)].

Introducing next the expression (3.66) into the wave equation (2.67) leads to an equation for the vectors $a_{\tilde{p}}$ and $a_{\tilde{q}}$. Here, we work out only the detailed operations for $a_{\tilde{p}} \exp(2\pi i\, S_{\tilde{p}}/\tilde{\lambda}_0)$; those for the second wave may be derived similarly

$$\nabla^2 \left[ a_{\tilde{p}} \exp\left( \frac{2\pi i}{\tilde{\lambda}_0} S_{\tilde{p}} \right) \right]$$

$$= \nabla \left[ \nabla \otimes \left( a_{\tilde{p}} \exp \frac{2\pi i}{\tilde{\lambda}_0} S_{\tilde{p}} \right) \right]$$

$$= \nabla \left[ (\nabla \otimes a_{\tilde{p}}) \exp\left( \frac{2\pi i}{\tilde{\lambda}_0} S_{\tilde{p}} \right) + \frac{2\pi i}{\tilde{\lambda}_0} \nabla S_{\tilde{p}} \otimes a_{\tilde{p}} \exp\left( \frac{2\pi i}{\tilde{\lambda}_0} S_{\tilde{p}} \right) \right]$$

$$= \left\{ \left( \frac{2\pi i}{\tilde{\lambda}_0} \right)^2 (\nabla S_{\tilde{p}})^2 a_{\tilde{p}} + \frac{2\pi i}{\tilde{\lambda}_0} \left[ 2\nabla S_{\tilde{p}}(\nabla \otimes a_{\tilde{p}}) + \nabla^2 S_{\tilde{p}} a_{\tilde{p}} \right] \right.$$

$$\left. + \nabla(\nabla \otimes a_{\tilde{p}}) \right\} \exp\left( \frac{2\pi i}{\tilde{\lambda}_0} S_{\tilde{p}} \right) \ ,$$

$$\nabla\left\{\nabla\cdot\left[a_{\tilde{p}}\exp\left(\frac{2\pi i}{\tilde{\lambda}_0}S_{\tilde{p}}\right)\right]\right\}$$

$$=\nabla\left[(\nabla\cdot a_{\tilde{p}})\exp\left(\frac{2\pi i}{\tilde{\lambda}_0}S_{\tilde{p}}\right)+\frac{2\pi i}{\tilde{\lambda}_0}(\nabla S_{\tilde{p}}\cdot a_{\tilde{p}})\exp\left(\frac{2\pi i}{\tilde{\lambda}_0}S_{\tilde{p}}\right)\right]\;.$$

Using the orthogonality requirements of (3.70), we obtain

$$\nabla\left\{\nabla\cdot\left[a_{\tilde{p}}\exp\left(\frac{2\pi i}{\tilde{\lambda}_0}S_{\tilde{p}}\right)\right]\right\}=\left\{\frac{2\pi i}{\tilde{\lambda}_0}[\nabla S_{\tilde{p}}(\nabla\cdot a_{\tilde{p}})]+\nabla^2 a_{\tilde{p}}\right\}\exp\left(\frac{2\pi i}{\tilde{\lambda}_0}S_{\tilde{p}}\right)\;.$$

Referring to the auxiliary relation (3.68) and multiplying the wave equation (2.67) by $\exp(-2\pi i\,S_{\tilde{p}}/\tilde{\lambda}_0)$, we get

$$\frac{4\pi^2}{\tilde{\lambda}_0^2}\left\{[(\nabla S_{\tilde{p}})^2-\tilde{n}_0^2]a_{\tilde{p}}+[(\nabla S_{\tilde{q}})^2-\tilde{n}_0^2]a_{\tilde{q}}\exp\left[-\frac{2\pi i}{\tilde{\lambda}_0}(S_{\tilde{p}}-S_{\tilde{q}})\right]\right\}$$

$$=\frac{2\pi i}{\tilde{\lambda}_0}\left(\left[\nabla^2 S_{\tilde{p}}a_{\tilde{p}}+2\nabla S_{\tilde{p}}(\nabla\otimes a_{\tilde{p}})-\nabla S_{\tilde{p}}(\nabla\cdot a_{\tilde{p}})\right.\right.$$

$$\left.-2i\tilde{n}_0\kappa_1\exp\left(-\frac{2\pi i}{\tilde{\lambda}_0}\Delta S\right)a_{\tilde{q}}\right]+\left[\nabla^2 S_{\tilde{q}}a_{\tilde{q}}+2\nabla S_{\tilde{q}}(\nabla\otimes a_{\tilde{q}})\right.$$

$$\left.-\nabla S_{\tilde{q}}(\nabla\cdot a_{\tilde{q}})-2i\tilde{n}_0\kappa_1\exp\left(\frac{2\pi i}{\tilde{\lambda}_0}\Delta S\right)a_{\tilde{p}}\right]\exp\left[-\frac{2\pi i}{\tilde{\lambda}_0}(S_{\tilde{p}}-S_{\tilde{q}})\right]$$

$$-2i\tilde{n}_0\kappa_1\left\{a_{\tilde{p}}\exp\left[\frac{2\pi i}{\lambda_0}(S_p-S_q)\right]\right.$$

$$\left.\left.+a_{\tilde{q}}\exp\left[-\frac{2\pi i}{\lambda_0}(S_p-S_q)-\frac{2\pi i}{\tilde{\lambda}_0}(S_{\tilde{p}}-S_{\tilde{q}})\right]\right\}\right)+O(1)\;.\qquad(3.71)$$

In this expression, we have used the deviation from the phase matching condition, i.e., the small mismatch: (always with $\tilde{\phi}=\tilde{\psi}=\phi=\psi=0$)

$$\frac{1}{\tilde{\lambda}_0}\Delta S=\frac{1}{\tilde{\lambda}_0}(S_{\tilde{p}}-S_{\tilde{q}})-\frac{1}{\lambda_0}(S_p-S_q)\;.\qquad(3.72)$$

Separating the slowly varying from steeply varying terms and first taking the terms of order $1/\tilde{\lambda}_0^2$ and then those of order $1/\tilde{\lambda}_0$, we obtain the four equations classified in the discussion which follows.

From the slowly varying terms of order $1/\tilde{\lambda}_0^2$, we deduce that

$$|\nabla S_{\tilde{p}}|^2-\tilde{n}_0^2=0\;,$$
$$|\nabla S_{\tilde{q}}|^2-\tilde{n}_0^2=0\;,\qquad(3.73)$$

where the latter is obtained when the full equation (3.71) is multiplied by $\exp\left[2\pi i\,(S_{\tilde{p}} - S_{\tilde{q}})/\tilde{\lambda}_0\right]$. These two equations are the well-known equations of the eikonal, see (2.82). In the same way, slowly varying terms of order $1/\tilde{\lambda}_0$ give

$$2\,\nabla S_{\tilde{p}}(\nabla \otimes \boldsymbol{a}_{\tilde{p}}) + \nabla^2 S_{\tilde{p}} \boldsymbol{a}_{\tilde{p}} - \nabla S_{\tilde{p}}(\nabla \cdot \boldsymbol{a}_{\tilde{p}}) = 2i\tilde{n}_0\,\kappa_1 \boldsymbol{a}_{\tilde{q}} \exp\left(-\frac{2\pi i}{\tilde{\lambda}_0}\varDelta S\right)$$

$$2\,\nabla S_{\tilde{q}}(\nabla \otimes \boldsymbol{a}_{\tilde{q}}) + \nabla^2 S_{\tilde{q}} \boldsymbol{a}_{\tilde{q}} - \nabla S_{\tilde{q}}(\nabla \cdot \boldsymbol{a}_{\tilde{q}}) = 2i\tilde{n}_0\,\kappa_1 \boldsymbol{a}_{\tilde{p}} \exp\left(\frac{2\pi i}{\tilde{\lambda}_0}\varDelta S\right) \ .$$

$$(3.74)$$

The two first-order differential equations (3.74) for the vectors $\boldsymbol{a}_{\tilde{p}}$ and $\boldsymbol{a}_{\tilde{q}}$ are coupled. In order to clarify their geometrical meaning, we may modify them by using (2.83) and (2.85), which read:

$$\nabla S_{\tilde{p}} = \tilde{n}_0 \boldsymbol{l}_{\tilde{p}} \ , \qquad \frac{d\boldsymbol{l}_{\tilde{p}}}{ds_{\tilde{p}}} = \frac{\boldsymbol{m}_{\tilde{p}}}{\varrho_{\tilde{p}}} \ ,$$

$$\nabla S_{\tilde{q}} = \tilde{n}_0 \boldsymbol{l}_{\tilde{q}} \ , \qquad \frac{d\boldsymbol{l}_{\tilde{q}}}{ds_{\tilde{q}}} = \frac{\boldsymbol{m}_{\tilde{q}}}{\varrho_{\tilde{q}}} \ .$$

$$(3.75)$$

The increment $ds_{\tilde{p}}$ represents an infinitesimal length on the light ray with $\boldsymbol{l}_{\tilde{p}}$ as tangential unit vector. Here $\varrho_{\tilde{p}}$ is the radius or curvature and $\boldsymbol{m}_{\tilde{p}}$ the principal unit vector perpendicular to the ray, pointing towards the center of curvature. The first term of (3.74) then becomes

$$2\,\nabla S_{\tilde{p}}(\nabla \otimes \boldsymbol{a}_{\tilde{p}}) = 2\tilde{n}_0 \boldsymbol{l}_{\tilde{p}}(\nabla \otimes \boldsymbol{a}_{\tilde{p}}) = 2\tilde{n}_0 \frac{d\boldsymbol{a}_{\tilde{p}}}{ds_{\tilde{p}}} \ .$$

The second term of (3.74) is transformed with the relation $\nabla \cdot \boldsymbol{l}_{\tilde{p}} = \nabla_1 \cdot \boldsymbol{l}_{\tilde{p}} + (\boldsymbol{l}_{\tilde{p}} \otimes \boldsymbol{l}_{\tilde{p}})\nabla \cdot \boldsymbol{l}_{\tilde{p}} = \nabla_1 \cdot \boldsymbol{l}_{\tilde{p}} + \boldsymbol{l}_{\tilde{p}} \, d\boldsymbol{l}_{\tilde{p}}/ds_{\tilde{p}} = \nabla_1 \cdot \boldsymbol{l}_{\tilde{p}}$ in the following form

$$\nabla^2 S_{\tilde{p}} \boldsymbol{a}_{\tilde{p}} = (\nabla \cdot \nabla S_{\tilde{p}})\boldsymbol{a}_{\tilde{p}} = [(\nabla \tilde{n}_0) \cdot \boldsymbol{l}_{\tilde{p}} + \tilde{n}_0(\nabla \cdot \boldsymbol{l}_{\tilde{p}})]\boldsymbol{a}_{\tilde{p}}$$

$$= \tilde{n}_0 \left( \frac{1}{\tilde{n}_0} \frac{d\tilde{n}_0}{ds_{\tilde{p}}} + \nabla_1 \cdot \boldsymbol{l}_{\tilde{p}} \right)\boldsymbol{a}_{\tilde{p}} \ .$$

Therefore, after being divided by $2\tilde{n}_0$, the two equations (3.74) become

$$\frac{d\boldsymbol{a}_{\tilde{p}}}{ds_{\tilde{p}}} + \frac{1}{2}\left( \nabla_1 \cdot \boldsymbol{l}_{\tilde{p}} + \frac{1}{\tilde{n}_0} \frac{d\tilde{n}_0}{ds_{\tilde{p}}} \right)\boldsymbol{a}_{\tilde{p}} - \frac{1}{2}\boldsymbol{l}_{\tilde{p}}(\nabla \cdot \boldsymbol{a}_{\tilde{p}}) = i\kappa_1 \boldsymbol{a}_{\tilde{q}} \exp\left(-\frac{2\pi i}{\tilde{\lambda}_0}\varDelta S\right) \ ,$$

$$\frac{d\boldsymbol{a}_{\tilde{q}}}{ds_{\tilde{q}}} + \frac{1}{2}\left( \nabla_1 \cdot \boldsymbol{l}_{\tilde{q}} + \frac{1}{\tilde{n}_0} \frac{d\tilde{n}_0}{ds_{\tilde{q}}} \right)\boldsymbol{a}_{\tilde{q}} - \frac{1}{2}\boldsymbol{l}_{\tilde{q}}(\nabla \cdot \boldsymbol{a}_{\tilde{q}}) = i\kappa_1 \boldsymbol{a}_{\tilde{p}} \exp\left(+\frac{2\pi i}{\tilde{\lambda}_0}\varDelta S\right) \ .$$

$$(3.76)$$

In addition, the longitudinal parts of (3.76) are obtained by contraction of each equation by $\boldsymbol{l}_{\tilde{p}}$ and $\boldsymbol{l}_{\tilde{q}}$, respectively. Noting that $(d\boldsymbol{a}_{\tilde{p}}/ds_{\tilde{p}}) \cdot \boldsymbol{l}_{\tilde{p}} =$

$d/ds_{\bar{p}}(a_{\bar{p}} \cdot l_{\bar{p}}) - (dl_{\bar{p}}/ds_{\bar{p}}) \cdot a_{\bar{p}} = -(dl_{\bar{p}}/ds_{\bar{p}}) \cdot a_{\bar{p}}$, with respect to (3.75), we conclude that $(da_{\bar{p}}/ds_{\bar{p}}) \cdot l_{\bar{p}} = -1/\varrho_{\bar{p}}(m_{\bar{p}} \cdot a_{\bar{p}})$; therefore, we have

$$-\frac{1}{\varrho_{\bar{p}}}(m_{\bar{p}} \cdot a_{\bar{p}}) - \frac{1}{2}(\nabla \cdot a_{\bar{p}}) = i\kappa_1(a_{\bar{q}} \cdot l_{\bar{p}})\exp\left(-\frac{2\pi i}{\tilde{\lambda}_0}\Delta S\right) ,$$

$$-\frac{1}{\varrho_{\bar{q}}}(m_{\bar{q}} \cdot a_{\bar{q}}) - \frac{1}{2}(\nabla \cdot a_{\bar{q}}) = i\kappa_1(a_{\bar{p}} \cdot l_{\bar{q}})\exp\left(\frac{2\pi i}{\tilde{\lambda}_0}\Delta S\right) .$$

These two equations contain the radii of curvature of the rays and could be used to alternatively determine the divergences $(\nabla \cdot a_{\bar{p}})$ and $(\nabla \cdot a_{\bar{q}})$.

Apparently (3.76) represent a system of $2 \times 3$ differential equations for the $2 \times 3$ components of the vectors $a_{\bar{p}}$ and $a_{\bar{q}}$. However, from (3.70) we know that these vectors are perpendicular to their corresponding rays. Since an interdependence exists, we must consider mainly the lateral parts of (3.76). Projecting the vectorial equations (3.76) by the operators $L_{\bar{p}} = I - l_{\bar{p}} \otimes l_{\bar{p}}$ and $L_{\bar{q}} = I - l_{\bar{q}} \otimes l_{\bar{q}}$, respectively, we find

$$L_{\bar{p}}\frac{da_{\bar{p}}}{ds_{\bar{p}}} + \frac{1}{2}\left(\nabla_{\bar{1}} \cdot l_{\bar{p}} + \frac{1}{\tilde{n}_0}\frac{d\tilde{n}_0}{ds_{\bar{p}}}\right)a_{\bar{p}} = L_{\bar{p}}a_{\bar{q}}i\kappa_1 \exp\left(-\frac{2\pi i}{\tilde{\lambda}_0}\Delta S\right)$$

$$L_{\bar{q}}\frac{da_{\bar{q}}}{ds_{\bar{q}}} + \frac{1}{2}\left(\nabla_{\bar{1}} \cdot l_{\bar{q}} + \frac{1}{\tilde{n}_0}\frac{d\tilde{n}_0}{ds_{\bar{q}}}\right)a_{\bar{q}} = L_{\bar{q}}a_{\bar{p}}i\kappa_1 \exp\left(\frac{2\pi i}{\tilde{\lambda}_0}\Delta S\right) . \tag{3.77}$$

In the coupled terms of these equations, only the projection of the polarization vector of one wave onto the plane perpendicular to the propagation direction of the second wave appears [Ref. 3.185, p. 2945]. In analogy to Sect. 2.2.1, we separate $a_{\bar{p}}$ multiplicatively into a real scalar $a_{\bar{p}}'$ and a complex unit vector $v_{\bar{p}}$, which is always perpendicular to $l_{\bar{p}}$ ($a_{\bar{q}}$ is likewise separated with its respective terms):

$$a_{\bar{p}} = a_{\bar{p}}' v_{\bar{p}} , \quad a_{\bar{q}} = a_{\bar{p}}' v_{\bar{q}} . \tag{3.78}$$

Rewriting, for example, (3.77a) with this separation gives

$$\left[\frac{da_{\bar{p}}'}{ds_{\bar{p}}} + \frac{1}{2}\left(\nabla_{\bar{1}} \cdot l_{\bar{p}} + \frac{1}{\tilde{n}_0}\frac{d\tilde{n}_0}{ds_{\bar{p}}}\right)a_{\bar{p}}'\right]v_{\bar{p}} + a_{\bar{p}}'L_{\bar{p}}\frac{dv_{\bar{p}}}{ds_{\bar{p}}}$$

$$= L_{\bar{p}}v_{\bar{q}}\left[a_{\bar{q}}'i\kappa_1\exp\left(-\frac{2\pi i}{\tilde{\lambda}_0}\Delta S\right)\right] . \tag{3.79}$$

the derivative $da_{\bar{p}}'/ds_{\bar{p}}$ describes the amplitude variation along the ray, whereas the projection, $L_{\bar{p}} dv_{\bar{p}}/ds_{\bar{p}}$, reveals the rotation of the polarization direction along the ray $s_{\bar{p}}$.

For a further separation, we first multiply (3.79) by the conjugate $v_{\bar{p}}^*(\text{Re}\{v_{\bar{p}}^* \cdot dv_{\bar{p}}/ds_{\bar{p}}\} = 0)$, and the corresponding equation to $a_{\bar{q}}$ by $v_{\bar{q}}^*$. We then get the following two real scalar coupled-wave equations:

$$\frac{da'_{\tilde{p}}}{ds_{\tilde{p}}} + \frac{1}{2}\left(\nabla_{\tilde{1}} \cdot l_{\tilde{p}} + \frac{1}{\tilde{n}_0}\frac{d\tilde{n}_0}{ds_{\tilde{p}}}\right)a'_{\tilde{p}} = a'_{\tilde{q}}\mathrm{Re}\left\{(v^*_{\tilde{p}} \cdot v_{\tilde{q}})i\kappa_1 \exp\left(-\frac{2\pi i}{\tilde{\lambda}_0}\Delta S\right)\right\}$$

$$\frac{da'_{\tilde{q}}}{ds_{\tilde{q}}} + \frac{1}{2}\left(\nabla_{\tilde{1}} \cdot l_{\tilde{q}} + \frac{1}{\tilde{n}_0}\frac{d\tilde{n}_0}{ds_{\tilde{q}}}\right)a'_{\tilde{q}} = a'_{\tilde{p}}\mathrm{Re}\left\{(v_{\tilde{p}} \cdot v^*_{\tilde{q}})i\kappa_1 \exp\left(\frac{2\pi i}{\tilde{\lambda}_0}\Delta S\right)\right\} .$$

$$(3.80)$$

Secondly, multiplying (3.79) by $E_{\tilde{p}}v_{\tilde{p}}$ [$E_{\tilde{p}}$ is the two-dimensional permutation tensor, see (2.34)] and the corresponding equation for $a_{\tilde{q}}$ by $E_{\tilde{q}}v_{\tilde{q}}$, we obtain two complex and two real scalar equations:

$$a'_{\tilde{p}}\left(\frac{dv_{\tilde{p}}}{ds_{\tilde{p}}} \cdot E_{\tilde{p}}v_{\tilde{p}}\right) = a'_{\tilde{q}}(v_{\tilde{q}} \cdot E_{\tilde{p}}v_{\tilde{p}})i\kappa_1 \exp\left(-\frac{2\pi i}{\tilde{\lambda}_0}\Delta S\right) ,$$

$$a'_{\tilde{q}}\left(\frac{dv_{\tilde{q}}}{ds_{\tilde{q}}} \cdot E_{\tilde{q}}v_{\tilde{q}}\right) = a'_{\tilde{p}}(v_{\tilde{p}} \cdot E_{\tilde{q}}v_{\tilde{q}})i\kappa_1 \exp\left(\frac{2\pi i}{\tilde{\lambda}_0}\Delta S\right) ,$$

$$a'_{\tilde{p}}\left(\frac{dv_{\tilde{p}}}{ds_{\tilde{p}}} \cdot v^*_{\tilde{p}}\right) = ia'_{\tilde{q}}\,\mathrm{Im}\left\{(v^*_{\tilde{p}} \cdot v_{\tilde{p}})i\kappa_1 \exp\left(-\frac{2\pi i}{\tilde{\lambda}_0}\Delta S\right)\right\} ,$$

$$a'_{\tilde{q}}\left(\frac{dv_{\tilde{q}}}{ds_{\tilde{q}}} \cdot v^*_{\tilde{q}}\right) = ia'_{\tilde{p}}\,\mathrm{Im}\left\{(v_{\tilde{p}} \cdot v^*_{\tilde{q}})i\kappa_1 \exp\left(\frac{2\pi i}{\tilde{\lambda}_0}\Delta S\right)\right\} .$$

$$(3.81)$$

Before resolving (3.80 and 81), the unit vector fields $l_{\tilde{p}}(x)$, $l_{\tilde{q}}(x)$ must be found from the integration of the eikonal equation (3.73) and (3.75), which then allows us to determine the rays $x(s_{\tilde{p}})$, $x(s_{\tilde{q}})$. We may restrict (3.80) to the most often encountered example in this chapter: point sources $\tilde{Q}$, $\tilde{P}$ (the existence of $\tilde{P}$ is assumed) are very closely set points, situated in a homogeneous material ($\tilde{n}_0 = $ constant, inside and outside). In this special case, the rays are straight lines, and with $\nabla_{\tilde{1}} \otimes l_{\tilde{p}} = (I - l_{\tilde{p}} \otimes l_{\tilde{p}})/\tilde{p}$, see (2.24), we have $\nabla_{\tilde{1}} \cdot l_{\tilde{p}}/2 = 1/\tilde{p}$. Therefore, the system (3.80) reduces to

$$\frac{da'_{\tilde{p}}}{ds_{\tilde{p}}} + \frac{1}{\tilde{p}}a'_{\tilde{p}} = a'_{\tilde{q}}\mathrm{Re}\left\{(v^*_{\tilde{p}} \cdot v_{\tilde{q}})i\kappa_1 \exp\left(-\frac{2\pi i}{\tilde{\lambda}_0}\Delta S\right)\right\} ,$$

$$\frac{da'_{\tilde{q}}}{ds_{\tilde{q}}} + \frac{1}{\tilde{q}}a'_{\tilde{q}} = a'_{\tilde{p}}\mathrm{Re}\left\{(v_{\tilde{p}} \cdot v^*_{\tilde{q}})i\kappa_1 \exp\left(\frac{2\pi i}{\tilde{\lambda}_0}\Delta S\right)\right\} .$$

$$(3.82)$$

It should be noted that (3.80, 81) cannot be used directly for a simple integration, such as with (3.65), since the increments $ds_{\tilde{p}}$ and $ds_{\tilde{q}}$ are not associated by a known relation. (In the one-dimensional case, such a relation is known beforehand; the homogeneity and the uniformity of the medium implying a unique dependence $a_{\tilde{p}}(z)$ and $a_{\tilde{q}}(z)$, so that $ds_{\tilde{p}}(\hat{n} \cdot \tilde{k}) = ds_{\tilde{q}}(\hat{n} \cdot \tilde{c}) = dz$.) Nevertheless, the systems (3.80 and 81) could be used in combination with some iteration.

## 3.2 Deformation and Distortion of the Image of an Object at the Reconstruction

We now return to the investigation of images formed by diffraction at *thin* holograms. The aberrations we have studied so far (Sect. 3.1.2) refer to the modifications of a single point "source" P → P̃. Here, we shall look at the modification of a whole neighborhood {P} of some object point P, which becomes another neighborhood {P̃} of P̃. In the terminology of optics, one often encounters the word distortion, when referring to a special kind of apparent deformation. Before treating the general deformation of such images, it seems advisable to first present some elements that link this field to the classical theory of Seidel aberrations as used in axi-symmetric lens imaging. Thereafter, we shall study general configurations with oblique incidence, in particular, the apparent projected deformation and, lastly, the corresponding deformation in space when the image is viewed stereoscopically by both eyes.

### 3.2.1 Primary Seidel Aberrations for a Rotational Symmetric Modification

The terminology of aberrations was originally adapted to the rotational symmetric case of lens imagery where the disturbing effects (with respect to simple Gaussian optics) caused by the *small* off-axis position of object points and image points are investigated. In the previous subsection, we have already encountered the notion of astigmatism in the case of a finite inclination of the image ray (direction $\tilde{k}$) with reference to the hologram normal (direction $\hat{n}$). If we want to study other effects of higher order, due to the modification of the image points influencing the apparent deformation of the whole image, it is helpful to first consider an example of rotational symmetry. Here, we can perform a development for a *small* inclination of the rays with respect to the hologram normal passing through a point of observation. This normal then plays the role of an optical axis. These kinds of problems (paraxial approximations) are treated in detail in [3.85, 135, 152, 263–286], and are very similar to the classical theory of Seidel aberrations in lens systems [Ref. 3.102, p. 211]. However, we introduce them here in a rather simple form which could lead to a possible generalization in nonsymmetric cases with finite oblicity of the rays.

Let us assume a modification due to a wavelength change ($\tilde{\lambda} \neq \lambda$) and, for simplicity, a collimated reference light at the recording ($q \rightarrow \infty$), as well as at the reconstruction ($\tilde{q} \rightarrow \infty$). Moreover, the two unit direction vectors $c$ and $\tilde{c}$, not necessarily parallel to $\hat{n}$, should be related to one another by the supplementary condition

$$\frac{1}{\tilde{\lambda}}\hat{N}\tilde{c} = \frac{1}{\lambda}\hat{N}c \ . \tag{3.83}$$

In other words, the wavelength change is somehow partially compensated by a geometrical change of the reference light direction. These conditions are very restrictive when compared to the case treated by *Meier* [3.263], but are necessary for our purpose inasmuch as a point on the optical axis remains on it at the reconstruction, as in classical lens systems. This permits us to interpret the variation of the off-axis distance of a point as a distortion in the sense of a proper image deformation $dr \to d\tilde{r}$, and not as a combination of lateral displacement and deformation. Furthermore, the Seidel coefficients are deduced from a Taylor series of the phase function with derivatives accompanied by increments; by contrast in the work of *Meier* [3.263] and *Mulak* [3.273], corresponding coefficients are deduced from an analogous series development of square roots, including the finite off-axis position of the rays from the sources. The present approach enables us to later make a separation between terms that are, in fact, constant, such as those indicating the positions of sources and terms varying in a neighborhood of the optical axis (e.g. the positions of hologram points and object points).

We now consider an object point P, situated on the normal to the hologram and passing through the center R of the observing instrument (Fig. 3.11), so that $k = \hat{n}$. Then, (3.15) and (3.83) show that we also have $\tilde{k} = \hat{n}$, i.e., the image $\tilde{P}$ is on the line $\check{R}P$. Because of the apparent symmetry of revolution for the set of neighboring rays through P and $\tilde{P}$ [expressed by (3.17, 83)], the image $\tilde{P} = \tilde{P}_0$ is stigmatic and from (3.33), we obtain Gauss' relation

$$\frac{1}{\tilde{\lambda}\tilde{p}_0} = \frac{1}{\tilde{\lambda}\tilde{p}} = \frac{1}{\lambda p} . \tag{3.84}$$

Let us also consider another neighboring object point $\overline{P}$ in the plane parallel to the hologram through P and separated from P by an increment $dr$. In fact, $dr$ will later be an incremental vector of arbitrary direction of the position vector $r$ on an opaque object. For the moment, however, $\overline{P}$ keeps the same distance from the hologram plane as P, and the unit vector $n$ normal to the object *remains* parallel to $\hat{n}$. Equation (3.15), applied at $H$ for the ray $\overline{P}H$, becomes

$$N\left(\frac{1}{\tilde{\lambda}}\tilde{k}' - \frac{1}{\lambda}\bar{k}'\right) = 0 , \tag{3.85}$$

which determines the image ray of direction $\tilde{\bar{k}}'$. Its intersection with the Gaussian image plane through $\tilde{P}_0$ is a point $\tilde{\bar{P}}$ which differs slightly from the Gaussian image $\tilde{\bar{P}}_0$. This latter point is found from the linearized equation (3.85)

$$\frac{d\tilde{r}_0}{\tilde{\lambda}\tilde{p}} = \frac{dr}{\lambda p} , \tag{3.86}$$

so that $d\tilde{r}_0 = dr$ because of (3.84). In this first-order approximation, a greater wavelength at the reconstruction ($\tilde{\lambda} > \lambda$), simply shifts the image from $P\overline{P}$ to $\tilde{P}_0\tilde{\bar{P}}_0$ without any change of size [3.85, 263].

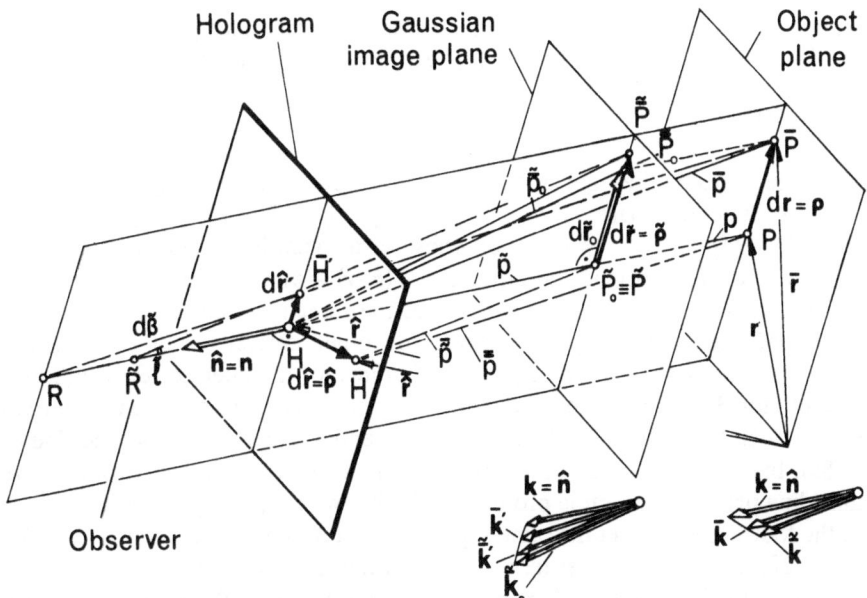

**Fig. 3.11.** Paraxial approximation around a hologram normal in the case of a particular "symmetric" optical modification. ($dr$, $d\bar{r}$: associated increments of the object at the recording and of the image at the reconstruction)

The wave aberration at the neighboring reference point $\bar{\bar{P}}_0$ is

$$\phi(\bar{\hat{r}}, \bar{r}) = \theta_0(\bar{\hat{r}}, \bar{\imath}) + \bar{\bar{\phi}}_0(\bar{r}) - \phi(\bar{r}) - \tilde{\psi} + \psi \ . \tag{3.87}$$

Contrary to (3.10), the function $\theta_0(\bar{\hat{r}}, \bar{r})$ is now a function of two variables $\bar{\hat{r}}$ and $\bar{r}$, the vector coordinates of the points $\bar{H}$ and $\bar{P}$, respectively, or the Seidel variables. A series development up to the fourth-order derivatives may thus be written in the form of a "triangle of Pascal" [3.273], with the abbreviations $d\hat{r} = \hat{\varrho}$, $dr = d\bar{r}_0 = \varrho$, $\partial_r \otimes \partial_r = \partial_{rr}$, $\partial_r \otimes \partial_r \otimes \partial_r = \partial_{rrr}$, ... in the form

$$\phi(\bar{\hat{r}}, \bar{r}) \simeq \phi(\hat{r}, r)$$

$$+ \hat{\varrho} \cdot \partial_{\hat{r}}\phi + \varrho \cdot \partial_r \phi$$

$$+ \frac{1}{2!}\hat{\varrho} \cdot \partial_{\hat{r}\hat{r}}\phi\hat{\varrho} + \frac{2}{2!}\hat{\varrho} \cdot \partial_{\hat{r}r}\phi\varrho + \frac{1}{2!}\varrho \cdot \partial_{rr}\phi\varrho$$

$$+ \frac{1}{3!}\hat{\varrho} \cdot (\hat{\varrho}\partial_{\hat{r}\hat{r}\hat{r}}\phi\hat{\varrho}) + \frac{3}{3!}\hat{\varrho} \cdot (\hat{\varrho}\partial_{\hat{r}\hat{r}r}\phi\varrho) + \frac{3}{3!}\varrho \cdot (\hat{\varrho}\partial_{\hat{r}rr}\phi\varrho) + \frac{1}{3!}\varrho \cdot (\varrho\partial_{rrr}\phi\varrho)$$

$$+ \frac{4}{4!}\hat{\varrho} \cdot (\hat{\varrho}\partial_{\hat{r}\hat{r}\hat{r}r}\phi\varrho)\hat{\varrho} + \frac{6}{4!}\hat{\varrho} \cdot (\hat{\varrho}\partial_{\hat{r}\hat{r}rr}\phi\varrho)\varrho + \frac{4}{4!}\varrho \cdot (\hat{\varrho}\partial_{\hat{r}rrr}\phi\varrho)\varrho$$

$$+ \frac{1}{4!}\hat{\varrho} \cdot (\hat{\varrho}\partial_{\hat{r}\hat{r}\hat{r}\hat{r}}\phi\hat{\varrho})\hat{\varrho} \qquad\qquad + \frac{1}{4!}\varrho \cdot (\varrho\partial_{rrrr}\phi\varrho)\varrho \ .$$

$$\tag{3.88}$$

Since the function $\tilde{\bar{\phi}}_0$ is defined as the phase of the reference point $\tilde{\bar{P}}_0$, and since the function $\theta_0$ is defined as in (3.10), because of (3.9), we have

$$\phi(\hat{r}, \bar{r}) = 0 . \tag{3.89}$$

The edge terms on the right side of the triangle represent the series development of this function, $\phi(\hat{r}, \bar{r}) = \varrho \cdot \partial_r \phi + \varrho \cdot \partial_{rr}\phi\varrho/2! + \varrho \cdot (\varrho\partial_{rrr}\phi\varrho)/3! + \varrho \cdot (\varrho\partial_{rrrr}\phi\varrho)\varrho/4! + \ldots = 0$, all of whose terms vanish simultaneously. In the literature, we find instead the difference $\phi(\bar{r}, \bar{r}) - \phi(\hat{r}, \bar{r}) = \theta_0(\bar{r}, \bar{r}) - \theta_0(\hat{r}, \bar{r})$, which corresponds to the development of the remaining left triangle of (3.88) [3.85, 152, 263, 271–273]. Performing the other derivatives of the function $\phi(\bar{r}, \bar{r})$, we notice that P and $P_0$ are fixed points for the operator $\partial_{\hat{r}}$, see (2.59), whereas for $\partial_r, \ldots$, H is the fixed point. Therefore, as a result of the first derivatives $\partial_r\phi$ and $\partial_{\hat{r}}\phi$, see (2.23), unit vectors $\hat{n}$ and $n$ appear in the second line of (3.88). In our special case, $\hat{n}$ is equivalent to $n$. The second derivative of $\phi$ in the third line produces the same projector $N$ in the two remaining terms, see (2.24). In the fourth line, we get the same superprojector $N = N \otimes n + n \otimes N + N \otimes n)^T$, see (2.27), for all third-order terms. In the fifth line (fourth derivatives), we have the same hyperprojector, $N = n \otimes N + {}^T(n \otimes N + N \otimes n)^T + N \otimes n - N \otimes N - N \otimes N)^T - (N \otimes N)^T$, see (2.28). The scalars obtained by contraction of the above tensors by the vectors $\varrho$ and $\hat{\varrho}$ vanish when containing $n$, which is the case for all of the terms in the second and fourth lines (usually, it is said by reason of symmetry).

Furthermore, with (3.10), the second-order terms (third line) become

$$\frac{1}{2!}2\pi \left( \frac{1}{\tilde{\lambda}\bar{p}} - \frac{1}{\lambda p} \right) \hat{\varrho} \cdot N\hat{\varrho} - \frac{2}{2!}2\pi \left( \frac{1}{\tilde{\lambda}\bar{p}} - \frac{1}{\lambda p} \right) \hat{\varrho} \cdot N\varrho = 0 , \tag{3.90}$$

and vanish because of Gauss' relation (3.84). Thus, only the four fourth-order terms remain, which all contain as a common factor the fourth-order tensor, again see (2.28):

$$\mp 2\pi \left( \frac{1}{\tilde{\lambda}\bar{p}^3} - \frac{1}{\lambda p^3} \right) [N \otimes N + N \otimes N)^T + (N \otimes N)^T] .$$

Note that an alternation of sign occurs since an exchange of the collineation center intervenes in the mixed derivatives, such as $\partial_{r\hat{r}}(\ldots)$. This alternation is due to the opposite position of the collineation centers involved. The order of the terms is named according to the order of the derivative. That is why the Gaussian imaging is represented by the second-order terms (second derivative), whereas here the primary Seidel aberrations are the fourth-order terms (fourth derivative), named with this order instead of third order as in classical aberrations. We choose this classification according to the order of derivatives so as to be consistent with Sects. 3.1.2 and 3.2.2, 3, where point aberrations and image deformations are treated in a general holographic configuration.

Performing then the contractions in accordance with the rule for quadriadics, see (2.12), eq. (3.88) reduces to the simpler form:

$$\frac{\tilde{\lambda}}{2\pi}\phi(\tilde{r},\,\tilde{r}) \simeq -\tilde{\lambda}\left(\frac{1}{\tilde{\lambda}\tilde{p}^3} - \frac{1}{\lambda p^3}\right) \cdot \left\{\frac{3}{4!}\,(\varrho \cdot \varrho)(\varrho \cdot \varrho) - \frac{3 \cdot 4}{4!}\,(\varrho \cdot \varrho)(\varrho \cdot \varrho)\right.$$

$$\left. + \frac{6}{4!}\left[(\varrho \cdot \varrho)(\varrho \cdot \varrho) + 2(\varrho \cdot \varrho)(\varrho \cdot \varrho)\right] - \frac{3 \cdot 4}{4!}\,(\varrho \cdot \varrho)(\varrho \cdot \varrho)\right\}$$

$$= \tilde{\lambda}\left(\frac{1}{\tilde{\lambda}\tilde{p}^3} - \frac{1}{\lambda p^3}\right)\left[-\frac{1}{8}\varrho^4 + \frac{1}{2}\varrho^2(\varrho \cdot \varrho) - \frac{1}{4}\varrho^2\varrho^2 - \frac{1}{2}\,(\varrho \cdot \varrho)^2\right.$$

$$\left. + \frac{1}{2}\,(\varrho \cdot \varrho)\varrho^2\right]. \tag{3.91}$$

The five terms in the bracket, multiplied by the characteristic factor

$$\tilde{\lambda}\left(\frac{1}{\tilde{\lambda}\tilde{p}^3} - \frac{1}{\lambda p^3}\right) = \chi \,,$$

describe the primary Seidel aberrations in the above order from left to right: the *spherical aberration*, the *coma*, the *field curvature*, the *astigmatism*, and the *distortion* [3.20, 84, 85, 102, 263, 273]. Some additional explanations are now necessary:

a) The astigmatism, here defined by the fourth term in (3.91), is consistent with the definition given by (3.34–36). With a similar development to (3.86) $(d\tilde{r}_0 = \varrho = dr)$

$$N\bar{k}' = -\frac{1}{p}\varrho \,, \qquad N\bar{k}_0 = -\frac{1}{\tilde{p}}\varrho \,, \tag{3.92}$$

we get

$$-\frac{1}{2}\chi(\varrho \cdot \varrho)^2 = -\frac{\tilde{\lambda}}{2}\left(\frac{1}{\tilde{\lambda}\tilde{p}^3} - \frac{1}{\lambda p^3}\right)\varrho \cdot (\varrho \otimes \varrho)\varrho$$

$$= -\frac{\tilde{\lambda}}{2}\varrho \cdot \left[\frac{1}{\tilde{\lambda}\tilde{p}}\,(\bar{k}_0 \otimes \bar{k}_0) - \frac{1}{\lambda p}(\bar{k}' \otimes \bar{k}')\right]\varrho$$

$$= -\frac{1}{2}\varrho \cdot N\tilde{S}N\varrho + O(\varrho^4) \,. \tag{3.93}$$

Moreover, because of symmetry, we have the special case of one eigenvector situated in the meridian plane and the other perpendicular to it. With $1/\tilde{\bar{p}}_2 = 1/\tilde{\bar{p}}_0$ and (3.34), we obtain

$$\bar{\bar{S}} = \frac{1}{2} \left( \frac{1}{\bar{p}_1} - \frac{1}{\bar{p}_2} \right) \bar{\bar{K}} + \frac{1}{2} \left( \frac{1}{\bar{p}_1} - \frac{1}{\bar{p}_2} \right) \bar{\bar{D}} \ , \tag{3.94}$$

from which we conclude that (3.93) contains only the *astigmatism*.

b) If $\hat{\varrho}$ is parallel to $\varrho$, the third term in (3.91) is half of the fourth term; hence for this isotropic term we may write a function of an invariant of $S$:

$$-\frac{1}{4} \chi \hat{\varrho}^2 \varrho^2 = -\frac{\tilde{\lambda}}{4} \left( \frac{1}{\tilde{\lambda}\bar{p}^3} - \frac{1}{\lambda p^3} \right) \varrho^2 \hat{\varrho}^2 = -\frac{1}{4} \hat{\varrho} \cdot N\bar{\bar{S}}N\hat{\varrho} \bigg|_{\hat{\varrho}\|\varrho} = -\frac{1}{4} \left( \frac{1}{\bar{p}_1} - \frac{1}{\bar{p}_0} \right) \hat{\varrho}^2$$

$$= -\frac{1}{2} \left[ \frac{1}{2} \left( \frac{1}{\bar{p}_1} + \frac{1}{\bar{p}_2} \right) - \frac{1}{\bar{p}_0} \right] \hat{\varrho}^2 + \dots \ . \tag{3.95}$$

We see that this result is now interpreted to depend only on the *field curvature*. Alternatively, with

$$\varrho^2/p^2 \cong 2(1 - \boldsymbol{n} \cdot \bar{\boldsymbol{k}}') \quad \text{and} \quad \varrho^2/\bar{p}^2 \cong 2(1 - \boldsymbol{n} \cdot \bar{\boldsymbol{k}}_0)$$

we may write

$$-\frac{1}{4} \chi \hat{\varrho}^2 \varrho^2 = -\frac{\tilde{\lambda}}{2} \left[ \frac{1}{\tilde{\lambda}\bar{p}} (1 - \boldsymbol{n} \cdot \bar{\boldsymbol{k}}_0) - \frac{1}{\lambda p} (1 - \boldsymbol{n} \cdot \bar{\boldsymbol{k}}') \right] \hat{\varrho}^2$$

$$= \frac{\tilde{\lambda}}{2} \left( \frac{1}{\tilde{\lambda}\bar{p}_0} - \frac{1}{\lambda \bar{p}} \right) \hat{\varrho}^2 \ , \tag{3.96}$$

which illustrates the change of curvature, $1/\bar{p}_0 - \tilde{\lambda}/\lambda\bar{p}$, compared to a simple magnification $\tilde{\lambda}/\lambda\bar{p}$.

c) In order to explain the fifth term of distortion in (3.91), a refined development is needed, namely: $(\tilde{\varrho} = d\bar{r})$

$$N\bar{\boldsymbol{k}}' = -\frac{1}{p} \varrho + \frac{1}{2p^3} \varrho(\varrho \cdot \varrho) \ , \qquad N\tilde{\bar{\boldsymbol{k}}}' = -\frac{1}{\bar{p}} \tilde{\varrho} + \frac{1}{2\bar{p}^3} \varrho(\varrho \cdot \varrho)$$

$$\tag{3.97}$$

$$N\tilde{\bar{\boldsymbol{k}}}_0 = -\frac{1}{\bar{p}} \varrho + \frac{1}{2\bar{p}^3} \varrho(\varrho \cdot \varrho) \ ,$$

A combination of these relations with (3.84 and 85) gives for the product of the difference $N(\bar{\boldsymbol{k}}' - \bar{\boldsymbol{k}}_0)$ with $\hat{\varrho}$:

$$\hat{\varrho} \cdot N(\bar{\boldsymbol{k}}' - \bar{\boldsymbol{k}}_0) = \hat{\varrho} \cdot \left( \frac{\varrho}{\bar{p}} - \frac{\tilde{\varrho}}{\bar{p}} \right) = -\frac{\tilde{\lambda}}{2} \left( \frac{1}{\tilde{\lambda}\bar{p}^3} - \frac{1}{\lambda p^3} \right) \hat{\varrho} \cdot \varrho(\varrho \cdot \varrho)$$

$$= -\frac{\chi}{2} (\hat{\varrho} \cdot \varrho) \varrho^2 \ . \tag{3.98}$$

This term is, therefore, due to the fact that $\tilde{\bar{P}}_0$ does not coincide with the point $\tilde{\bar{P}}$ and vanishes if $\hat{\varrho}$ is perpendicular to $\varrho$. However, if we choose $\hat{\varrho} = \hat{\varrho}' = d\hat{r}'$ parallel to $\varrho$, so that point $\tilde{\bar{P}}$ in the Gaussian plane and point $\overline{H}'$ on the hologram lie nearly on a line of sight through the observer's position $\tilde{R}$ on the axis, with an angular increment $d\tilde{\beta}$ and a distance $\tilde{l} \ll p$ (for the moment, we assume here the observer to be relatively close to the hologram), we may write

$$- \hat{\varrho} \cdot N(\tilde{\bar{k}}' - \tilde{\bar{k}}_0) = + \tilde{l}\, d\tilde{\beta} |\tilde{\bar{k}}' - \tilde{\bar{k}}_0| \ .$$

Therefore, the *distortion* (relative to the observer's distance from the hologram) describes the lateral refinement of the off-axis position of $\tilde{\bar{P}}_0$. Since (3.92) are linear in $\varrho$, whereas (3.98) is nonlinear in $\varrho$, a rectangular grid in the vicinity of P becomes an apparently curved grid in the vicinity of $\tilde{\bar{P}}_0$. This phenomenon is well known from classical lens optics (Fig. 3.12).

d) The first term in the bracket of (3.91) is independent of $\varrho$ and, therefore, also present on the axis. We now use an analogue to or dual development of (3.92) (Fig. 3.11)

$$N\bar{k} \ = \frac{1}{p}\, \hat{\varrho} \ , \qquad N\tilde{\bar{k}} = \frac{1}{\tilde{p}}\, \hat{\varrho} \ , \tag{3.99}$$

so that this term of *spherical aberration* is similar to (3.96):

$$\frac{\tilde{\lambda}}{2\pi 4!} \ \hat{\varrho} \cdot (\hat{\varrho}\partial_{\textit{ffff}}\varphi\,\hat{\varrho})\,\hat{\varrho} \ = -\frac{\chi}{8}\,\hat{\varrho}^4 = -\frac{\tilde{\lambda}}{8}\left(\frac{1}{\tilde{\lambda}\tilde{p}^3} - \frac{1}{\lambda p^3}\right)\hat{\varrho}^4$$

$$= \frac{\tilde{\lambda}}{4}\left(\frac{1}{\tilde{\lambda}\tilde{p}} - \frac{1}{\lambda\bar{p}}\right)\hat{\varrho}^2 \ . \tag{3.100}$$

A spherical wave, starting from point P at the recording and having the curvature $1/\bar{p}$ at $\overline{H}$, is not simply "magnified" at the reconstruction by the factor $\tilde{\lambda}/\lambda$. On the contrary, the neighboring rays to the axis have a change of curvature $1/\tilde{\bar{p}} - \tilde{\lambda}/\lambda\bar{p} \neq 0$. In this case of symmetry of revolution, we also can easily apply

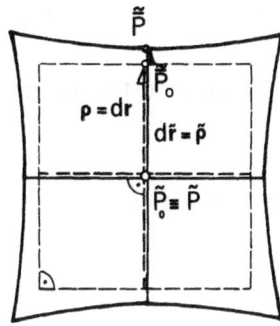

Fig. 3.12. Distortion of a rectangular grid in the vicinity of the axis

Hologram

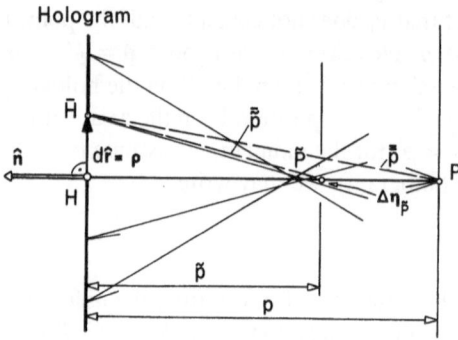

Fig. 3.13. Spherical aberration: Improper location of the image $\tilde{P}$ of P on the axis

(3.21), containing the transverse ray aberration. Here, with $\phi|_{\varrho=0} = \theta_0 + \tilde{\phi}_0 - \phi - \tilde{\psi} + \psi$ on the axis at $\tilde{P}$, we first have

$$d\eta_{\tilde{P}} = \frac{\tilde{\lambda}\tilde{p}}{2\pi} d\tilde{r} \partial_{\tilde{r}\tilde{r}} \phi = 0 , \quad d^2\eta_{\tilde{P}} = 0 , \quad \text{but}$$

$$d^3\eta_{\tilde{P}} = \frac{\tilde{\lambda}\tilde{p}}{2\pi} d\tilde{r}(d\tilde{r} \, \partial_{\tilde{r}\tilde{r}\tilde{r}}\phi d\tilde{r}) \neq 0 .$$

Therefore, the refined aberration with (3.99) is

$$\varDelta \eta_{\tilde{P}} = d\eta_{\tilde{P}} + \frac{1}{2!}d^2\eta_{\tilde{P}} + \frac{1}{3!}d^3\eta_{\tilde{P}} = \frac{1}{3!}d^3\eta_{\tilde{P}} = \frac{\tilde{\lambda}\tilde{p}}{2\pi 3!} \, \varrho(\varrho\partial_{\tilde{r}\tilde{r}\tilde{r}}\phi\varrho) ,$$

or

$$\varDelta \eta_{\tilde{P}} = -\frac{\tilde{\lambda}\tilde{p}}{2} \left(\frac{1}{\tilde{\lambda}\tilde{p}^3} - \frac{1}{\lambda p^3}\right) \hat{\varrho}^2\varrho = \tilde{\lambda}\tilde{p} \left(\frac{1}{\tilde{\lambda}\tilde{p}} - \frac{1}{\lambda\tilde{p}}\right) \varrho .$$

(3.101)

The spherical aberration causes a lateral refinement of rays, from which, in case of a large aperture, results an improper longitudinal location of $\tilde{P}$ (Fig. 3.13).

e) Finally, as for the second term in (3.91), and because of symmetry, we may write $\partial_r\phi_0(\tilde{r})|_{\varrho=0} = 0$, so that with the differential of the transverse ray aberration (relative to the axis) $d^2\eta = (\tilde{\lambda}\tilde{p}/2\pi)N \, d\tilde{r} \, \partial_{\tilde{r}\tilde{r}\tilde{r}}\theta \, d\tilde{r}$, we obtain the increment

$$dr(\partial_r \otimes d^2\eta) = \frac{\tilde{\lambda}\tilde{p}}{2\pi} dr(d\tilde{r} \, \partial_{\tilde{r}\tilde{r}\tilde{r}}\phi \, d\tilde{r}) , \quad dr(\partial_r \otimes d\eta) = 0 .$$

(3.102)

Therefore, the variation of the refined transverse ray aberration for the ray between $\overline{P}$ and $\overline{H}$ (!) is

$$dr \cdot (\partial_r \otimes \varDelta\eta)\varrho = \frac{1}{2!} dr \cdot (\partial_r \otimes d^2\eta) \varrho = \frac{\tilde{\lambda}\tilde{p}}{2\pi 2!} \varrho \cdot (\varrho\partial_{\tilde{r}\tilde{r}\tilde{r}}\phi\varrho) \varrho$$

$$= \frac{3\tilde{\lambda}\tilde{p}}{2} \left(\frac{1}{\tilde{\lambda}\tilde{p}^3} - \frac{1}{\lambda p^3}\right)(\varrho \cdot \varrho)\hat{\varrho}^2 .$$

(3.103)

Consequently, at the neighboring rays of the axis ($\varrho \neq 0$), the transverse ray aberration ($\hat{\varrho} \neq 0$) becomes a lateral refinement, resulting in an improper lateral location of $\tilde{\tilde{P}}$, called the *coma*. Let us add that, when the vectors $\varrho$ and $\hat{\varrho}$ are simply exchanged, the term describing the phenomenon of coma is dual to the term describing the distortion.

We conclude this paragraph by adding some general remarks. The five primary Seidel aberrations we have discussed here are all of the same order (fourth), if $\varrho$ and $\hat{\varrho}$ are of the same order of magnitude $|\varrho = O(\hat{\varrho})|$. In classical lens imaging, in which all of the rays going through an optical instrument are only slightly inclined with respect to the axis, this is always the case. (There, the Seidel variables are not the true "distance" from the axis, but their definition contains a magnification factor so that they are, in fact, "angular" variables). In holography, however, the inclination of the rays with respect to the hologram normal is generally large. Therefore, we must develop around another ray of reference (i.e., the line of sight), i.e. we perform the non-paraxial development similar to what we provisionally have done in Sect. 3.1.

The foregoing calculations with the "triangle" of Pascal, (3.88), are relatively simple because many terms in the diadics, triadics, and quadriatics vanish immediately due to symmetry of revolution. In particular, the astigmatism and the coma are zero on the axis and the Seidel aberrations are, in fact, their off-axis "perturbations from zero". Around the "axis", in the non-symmetrical case, one could use the polynomials of *Zernike* [3.287, 288]. If, in the non-paraxial case, we perform a development with respect to $d\hat{r}$ alone, the appearing astigmatism on the reference line ($dr = 0$) is different from zero and is of second order. The term which causes the lateral refinement, still on the line of reference, is of third order and thereby corresponds to the coma. Finally, the term which causes the longitudinally improper location of $\tilde{P}$, also on the line of reference, is of fourth order and corresponds to the spherical aberration. The terminology just defined will apply to the following discussion.

A peculiar feature of this approach may be noted here. Although the Seidel aberrations all come from the same development and are of the same order, thus somehow equivalent, their names describe somewhat different phenomena. For instance, "coma" indicates an improper formation of an off-axis *image point* $\tilde{\tilde{P}}$. Contrary to this, the word "distortion" suggests a type of deformation of the *whole vicinity* $\{\tilde{\tilde{P}}\}$ of some axial point $\tilde{P}$ (the undeformed configuration being $\{\overline{P}\}$ around P). Therefore, we shall distinguish between investigations concerning a point-aberration and those concerning the deformation of a point neighborhood. In a first step, as in Sect. 3.1.2, if we set $dr = 0$, $d\hat{r}$ is then the independent variable. In a second step, we assure $dr \neq 0$ as before; $d\hat{r}$ is then related to another $d\hat{r}'$ (corresponding, in fact, to $\overline{P}$) by the collineation with center $\tilde{R}$, which here is the center of the observing instrument on the line of sight at the reconstruction. Since we seek the mapping $dr \to d\tilde{r}$, it is convenient to introduce a "point" R, the origin of $\tilde{R}$ at the recording which relates $dr$ and $d\hat{r}'$, Figure 3.11 illustrates the case with this alternate collineation center on the axis.

## 3.2.2 Apparent Deformation of an Image and Duality in Holography

With the previous derivations, let us now investigate the deformation around a general point P caused by a nonsymmetrical modification at the reconstruction [3.289, 290]. In fact, we seek the mapping $dr \rightarrow d\tilde{r} = \tilde{F}^{-1} dr$ of corresponding increments on the object and image surfaces; but first, we shall study only a sort of apparent, projected deformation. As in Sect. 3.1, we begin with the simple case of a wavelength change $\lambda \rightarrow \tilde{\lambda}$, combined with an arbitrary shift $Q \rightarrow \tilde{Q}$ and a phase change $\psi \rightarrow \tilde{\psi}$ of the reference source. Although the image surface is not really a proper surface, due to the astigmatism of its points, we can at first imagine that $\tilde{P}$ and $\bar{P}$ are any pair of points in two corresponding astigmatic intervals, e.g., associated with the same vector $\tilde{m}$ and observed at the reconstruction from a fixed point $\tilde{R}$. Point $\tilde{R}$ is the center of the observing instrument.

Applying (3.15, 17) at the two neighboring intersection points H, $\overline{H}'$, respectively, of the lines of sight with the hologram (separated by the increment $d\tilde{r}'$), we have

$$\hat{N}\left[\frac{1}{\lambda}(k - c) - \frac{1}{\tilde{\lambda}}(\tilde{k} - \tilde{c})\right] = 0 ,$$

$$\hat{N}\left[\frac{1}{\lambda}(\bar{k}' - \bar{c}') - \frac{1}{\tilde{\lambda}}\tilde{k}' - \tilde{c}')\right] = 0 . \tag{3.104}$$

Contrary to Sect. 3.1, the unknown vectors here are $k$, $\bar{k}'$ at the recording, whereas $\tilde{k}$, $\tilde{k}'$ at the reconstruction are given unit vectors. Therefore, $k$ and $\bar{k}'$ define two *skewed* rays of a rectilinear congruence which has an "astigmatic" interval $\langle R_1, R_2 \rangle$ of points $\{R\}$ (Fig. 3.15). This situation is exactly analogous to that of Fig. 3.14, for which appears a *duality* [3.147, 291] between two configurations:

$$P \leftrightarrow \tilde{R} , \qquad Q \leftrightarrow \tilde{Q} ,$$
$$\{\tilde{P}\} \leftrightarrow \{R\} , \qquad \tilde{Q} \leftrightarrow Q .$$

Roughly stated, the roles of object point and observer's center are exchanged simultaneously with the roles of recording and reconstruction. Therefore, in addition to the phase difference

$$\theta_{\tilde{P}}(\tilde{r}) = \frac{2\pi}{\tilde{\lambda}}(\tilde{p} - \tilde{q}) - \frac{2\pi}{\lambda}(p - q) , \tag{3.105}$$

we may introduce a dual function (the indices $\tilde{P}$ and R mark the sought after points)

$$\theta_R(\tilde{r}) = \frac{2\pi}{\lambda}(l + q) - \frac{2\pi}{\tilde{\lambda}}(\tilde{l} + \tilde{q}) , \tag{3.106}$$

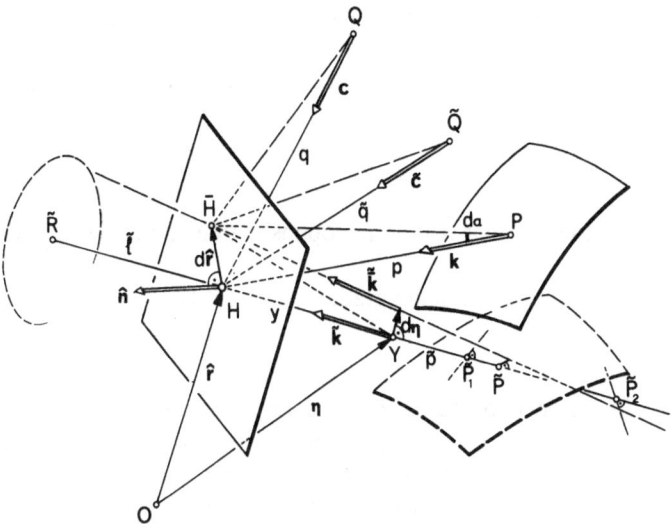

**Fig. 3.14.** Modification at the reconstruction by a shift of the reference source $Q \rightarrow \tilde{Q}$ and by a wavelength change $\lambda \rightarrow \tilde{\lambda}$, illustrating the point aberration. ($\{\tilde{P}\}$: image set of the object point P; $\langle \tilde{P}_1 \tilde{P}_2 \rangle$: astigmatic interval)

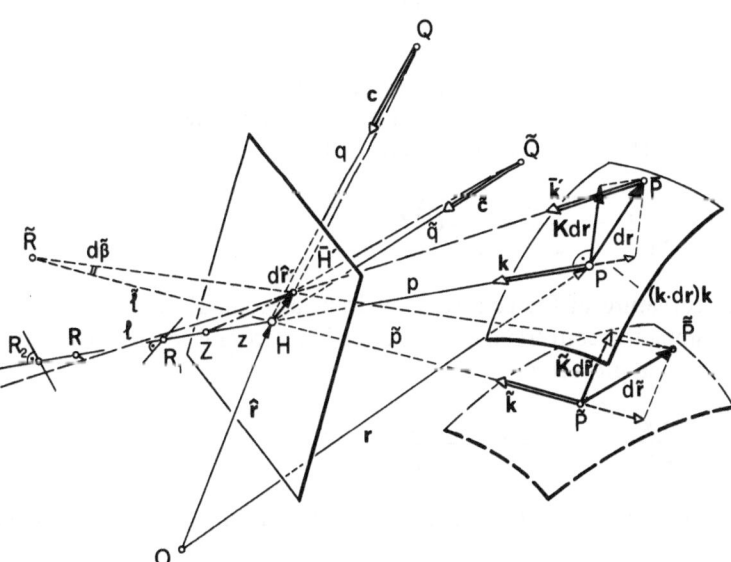

**Fig. 3.15.** The same modification as in Fig. 3.14, but illustrating the apparent deformation of a point neighborhood. (*dr*, *dr̃*: associated increments on the object surface and on the image "surface"; $\tilde{R}$: center of the observing instrument at the reconstruction. $\langle R_1 R_2 \rangle$: corresponding astigmatic interval at the recording). This figure is the counterpart of Fig. 3.14

where $\tilde{l}$ denotes the given distance $\tilde{R}H$ and where $l$ is thus far defined as the unknown distance from the origin R of $\tilde{R}$ at the recording to the same hologram point H. Therefore, we have a correspondence between the distances and between the wavelengths

$$\tilde{l} \leftrightarrow p \ , \quad l \leftrightarrow \tilde{p} \ , \quad q \leftrightarrow -\tilde{q} \ , \quad \lambda \leftrightarrow \tilde{\lambda} \ . \tag{3.107}$$

The minus sign is due to the fact that $\tilde{R}$ is on the opposite side of point P relative to the hologram.

Both function $\theta_R$ and $\theta_{\tilde{P}}$ have the same gradient at point H,

$$\nabla_{\hat{n}} \theta_R = \nabla_{\hat{n}} \theta_{\tilde{P}} \ , \tag{3.108}$$

since the derivative of a distance to a fixed point results in a unit vector, independently of the location of this fixed point on the line in question. Therefore, as with the stationary behavior of $\theta_{\tilde{P}}$, the stationary behavior of the function $\theta_R$ at point H also leads to the first equation of (3.104), see (3.15). This formally justifies the introduction of the function $\theta_R$ in addition to the function $\theta_{\tilde{P}}$.

On a similar line of thought, we may now repeat the principles developed in Sect. 3.1.2 and calculate, for instance, the second derivative $\nabla_{\hat{n}} \otimes \nabla_{\hat{n}} \theta_R$ in order to find $l$, the wave aberration and the transverse ray aberration. Every equation in Sect. 3.1.2 has a counterpart. For instance, the characteristic tensor defined by (3.31)

$$\tilde{T} = \tilde{\lambda} \tilde{M} \left( \frac{1}{\tilde{\lambda}\tilde{q}} \tilde{C} + \frac{1}{\lambda p} K - \frac{1}{\lambda q} C \right) \tilde{M}^T$$

describes, as we already said, the curvature of the convex wavefront at point H at the reconstruction, which travels along diverging rays in the direction $\tilde{k}$ from $\langle \tilde{P}_1, \tilde{P}_2 \rangle$ to $\tilde{R}$. In return, the dual characteristic tensor

$$T = \lambda \hat{M} \left( -\frac{1}{\lambda q} C + \frac{1}{\tilde{\lambda}\tilde{l}} \tilde{K} + \frac{1}{\tilde{\lambda}\tilde{q}} \tilde{C} \right) \hat{M}^T \tag{3.109}$$

describes the curvature of the concave wavefront at point H at the recording, which travels along the converging rays in direction $k$ from P to $\langle R_1, R_2 \rangle$. Note that the two oblique projectors

$$\hat{M} = I - \frac{\hat{n} \otimes k}{\hat{n} \cdot k} \ , \quad \tilde{M} = I - \frac{\hat{n} \otimes \tilde{k}}{\hat{n} \cdot \tilde{k}}$$

are dual. The eigenvalues $1/l_1$, $1/l_2$ of $T$ determine the "focal" lines of the astigmatic interval $\langle R_1, R_2 \rangle$.

We also may choose an arbitrary point Z on the reference ray HP at a distance $z$ from H and write the dual of (3.26)

$$d\tilde{k} = z \ dk \ T \ . \tag{3.110}$$

The vectors $dk$, $d\bar{k}$ here are increments of the direction unit vector $k$ relative to $\overline{H'Z}$, $\overline{PH'}$, respectively (Fig. 3.15). Moreover, the normal projection $K \, dr$ of the object increment $dr$ is geometrically related to the other increments by the form

$$K \, dr = K \, d\hat{r}' - p \, d\bar{k} = - (p \, d\bar{k} + z \, dk) \; . \tag{3.111}$$

Using (3.110), this relation becomes

$$K \, dr = - zp \left( T + \frac{1}{p} K \right) dk \; ,$$

or, in particular, for $z = l$, (the increment $dk$ is from now on related to the point R and not to Z):

$$K \, dr = - lp \left( T + \frac{1}{p} K \right) dk \; . \tag{3.112}$$

In order to afterwards perform the passage to the normal projection $\tilde{K} \, d\tilde{r}$ of the image increment $d\tilde{r}$, we use (3.23), describing the affine connection for $d\hat{r}'$, and with $y = - \tilde{l}$, we write

$$d\hat{r}' = - \tilde{l}\tilde{M}^T d\tilde{k} = - \tilde{l}\tilde{M}^T \tilde{m} \, d\tilde{\beta} \; ; \tag{3.113}$$

together with $z = l$, its counterpart is

$$d\hat{r}' = - l\hat{M}^T dk = - l\hat{M}^T m \, d\beta \; , \tag{3.114}$$

with unit vectors $m$, $\tilde{m}$ and corresponding angular increments $d\beta$, $d\tilde{\beta}$. Elimination of $d\hat{r}'$ gives

$$l\hat{M}^T dk = \tilde{l}\tilde{M}^T d\tilde{k} \quad \text{or} \quad dk = \frac{\tilde{l}}{l} K\tilde{M}^T d\tilde{k} \; . \tag{3.115}$$

But since $\tilde{R}$ is a collineation center, we also have

$$d\tilde{k} = - \frac{1}{\tilde{l} + \tilde{p}} \tilde{K} \, d\tilde{r} \; . \tag{3.116}$$

Therefore, it is possible to make $\tilde{K} \, d\tilde{r}$ appear in (3.112) by combining (3.115) with (3.116). It follows that

$$K \, dr = \frac{\tilde{l}p}{\tilde{l} + \tilde{p}} \left( T + \frac{1}{p} K \right) \tilde{M}^T \tilde{K} \, d\tilde{r} \; , \tag{3.117}$$

which provisionally describes the inverse mapping $\tilde{K} \, d\tilde{r} \rightarrow K \, dr$ of the apparent (projected) image deformation due to the optical modification (see also the example of Figs. 3.16a, b).

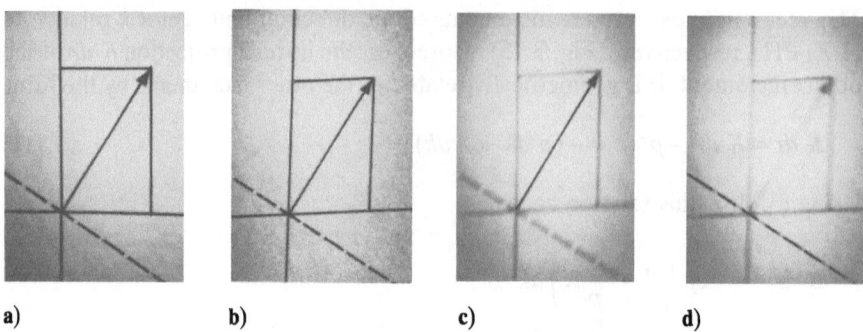

a)                    b)                    c)                    d)

**Fig. 3.16a–d.** Illustration of simultaneous deformation and astigmatism according to duality: (a) Reconstruction of a small rectangle containing an arrow and a dashed line without any modification. (b) Reconstruction of the same image as in (a), but after an optical modification i.e., a relative shift of the reference source. The rectangle here becomes a parallelogram (apparent deformation). (c) Reconstruction of the same image as in (b), but the camera has a larger aperture and is focused on $\check{P}_1$. The arrow is sharp, but the normal dashed line is blurred. (d) Reconstruction of the same image as in (b), but the camera, again with a large aperature, is focused on $\check{P}_2$. The arrow is now blurred and the dashed line is clear

As compared with (2.138), the tensor on the right side of (3.117) and operating on $\tilde{K}\, d\bar{r}$ is a mixed projection of type $K\tilde{F}\tilde{K}$. Since $\hat{M}\hat{M} = \hat{M}$, $\hat{M}^\mathrm{T}\tilde{M}^\mathrm{T} = \tilde{M}^\mathrm{T}$, we may alternatively write the dual form

$$\tilde{\lambda}\left(T + \frac{1}{p}K\right)\tilde{M}^\mathrm{T} = \tilde{\lambda}\lambda\hat{M}\left(\frac{1}{\lambda p}K - \frac{1}{\lambda q}C + \frac{1}{\tilde{\lambda}\tilde{l}}\tilde{K} + \frac{1}{\tilde{\lambda}\tilde{q}}\tilde{C}\right)\tilde{M}^\mathrm{T}$$

$$= \lambda\hat{M}\left(\tilde{T} + \frac{1}{\tilde{l}}\tilde{K}\right), \tag{3.118}$$

so that (3.117) is also [Ref. 3.292, Eq. (12)]:

$$\left(\frac{1}{\tilde{l}} + \frac{1}{\tilde{p}}\right)K\, dr = \frac{p\lambda}{\tilde{p}\tilde{\lambda}}\hat{M}\left(\tilde{T} + \frac{1}{\tilde{l}}\tilde{K}\right)\tilde{K}\, d\bar{r}\ . \tag{3.119}$$

Finally, with a decomposition similar to (3.34),

$$\tilde{T} + \frac{1}{\tilde{l}}\tilde{K} = \left(\frac{1}{\tilde{l}} + \frac{1}{\tilde{p}}\right)\tilde{K} + \left[\frac{1}{2}\left(\frac{1}{\tilde{p}_1} + \frac{1}{\tilde{p}_2}\right) - \frac{1}{\tilde{p}}\right]\tilde{K} + \frac{1}{2}\left(\frac{1}{\tilde{p}_1} - \frac{1}{\tilde{p}_2}\right)\tilde{D}\ ,$$

and (3.119) becomes

$$K\, dr = \frac{p\lambda}{\tilde{p}\tilde{\lambda}}\hat{M}\tilde{K}\, d\bar{r} + \frac{p\tilde{l}}{\tilde{l} + \tilde{p}}\frac{\lambda}{\tilde{\lambda}}\left\{\left[\frac{1}{2}\left(\frac{1}{\tilde{p}_1} + \frac{1}{\tilde{p}_2}\right) - \frac{1}{\tilde{p}}\right]\hat{M}\tilde{K}\right.$$

$$\left. + \frac{1}{2}\left(\frac{1}{\tilde{p}_1} - \frac{1}{\tilde{p}_2}\right)\hat{M}\tilde{D}\right\}\tilde{K}\, d\bar{r}\ . \tag{3.120}$$

Figures 3.16c, d illustrate the astigmatism which simultaneously appears with the deformation. When the vector $\tilde{K}\,d\tilde{r}$ lies parallel to the principal directions 1, 2 of the tensor $\tilde{T}$, the last term disappears and we have:

$$\frac{1}{\lambda p}\,(K\,dr)_1 = \frac{1}{\tilde{\lambda}\tilde{p}_1}\hat{M}(\tilde{K}\,d\tilde{r})_1 \ , \qquad \frac{1}{\lambda p}\,(K\,dr)_2 = \frac{1}{\tilde{\lambda}\tilde{p}_2}\hat{M}(\tilde{K}\,d\tilde{r})_2 \ . \qquad (3.121)$$

It follows that this result is independent of the observer's position along the line of sight. On the other hand, when the image point is stigmatic, i.e., when there is no astigmatism and no "difference" of field curvature, the first term remains alone for any direction of $\tilde{K}\,d\tilde{r}$. As one might suspect, this term contains simply a magnification factor and an oblique projection. In particular, the case of Sect. 3.2.1, with Gauss' relation (3.84), gives $K\,dr = \tilde{K}\,d\tilde{r}$, as we have already found. In general, however, we may say that the "aberration" from the first term of the apparent deformation of the point neighborhood (pure obliquity term) depends directly on the astigmatism and the field curvature of this point.

Let us add that, if we apply $d\tilde{r}'$ twice to (3.118), once on each side, and contract, and if, in addition, we make use of (3.113, 114), of the relations $m \cdot Tm = 1/l$, $\tilde{m} \cdot \tilde{T}\tilde{m} = 1/\tilde{p}$, and of the combined projections $\tilde{M}^T\hat{M}^T = \hat{M}^T$ and $\tilde{M}\hat{M} = \hat{M}$, we find the scalar equation (Fig. 3.15)

$$\frac{l^2}{\lambda}\left(\frac{1}{l}+\frac{1}{p}\right)(d\beta)^2 = \frac{\tilde{l}^2}{\tilde{\lambda}}\left(\frac{1}{\tilde{p}}+\frac{1}{\tilde{l}}\right)(d\tilde{\beta})^2 \ , \qquad (3.122)$$

which reminds us of some relationship encountered in elementary optics. Similarly, with angular increments $da$ at point P and $d\tilde{a}$ at point $\tilde{P}$ (Fig. 3.14), we could deduce the dual relation

$$\frac{\tilde{p}^2}{\tilde{\lambda}}\left(\frac{1}{\tilde{p}}+\frac{1}{\tilde{l}}\right)(d\tilde{a})^2 = \frac{p^2}{\lambda}\left(\frac{1}{l}+\frac{1}{p}\right)(da)^2 \ . \qquad (3.123)$$

which will be used in the following subsection.

Of course, the duality concept is not limited to special cases of wavelength changes and shifts of reference sources. If we recall the image point formation in the case of a general modification, (Sect. 3.1.2), we may, by means of (3.39), first write (3.44) in the alternative form (Fig. 3.17):

$$\frac{1}{\tilde{\lambda}}\hat{N}\hat{F}^T(\tilde{k} - \tilde{c}) - \frac{1}{\lambda}\hat{N}(k-c) = 0 \ . \qquad (3.124)$$

This form should suggest that, as in (3.104), it determines the direction $k$ to the origin R of the observer's point $\tilde{R}$, when the direction $\tilde{k}$ of the ray $\tilde{P}\tilde{R}$ at the reconstruction is given. Next, from (3.48), we have, i.e.,

$$\tilde{T} = \tilde{\lambda}\hat{M} \left\{ \frac{1}{\tilde{\lambda}}\left[\frac{1}{\tilde{q}}\tilde{C} - \hat{B}(\hat{n} \cdot \tilde{k} - \hat{n} \cdot \tilde{c})\right] \right.$$

$$\left. + \frac{1}{\lambda}\hat{F}^T\left[\frac{1}{p}K - \frac{1}{q}C + \hat{B}(\hat{n} \cdot k - \hat{n} \cdot c)\right]\hat{F} \right\}\hat{M}^T \ ,$$

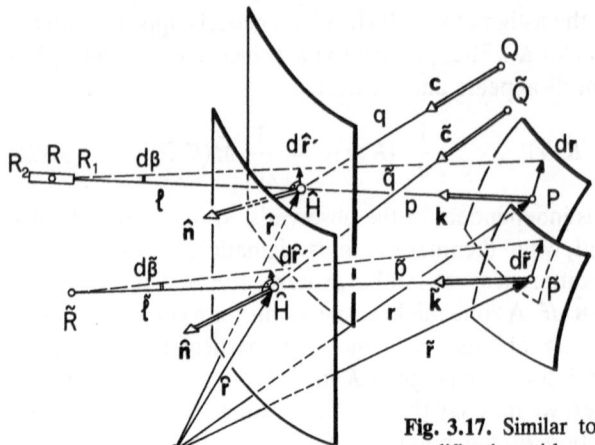

**Fig. 3.17.** Similar to Fig. 3.15, but for a general modification with a curved hologram. This figure is the counterpart of Fig. 3.7

and, thereby, find the dual tensor of the wavefront at point H for the ray from P to R at the recording

$$T = \lambda \hat{M} \left\{ \frac{1}{\lambda} \left[ -\frac{1}{q}C + \hat{B}(\hat{n} \cdot k - \hat{n} \cdot c) \right] \right.$$

$$\left. + \frac{1}{\tilde{\lambda}}\hat{F}^{T} \left[ \frac{1}{\tilde{l}}\tilde{K} + \frac{1}{\tilde{q}}\tilde{C} - \hat{B}(\hat{n} \cdot \tilde{k} - \hat{n} \cdot \tilde{c}) \right]\hat{F} \right\} \hat{M}^{T} , \qquad (3.125)$$

where the projector $\hat{M} = I - \hat{n} \otimes k/\hat{n} \cdot k$ is now dual to the projector $\hat{M} = I - \hat{n} \otimes \tilde{k}/\hat{n} \cdot \tilde{k}$ and not to $\tilde{M}$.

Further, the development concerning the ray tracing, leading to (3.26), is exactly the same; but the affine connections (3.113 and 114) must be replaced by the relations

$$d\hat{r}' = -\tilde{l}\hat{M}^{T}d\tilde{k} , \qquad d\hat{r}' = \hat{F} \, d\hat{r}' , \qquad d\hat{r}' = -l\hat{M}^{T}dk , \qquad (3.126)$$

so that instead of (3.115) with $\hat{F} = \hat{F}^{-1}$, we obtain the "bridging" relation in the form:

$$l\hat{M}^{T}dk = \tilde{l}\hat{F}\hat{M}^{T}d\tilde{k} . \qquad (3.127)$$

The apparent image deformation, thereafter, with (3.112 and 116) is

$$K \, dr = \frac{\tilde{l}p}{\tilde{l} + \tilde{p}} \left( T + \frac{1}{p}K \right)\hat{F}\hat{M}^{T}\tilde{K} \, d\tilde{r} . \qquad (3.128)$$

This is the generalization of (3.117). The tensor $\hat{F}$ in (3.128), describing the rotation and deformation of the holographic plate, may be used to adjust and modify the apparent deformation $K\,dr \to \tilde{K}\,d\tilde{r}$. In an orthoscopic angular magnification [3.292], the tensor $(T + 1/p\,K)\hat{F}\hat{M}^{\mathrm{T}}$ must be a mixed projection $K\tilde{Q}\tilde{K}$ of an orthogonal tensor $\hat{Q}$. This may be realized by changing the tensor $\hat{F}$. Such study of invariance properties in holography may also be applied, for instance, to the relative location of image points [3.293, 294]. From the following auxiliary relations for an obliquely projected deformation gradient

$$\hat{F}\hat{M}^{\mathrm{T}}\hat{F}\hat{M}^{\mathrm{T}} = \hat{F}\hat{M}^{\mathrm{T}}\hat{F}\hat{N}\hat{M}^{\mathrm{T}} = \hat{F}\hat{M}^{\mathrm{T}}\hat{N}\hat{F}\hat{N}\hat{M}^{\mathrm{T}} = \hat{F}\hat{F}\hat{N}\hat{M}^{\mathrm{T}} = \hat{M}^{\mathrm{T}}\ldots\;, \tag{3.129}$$

we also find the equation of duality

$$\tilde{\lambda}\left(T + \frac{1}{p}K\right)\hat{F}\hat{M}^{\mathrm{T}} \;=\; \tilde{\lambda}\lambda\hat{M}\left\{\left[\frac{1}{\lambda p}K \;-\;\frac{1}{\lambda q}C + \frac{1}{\lambda}\hat{B}(\hat{n}\cdot k - \hat{n}\cdot c)\right]\hat{F}\right. \tag{3.130}$$

$$+\; \hat{F}^{\mathrm{T}}\left.\left[\frac{1}{\tilde{\lambda}\tilde{l}}\tilde{K} + \frac{1}{\tilde{\lambda}\tilde{q}}\tilde{C} \;-\;\frac{1}{\tilde{\lambda}}\hat{B}(\hat{n}\cdot\tilde{k}-\hat{n}\cdot\tilde{c})\right]\right\}\hat{M}^{\mathrm{T}} = \lambda\hat{M}\hat{F}^{\mathrm{T}}\left(\tilde{T} + \frac{1}{\tilde{l}}\tilde{K}\right),$$

so that, as (3.119), eq. (3.128) alternatively becomes

$$\left(\frac{1}{\tilde{l}} + \frac{1}{\tilde{p}}\right)K\,dr = \frac{p\lambda}{\tilde{p}\tilde{\lambda}}\hat{M}\hat{F}^{\mathrm{T}}\left(\tilde{T} + \frac{1}{\tilde{l}}\tilde{K}\right)\tilde{K}\,d\tilde{r}\;. \tag{3.131}$$

### 3.2.3 Deformation of an Image in Space

The preceeding subsection comprises an investigation of the apparent projected increments and their mapping $K\,dr \to \tilde{K}\,d\tilde{r}$ for any modification of the optical set-up at the reconstruction. However, this constitutes only a part of the solution of the problem $dr \to d\tilde{r}$ in space. Such increments $dr$, $d\tilde{r}$ between two neighboring points P, $\bar{\mathrm{P}}$ and their images $\tilde{\mathrm{P}}$, $\check{\mathrm{P}}$ may, of course, be recognized only by stereoscopic viewing with both eyes [3.295, 296]. Therefore, it will be convenient to later work with two collineation centers $\tilde{\mathrm{R}}$, $\tilde{\mathrm{S}}$. For the moment, let us only remark that we should, in addition to (3.117 or 128), investigate the mapping $k\cdot dr \to \tilde{k}\cdot d\tilde{r}$ of the longitudinal parts of the increments. It is now quite a general feature that lateral and longitudinal components of any quantity arc often related to successive derivatives. Since the lateral parts $K\,dr$, $\tilde{K}\,d\tilde{r}$ are related to the second derivative of the phase difference function, i.e. the second-order wave aberration, we therefore expect that the longitudinal parts are related to the third derivative of the phase; or, in other words, to the third-order wave aberration. On the other hand, it has already been mentioned at the end of Sect. 3.2.1, during the discussion of the primary Seidel aberrations, that in a Taylor development of the phase in the neighborhood of a hologram point, the third-order terms must show the effect of the coma. Indeed, similarly to the relationship (3.103), we expect that a third-order wave aberration produces a second-order transverse ray aberration. When the increments are varied, the latter is a correction of the first-order transverse ray aberration and, therefore, causes an improper lateral location of $\tilde{P}$. This phenomenon of image point aberration will first be studied, but our considerations will be limited to the relatively simple case of a wavelength change and of a shift of the reference source (Sect. 3.1.2).

Let us first refer to (3.17) and develop the three unit vectors $\bar{k}$, $\tilde{c}$, $\bar{\tilde{c}}$ up to second-order terms. For instance, for the vector $\bar{k}$ we have

$$\bar{k} = k + \Delta k = k + dk + \frac{1}{2}d^2k$$

$$= k + \left(d\hat{r} + \frac{1}{2}d^2\hat{r}\right)\nabla \otimes k + \frac{1}{2}d\hat{r}[d\hat{r}(\nabla \otimes \nabla \otimes k)] \quad (3.132)$$

$$= k + \left(d\hat{r} + \frac{1}{2}d^2\hat{r}\right)\left(\frac{1}{p}K\right) + \frac{1}{2}d\hat{r}\left[d\hat{r}\left(-\frac{1}{p^2}K\right)\right],$$

with the already defined superprojector $K = K \otimes k + k \otimes K + K \otimes k)^{\mathrm{T}}$. Since here we assume a plane hologram, the second differential $d^2\hat{r}$ necessarily lies in the hologram plane. For simplicity, we may even place it in the direction of the first differential $d\hat{r}$. It should be noted that, if $\hat{r}$ was chosen as the independent variable, $d^2\hat{r}$ would be zero. But here, instead of $\hat{r}$, we shall take the vector $\bar{k}$ as the "independent" variable, thereby making $d^2\hat{r} \neq 0$. Thus by taking the difference between (3.17 and 15), and by separating first-order from second-order terms, we find an extension to (3.19) ($C = C \otimes c + c \otimes C + C \otimes c)^{\mathrm{T}}$)

$$\hat{N}\,d\bar{\tilde{k}} = \tilde{\lambda}\,d\hat{r}\,\hat{N}\left(\frac{1}{\tilde{\lambda}\tilde{q}}\tilde{C} + \frac{1}{\lambda p}K - \frac{1}{\lambda q}C\right)\hat{N}, \quad (3.133)$$

$$\hat{N}\,d^2\bar{\tilde{k}} = \tilde{\lambda}\,d^2\hat{r}\,\hat{N}\left(\frac{1}{\tilde{\lambda}\tilde{q}}\tilde{C} + \frac{1}{\lambda p}K - \frac{1}{\lambda q}C\right)\hat{N}$$

$$- \tilde{\lambda}\,d\hat{r}\,\hat{N}\left[d\hat{r}\,\hat{N}\left(\frac{1}{\tilde{\lambda}\tilde{q}^2}\tilde{C} + \frac{1}{\lambda p^2}K - \frac{1}{\lambda q^2}C\right)\hat{N}\right], \quad (3.134)$$

where the normal projectors $\hat{N}$ remind us of the two-dimensional character of the tensors involved. With reference to any fixed Y on the ray through H, we have, as previously, the affine connection at the reconstruction

$$d\hat{r} = y\tilde{M}^{\mathrm{T}}d\tilde{k}, \quad (3.135)$$

where we repeat that $d\tilde{k}$ should not be confused with $d\bar{\tilde{k}}$. The difference between these two increments is due to the fact that the rays through $\overline{H}$ and H are skew. The second differential thus yields

$$d^2\hat{r} = d(d\hat{r}) = dy\,\tilde{M}^{\mathrm{T}}d\tilde{k} + y\,d\tilde{M}^{\mathrm{T}}d\tilde{k} + y\tilde{M}^{\mathrm{T}}d^2\tilde{k}. \quad (3.136)$$

We need here the differential of the oblique projector $\tilde{M}^T = I - \tilde{k} \otimes \hat{n}/\tilde{k} \cdot \hat{n}$, which reads with $\tilde{k}$ variable and $\hat{n}$ constant:

$$d\tilde{M}^T = \frac{-d\tilde{k} \otimes \hat{n}}{\tilde{k} \cdot \hat{n}} + \frac{(d\tilde{k} \cdot \hat{n})(\tilde{k} \otimes \hat{n})}{(\tilde{k} \cdot \hat{n})^2} = \frac{-1}{\tilde{k} \cdot \hat{n}} d\tilde{k}(\tilde{M} \otimes \hat{n}) \ . \tag{3.137}$$

The second term of (3.136) then becomes

$$y \, d\tilde{M}^T d\tilde{k} = y \, d\tilde{k} \, \tilde{M}\left(-\frac{\hat{n} \cdot d\tilde{k}}{\tilde{k} \cdot \hat{n}}\right) = y\tilde{M}^T d\tilde{k}(\tilde{k}\tilde{M}^T \cdot d\tilde{k}) \ . \tag{3.138}$$

Furthermore, we have $dy = y\tilde{k} \cdot \tilde{M}^T d\tilde{k}$ and, additionally, $\tilde{M}^T d^2\tilde{k} = 0$, since $d^2\tilde{k}$ here is parallel to $\tilde{k}$. Consequently, with (3.138), we obtain a shorter form for the vector $d^2\hat{r}$:

$$d^2\hat{r} = 2y(\tilde{k} \cdot \tilde{M}^T d\tilde{k})\tilde{M}^T d\tilde{k} \ .$$

As far as the vector $\Delta\hat{r}$ is concerned, we find

$$\Delta\hat{r} = d\hat{r} + \tfrac{1}{2!}d^2\hat{r} = y(1 + \tilde{k} \cdot \tilde{M}^T d\tilde{k})\tilde{M}^T d\tilde{k} \ , \tag{3.139}$$

which may also be recognized from Fig. 3.18.

By means of the curvature tensor $-\tilde{T}$

$$\tilde{T} = \tilde{\lambda}\tilde{M}\left(\frac{1}{\tilde{\lambda}\tilde{q}}\tilde{C} + \frac{1}{\lambda p}K - \frac{1}{\lambda q}C\right)\tilde{M}^T \ , \tag{3.140}$$

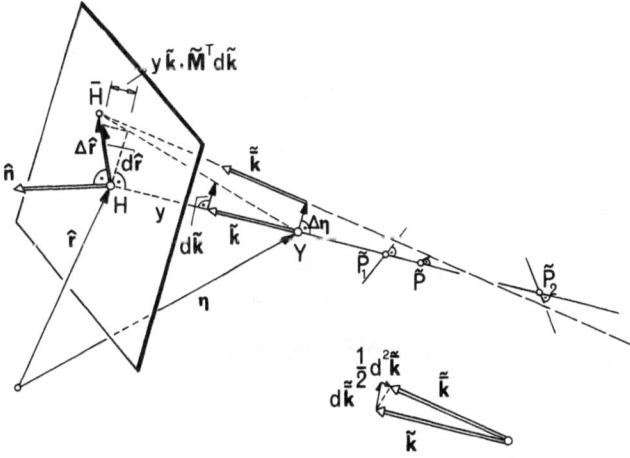

**Fig. 3.18.** Transverse ray aberration $\Delta\eta$ including second-order terms ($\tilde{k}$, $\check{k}$: unit direction vectors)

of the wave front at point H and the corresponding third-order tensor, defined at the same point H,

$$\tilde{T} = -\tilde{\lambda}\tilde{M} \lrcorner \tilde{M}\left(\frac{1}{\tilde{\lambda}\tilde{q}^2}\tilde{C} + \frac{1}{\lambda p^2}K - \frac{1}{\lambda q^2}C\right)\tilde{M}^{\mathrm{T}} , \qquad (3.141)$$

we may rewrite (3.133 and 134), respectively, after an oblique projection in the forms

$$d\tilde{\tilde{k}} = y\, d\tilde{k}\, \tilde{T} , \qquad (3.142)$$

$$\tilde{M}\, d^2\tilde{\tilde{k}} = 2y(d\tilde{k} \cdot \tilde{M}\tilde{k})d\tilde{k}\tilde{T} + y^2 d\tilde{k}(d\tilde{k}\, \tilde{T}) . \qquad (3.143)$$

Note that in (3.141), the first oblique projector $\tilde{M}$ acts from the middle (which is marked by the sign $\lrcorner$), the second from the left, and the third from the right. The last projector is, therefore, written in the transpose form when it is placed at the right side of the triadics $\tilde{C}$, $K$, and $C$.

The oblique projection $\tilde{M}\, d^2\tilde{\tilde{k}}$ of $d^2\tilde{\tilde{k}}$ in (2.143) is not sufficient for determining the second-order transverse ray aberration which we will determine. In fact, we need the true vector $d^2\tilde{k}$ in space, which may be decomposed in two parts (Fig. 3.18): namely,

$$d^2\tilde{k} = \tilde{M}\, d^2\tilde{\tilde{k}} + \frac{\hat{n}}{\hat{n} \cdot \tilde{k}}\, \tilde{k} \cdot d^2\tilde{k} , \qquad (3.144)$$

where the, as yet, unknown part parallel to $\hat{n}$ may be found from the auxiliary condition $|\tilde{k}| = 1$. Indeed, we have

$$1 = (\tilde{k} + d\tilde{k} + \tfrac{1}{2}d^2\tilde{k})^2 \simeq 1 + |d\tilde{k}|^2 + \tilde{k} \cdot d^2\tilde{k} , \quad \text{or}$$

$$\tilde{k} \cdot d^2\tilde{k} = -|d\tilde{k}|^2 .$$

Therefore, for (3.144), together with (3.143), we get

$$d^2\tilde{k} = \tilde{M}\, d^2\tilde{\tilde{k}} - \frac{\hat{n}}{\hat{n} \cdot \tilde{k}}\, |d\tilde{k}|^2$$

$$= 2y(d\tilde{k} \cdot \tilde{M}\tilde{k})d\tilde{k}\tilde{T} + y^2 d\tilde{k}\left[d\tilde{k}\left(\tilde{T} - \tilde{T}^2 \otimes \frac{\hat{n}}{\hat{n} \cdot \tilde{k}}\right)\right] . \qquad (3.145)$$

Now, the transverse ray aberration at point Y is (Fig. 3.18)

$$\Delta\eta = y\tilde{k} + \Delta\hat{r} - \tilde{y}\tilde{\tilde{k}}$$

which, when projected perpendicularly and along $\tilde{k}$, becomes

$$\Delta\eta = \tilde{K}(\Delta\hat{r} - \tilde{y}\tilde{\tilde{k}}) , \quad 0 = y + \Delta\hat{r} \cdot \tilde{k} - \tilde{y}\tilde{k} \cdot \tilde{\tilde{k}} .$$

Then, eliminating the unknown distance $\bar{y}$, we obtain the equation used by *Miles*
[Ref. 3.130, Eqs. (21, 22)]

$$\Delta\eta \;=\; \tilde{K}\Delta\hat{r} \;-\; \frac{y + \Delta\hat{r}\cdot\tilde{k}}{\tilde{k}\cdot\tilde{k}}\,\tilde{K}\tilde{\tilde{k}}\;. \tag{3.146}$$

This relation will here be developed up to second-order terms which gives

$$\Delta\eta = y(1 + d\tilde{k}\cdot\tilde{M}\tilde{k})d\tilde{k} - y[1 + (1 + d\tilde{k}\cdot\tilde{M}\tilde{k})(d\tilde{k}\cdot\tilde{M}\tilde{k})](d\tilde{\tilde{k}} + \tfrac{1}{2}\tilde{K}\,d^2\tilde{k})\;.$$

Using (3.142, 143, 145) and the auxiliary relation $\tilde{K}\hat{n}/\hat{n}\cdot\tilde{k} = -\tilde{M}\tilde{k}$, we then find
for the transverse ray aberration:

$$\begin{aligned}
\Delta\eta \;=\;& y\,d\tilde{k}(\tilde{K} - y\tilde{T}) + y\,d\tilde{k}(d\tilde{k}\cdot\tilde{M}\tilde{k})(\tilde{K} - 2y\tilde{T})\\
&- \frac{y^3}{2}\,d\tilde{k}[d\tilde{k}(\tilde{T} + \tilde{T}^2 \otimes \tilde{M}\tilde{k})]\;.
\end{aligned} \tag{3.147}$$

The first term on the right side of this equation is the linear transverse ray
aberration $d\eta$, see (3.27). The remaining part represents half of the second-order
transverse ray aberration $d^2\eta/2$, so that the vector $d^2\eta$ reduces to

$$d^2\eta = 2y\,d\tilde{k}(d\tilde{k}\cdot\tilde{M}\tilde{k})(\tilde{K} - 2y\tilde{T}) - y^3 d\tilde{k}[d\tilde{k}(\tilde{T} + \tilde{T}^2 \otimes \tilde{M}\tilde{k})]\;. \tag{3.148}$$

On the other hand, we may start with a development of the phase-difference $\theta$
around H, see (3.13), which with (3.14, 18, 25, 139 and 141) gives

$$\begin{aligned}
\Delta\theta \;=\;& \bar{\theta} - \theta = \tfrac{1}{2}(d\hat{r} + \tfrac{1}{2}d^2\hat{r})\cdot(\nabla_{\hat{n}} \otimes \nabla_{\hat{n}}\theta)(d\hat{r} + \tfrac{1}{2}d^2\hat{r})\\
&+ \tfrac{1}{6}d\hat{r}\cdot[d\hat{r}(\nabla_{\hat{n}} \otimes \nabla_{\hat{n}} \otimes \nabla_{\hat{n}}\theta)d\hat{r}]\\
\cong\;& \frac{y^2}{2}(1 + 2\,d\tilde{k}\cdot\tilde{M}\tilde{k})\frac{2\pi}{\tilde{\lambda}}d\tilde{k}\cdot\left(\frac{1}{y}\tilde{K} - \tilde{T}\right)d\tilde{k}\\
&- \frac{y^3}{6}\cdot\frac{2\pi}{\tilde{\lambda}}d\tilde{k}\cdot(d\tilde{k}\,\tilde{T}\,d\tilde{k}) - \frac{y}{6}\frac{2\pi}{\tilde{\lambda}}d\tilde{k}\,\tilde{M}\cdot(d\tilde{k}\,\tilde{M}\tilde{K}\tilde{M}^{\mathrm{T}}d\tilde{k})\;.
\end{aligned}$$

From the above equation, we deduce the following form for the third-order wave
aberration with $\tilde{K} = \tilde{K}\otimes\tilde{k} + \tilde{k}\otimes\tilde{K} + \tilde{K}\otimes\tilde{k})^{\mathrm{T}}$:

$$d^3\theta_y \;=\; \frac{2\pi}{\tilde{\lambda}}\,[3y(d\tilde{k}\cdot\tilde{M}\tilde{k})d\tilde{k}\cdot(\tilde{K} - 2y\tilde{T})d\tilde{k} - y^3 d\tilde{k}\cdot(d\tilde{k}\,\tilde{T}\,d\tilde{k})]\;. \tag{3.149}$$

Comparing (3.148) with (3.149), we see, that at a general point Y, in the special
case of normal incidence $(\tilde{k} = \hat{n})$ where $\tilde{M}\tilde{k} = 0$, there is a simple relationship

$$d^3\theta_y \;=\; \frac{2\pi}{\tilde{\lambda}y}\,d\hat{r}\cdot d^2\eta \tag{3.150}$$

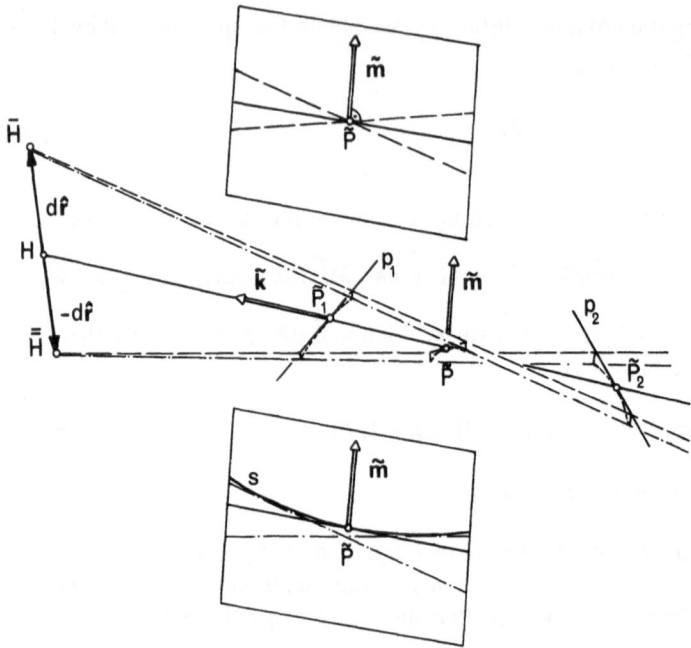

**Fig. 3.19.** Transverse ray aberration including second-order terms. The projected rays result in a sort of "caustic". ($\tilde{m}$: transversal unit vector)

between the third-order wave aberration and the second-order transverse ray aberration conforming with what is known in lens imagery [Ref. 3.102, p. 206]. For this specific case, the relationship is characterized only by the tensor $\tilde{T}$ and is analogous to the second-order relations (3.21 and 27), where the tensor $\tilde{T}$ intervenes.

Let us now recall (3.37) for the point $\tilde{P}$, with $d\tilde{k} = \tilde{m}\,d\tilde{a}$, written in the form

$$\tilde{m} \cdot d\boldsymbol{\eta}_{\tilde{P}} = 0 \; , \tag{3.151}$$

and join to it the scalar product of $\tilde{m}$ with (3.147) for $y = \tilde{p}$. With $\tilde{m} \cdot \tilde{T}\tilde{m}\tilde{p} = 1$, we find

$$\tilde{m} \cdot \Delta\boldsymbol{\eta}_{\tilde{P}} = \tfrac{1}{2}\tilde{m} \cdot d^2\boldsymbol{\eta}_{\tilde{P}}$$

$$= -(d\tilde{a})^2 \tilde{p} \left\{ (\tilde{m} \cdot \tilde{M}\tilde{k}) + \tfrac{1}{2}\tilde{p}^2 \tilde{m}[\tilde{m}(\tilde{T} + \tilde{T}^2 \otimes \tilde{M}\tilde{k})] \cdot \tilde{m} \right\} \neq 0 \; . \tag{3.152}$$

Equations (3.151 and 152) may be interpreted in the following manner [3.297]. For a given direction of $d\hat{r}$, i.e., a fixed $\tilde{m}$, all of the skewed rays around $\tilde{P}H$ (variable length $|d\hat{r}|$) considered within the first-order ray aberration intersect the normal to the plane $(\tilde{k}, \tilde{m})$ through $\tilde{P}$. We may also say that their projection onto this plane intersects at the same point $\tilde{P}'$. This leads to the well known astigmatic pencil of Fig. 3.6. By contrast, within the second-order ray aberration, the projection of these skewed rays onto the plane $(\tilde{k}, \tilde{m})$ pass near $\tilde{P}'$ and, since $(d\tilde{a})^2 > 0$ has always the same sign, it remains on the same side of this point. This may be compared to the effect of the coma. The projections of the rays are, in fact, tangent to a sort of caustics (Fig. 3.19).

A particular situation arises at the two endpoints $\tilde{P}_1$, $\tilde{P}_2$ of the astigmatic interval. In this case, we do not need to contract with $\tilde{m}$, but with $\tilde{m}_a \tilde{T} \tilde{p}_a = \tilde{m}_a$, $\alpha = 1, 2$, we get directly from (3.147) that

$$\Delta \boldsymbol{\eta}_{\tilde{P}_a} = \tfrac{1}{2} d^2 \boldsymbol{\eta}_{\tilde{P}_a} = -\tfrac{1}{2} (d\tilde{a}_a)^2 \tilde{p}_a \tilde{m}_a [\tilde{m}_a (\tilde{M} \lrcorner \tilde{M} \tilde{K} \tilde{M}^{\mathrm{T}} + \tilde{p}_a^2 \tilde{T})] \ , \tag{3.153}$$

which means that near $\tilde{P}_1$ or $\tilde{P}_2$ the rays within the first-order aberration really intersect at $\tilde{P}$, and the rays within the second-order aberration are almost tangent to a real caustic. Finally, if we contract (3.153) with $m_a$, we find

$$\tilde{m}_a \cdot d^2 \boldsymbol{\eta}_{\tilde{P}_a} = - (d\tilde{a}_a)^2 \tilde{p}_a [3 \tilde{m}_a \cdot \tilde{M} \tilde{k} + \tilde{p}_a^2 \tilde{m}_a \cdot (\tilde{m}_a \tilde{T} \tilde{m}_a)] \quad (\alpha = 1, 2) \ .$$

Comparing this relation with (3.149) for $y = \tilde{p}_a$, we see that in $\tilde{P}_1$ or $\tilde{P}_2$ (if $\tilde{k} \neq \hat{n}$ only at $\tilde{P}_1$ or $\tilde{P}_2$)

$$d^3 \theta_{\tilde{P}_a} = \frac{2\pi}{\tilde{\lambda} \tilde{p}_a} d\tilde{r}_a \cdot d^2 \boldsymbol{\eta}_{\tilde{P}_a} \quad (\alpha = 1, 2) \ , \tag{3.154}$$

holds similarly as in (3.150), i.e., the third-order wave aberration and the second-order transverse ray aberration are again related. Let us add that we could look to the dual of the coma, which would lead us to the distortion. However, here we shall rather look at the deformation from the standpoint of its linear approximation around an oblique reference ray.

We shall next move one step forward by passing to the already mentioned *deformation in space*, by first considering (3.117) for the apparent image deformation gradient. With the definition (3.140), the abbreviation $\tilde{l} + \tilde{p} = \tilde{L}_R$, and an additional projection with $\hat{N}$, (3.119) becomes

$$\frac{1}{p} \hat{N} K \, dr = \frac{\lambda}{\tilde{L}_R} \hat{N} \left[ \tilde{l} \left( \frac{1}{\tilde{\lambda} \tilde{q}} \tilde{C} + \frac{1}{\lambda p} K - \frac{1}{\lambda q} C \right) \tilde{M}^{\mathrm{T}} + \frac{1}{\tilde{\lambda}} \tilde{K} \right] d\tilde{r} \ , \tag{3.155}$$

where we have used the auxiliary relations for successive projectors $\hat{N} \hat{M} = \hat{N}$, $\tilde{M}^{\mathrm{T}} \tilde{K} = \tilde{M}^{\mathrm{T}}$, $\tilde{M} \tilde{K} = \tilde{K}$. The image point $\tilde{P}$ and its neighborhood $\{\tilde{P}\}$ are viewed here from the point $\tilde{R}$, the locus of one of the two eyes, for instance. It has already been stated that we need both eyes for stereoscopic viewing. That is why we now rewrite (3.155), with respect to a neighboring point $\tilde{S}$, separated from $\tilde{R}$ by an increment $d\tilde{\varrho}$ (Fig. 3.20):

$$\frac{1}{\tilde{p}} \hat{N} \tilde{K} \, dr = \frac{\lambda}{\tilde{L}_R} \hat{N} \left[ \tilde{l} \left( \frac{1}{\tilde{\lambda} \tilde{q}} \tilde{C} + \frac{1}{\lambda \tilde{p}} \tilde{K} - \frac{1}{\lambda \tilde{q}} \tilde{C} \right) \tilde{M}^{\mathrm{T}} + \frac{1}{\tilde{\lambda}} \tilde{K} \right] d\tilde{r} \ . \tag{3.156}$$

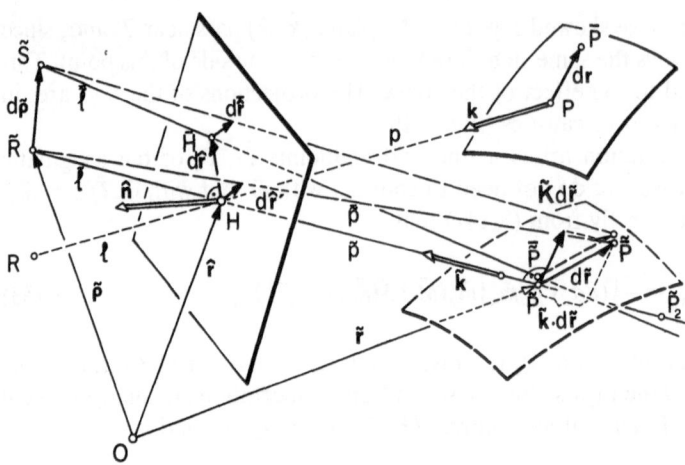

**Fig. 3.20.** Stereoscopic viewing of a modified image in space with eyes placed at $\tilde{R}$ and $\tilde{S}$, respectively. ($dr$, $d\tilde{r}$: associated increments of the object and the image surface)

It should be noted that not all of the quantities vary in (3.156). We have made $\tilde{\tilde{L}}_R = \tilde{L}_R$, assuming that the observer faces the region around $\tilde{P}$. Further, $d\tilde{\tilde{r}} = dr$, since both eyes should recognize the "same" materialized points $\tilde{P}$ and $\tilde{\tilde{P}}$, which is only possible when the region around $\tilde{P}$ has a typical appearance, permitting a correct association [3.296]. Finally, we choose $d\tilde{\tilde{r}} = d\tilde{r}$, since point $\tilde{S}$ is close to point $\tilde{R}$ as compared with distance $\tilde{L}_R$. Consequently, point $\tilde{\tilde{P}}$ is close to $\tilde{P}$ to some higher order than $d\tilde{r}$.

Now, as far as the oblique projector $\tilde{\tilde{M}}^T$ is concerned, we may proceed as in (3.137). The difference, however, is that $d\tilde{k}$ has to be replaced by $d\tilde{\tilde{k}}$ since the rays $\tilde{R}\tilde{P}$ and $\tilde{S}\tilde{\tilde{P}}$ are skew. Thus, with (3.142), we have

$$\tilde{\tilde{M}}^T = \tilde{M}^T + d\tilde{M}^T = \tilde{M}^T - \frac{1}{\hat{n} \cdot \tilde{k}} d\tilde{\tilde{k}}(\tilde{M} \otimes \hat{n})$$

$$= \tilde{M}^T\left(\tilde{K} - \tilde{p}\, d\tilde{k}\, \tilde{T} \otimes \frac{\hat{n}}{\hat{n} \cdot \tilde{k}}\right) . \tag{3.157}$$

The projector $\tilde{\tilde{K}}$ may be developed in a similar way

$$\tilde{\tilde{K}} = \tilde{K} + d\tilde{K} = \tilde{K} - d\tilde{\tilde{k}} \otimes \tilde{k} - \tilde{k} \otimes d\tilde{\tilde{k}}$$

$$= \tilde{K} - \tilde{p}\, d\tilde{k}\, \tilde{T}[\tilde{K} \otimes \tilde{k} + \tilde{K} \otimes \tilde{k})^T] . \tag{3.158}$$

All of the other projectors, including the corresponding inverse of the distance to the respective collineation center, as in $\bar{C}/\bar{q}$, for instance, must be developed into projectors and superprojectors as follows

$$\frac{1}{\tilde{p}}\tilde{K} = \frac{1}{p}K - \frac{1}{p^2} d\hat{r}\, K . \tag{3.159}$$

Finally, since $d\tilde{L}_R = 0$, we have $d\tilde{l} = -d\tilde{p} = -\tilde{p}\,d\tilde{k} \cdot \tilde{M}\tilde{k}$. Taking the difference between (3.156 and 155), and contracting the result with $d\hat{r}$, which is the increment relative to $d\tilde{\varrho}_R$, with P as the collineation center, we find the scalar equation

$$d\hat{r} \cdot \hat{N}\left[d\hat{r}\,\hat{N}\left(\frac{1}{p^2}K\,dr\right)\right]$$

$$= \frac{\lambda}{\tilde{L}_R}\left\{\tilde{l}\,d\hat{r} \cdot \hat{N}\left[d\hat{r}\,\hat{N}\left(\frac{1}{\tilde{\lambda}\tilde{q}^2}\tilde{C} + \frac{1}{\lambda p^2}K - \frac{1}{\lambda q^2}C\right)\tilde{M}^T\right]d\tilde{r}\right.$$

$$+ \tilde{l}\,d\hat{r} \cdot \hat{N}\left[\left(\frac{1}{\tilde{\lambda}\tilde{q}}\tilde{C} + \frac{1}{\lambda p}K - \frac{1}{\lambda q}C\right)\tilde{M}^T\tilde{p}\left(d\tilde{k}\,\tilde{T} \otimes \frac{\hat{n}}{\hat{n} \cdot \tilde{k}}\right)\right]d\tilde{r}$$

$$+ \tilde{p}\,d\hat{r} \cdot \hat{N}\left[(d\tilde{k}\,\tilde{M} \cdot \tilde{k})\left(\frac{1}{\tilde{\lambda}\tilde{q}}\tilde{C} + \frac{1}{\lambda p}K - \frac{1}{\lambda q}C\right)\tilde{M}^T\right]d\tilde{r}$$

$$+ \frac{\tilde{p}}{\lambda}\,d\hat{r} \cdot \hat{N}\left[\tilde{T}\,d\tilde{k}(\tilde{k} \cdot d\tilde{r}) + \tilde{k}(d\tilde{k} \cdot \tilde{T}\,d\tilde{r})\right]\right\} . \tag{3.160}$$

On the left side of this equation, we may isolate the longitudinal part $k \cdot dr$ of $dr$, by applying the affine connection at the recording ($\hat{M} = I - \hat{n} \otimes k/\hat{n} \cdot k$)

$$d\hat{r} = p\hat{M}^T m\,d\alpha ,$$

with a unit vector $m$ perpendicular to $k$ and an angular increment $d\alpha$ between the rays PH and P$\overline{\text{H}}$. With $K\,dr = k \otimes K\,dr + K(k \cdot dr) + K\,dr \otimes k$, we find

$$d\hat{r} \cdot \hat{N}\left[d\hat{r}\hat{N}\left(\frac{1}{p^2}K\,dr\right)\right] = (d\alpha)^2[(k \cdot dr) + 2(m \cdot \hat{M}k)(m \cdot K\,dr)] . \tag{3.161}$$

In a similar way, on the right side of (3.160), we use the affine connection at the reconstruction

$$d\hat{r} = \tilde{p}\tilde{M}^T\tilde{m}\,d\tilde{\alpha} ,$$

with the unit vector $\tilde{m}$ perpendicular to $\tilde{k}$ and an angular increment $d\tilde{\alpha}$ at $\tilde{P}$. Using the definitions (3.140, 141), the right side of (3.160) reduces to

$$(d\tilde{\alpha})^2\,\frac{\lambda\tilde{l}\tilde{p}^2}{\tilde{\lambda}\tilde{L}_R}\left\{\tilde{m} \cdot \left[\tilde{m}\left(-\tilde{l} + \tilde{T}^2 \otimes \frac{\hat{n}}{\hat{n} \cdot \tilde{k}}\right)\right]d\tilde{r} + \frac{1}{\tilde{l}}[(\tilde{k} \cdot d\tilde{r})\tilde{m} \cdot \tilde{T}\tilde{m}\right.$$

$$+ 2(\tilde{m} \cdot \tilde{M}\tilde{k})\tilde{m} \cdot \tilde{T}\,d\tilde{r}]\right\} . \tag{3.162}$$

If we separate $d\tilde{r}$ into $\tilde{K}\,d\tilde{r}$ and $\tilde{k}(\tilde{k} \cdot d\tilde{r})$, we have

$$\frac{\hat{n}}{\hat{n} \cdot \tilde{k}} \cdot d\tilde{r} = -(\tilde{M}\tilde{k}) \cdot \tilde{K}\,d\tilde{r} + \tilde{k} \cdot d\tilde{r} . \tag{3.163}$$

Further, $d\tilde{a}$ may be expressed by $da$, see (3.123), here written in the form:

$$(d\tilde{a})^2 = (da)^2 \frac{L_R \tilde{\lambda} \tilde{p} \tilde{l}}{\tilde{L}_R \lambda \tilde{p} l} . \tag{3.164}$$

With (3.161–164), we obtain for (3.160) the sought after relation

$$\frac{l}{L_R p} [k \cdot dr + 2(m \cdot \hat{M}k)(m \cdot K \, dr)]$$

$$= \frac{\tilde{p}\tilde{l}}{\tilde{L}_R^2} \{\tilde{m} \cdot (l\tilde{T} + \tilde{K})\tilde{T}\tilde{m}(\tilde{k} \cdot d\tilde{r}) - \tilde{l}\tilde{m} \cdot [\tilde{m}(\tilde{T} + \tilde{T}^2 \otimes \tilde{M}\tilde{k})]\tilde{K} \, d\tilde{r}$$

$$+ 2(\tilde{m} \cdot \tilde{M}\tilde{k})(\tilde{m} \cdot \tilde{T}\tilde{K} \, d\tilde{r})\} , \tag{3.165}$$

describing, together with (3.119), the mapping $k \cdot dr \to \tilde{k} \cdot d\tilde{r}$.

Indeed, we may express $K \, dr$ by (3.119), $m$ by $\tilde{m}$, and $da$ by $d\tilde{a}$ with (3.164), and isolate the term $(k \cdot dr)$ in (3.165):

$$\frac{l}{L_R p} (k \cdot dr) = \frac{\tilde{l}^2 \tilde{p}}{\tilde{L}_R^2} (\tilde{k} \cdot d\tilde{r}) \left[ \tilde{m} \cdot \left( \tilde{T} + \frac{1}{\tilde{l}} \tilde{K} \right) \tilde{T}\tilde{m} \right]$$

$$- 2 \frac{\tilde{l}^2 \tilde{p}}{\tilde{L}_R^2 p} (\tilde{m} \cdot \tilde{M}k) \left[ \tilde{m} \cdot \left( \tilde{T} + \frac{1}{\tilde{l}} \tilde{K} \right) \tilde{K} \, d\tilde{r} \right]$$

$$+ \frac{2\tilde{l}\tilde{p}}{\tilde{L}_R^2} (\tilde{m} \cdot \tilde{M}\tilde{k})(\tilde{m} \cdot \tilde{T}\tilde{K} \, d\tilde{r})$$

$$- \frac{\tilde{l}^2 \tilde{p}}{\tilde{L}_R^2} \tilde{m} \cdot [\tilde{m}(\tilde{T} + \tilde{T}^2 \otimes \tilde{M}\tilde{k})\tilde{K} \, d\tilde{r}] . \tag{3.166}$$

Taking the first principal direction $(\tilde{m} = \tilde{m}_1)$ of the tensor $\tilde{T}$ and, since $\tilde{T}\tilde{m}_1 = \tilde{m}_1/\tilde{p}_1$, the above equation simplifies to:

$$\frac{l}{L_R p} (k \cdot dr) = \frac{\tilde{l}}{\tilde{L}_R \tilde{p}_1} (\tilde{k} \cdot d\tilde{r})$$

$$+ \frac{2\tilde{l}}{\tilde{L}_R^2} (\tilde{m}_1 \cdot \tilde{K} \, d\tilde{r}) \left[ (\tilde{m}_1 \cdot \tilde{M}\tilde{k}) - \frac{L_R}{p} (\tilde{m}_1 \cdot \tilde{M}k) \right]$$

$$- \frac{\tilde{l}^2}{\tilde{L}_R^2 \tilde{p}_1} (\tilde{M}\tilde{k}) \cdot \tilde{K} \, d\tilde{r} - \frac{\tilde{l}^2 \tilde{p}_1}{\tilde{L}_R^2} (\tilde{m}_1 \tilde{T}\tilde{m}_1) \cdot \tilde{K} \, d\tilde{r} . \tag{3.167}$$

In the special case $q \to \infty$, $\tilde{q} \to \infty$ and $\hat{N}\tilde{c} = (\tilde{\lambda}/\lambda)\hat{N}c$, discussed in Sect. 3.2.1 where the primary Seidel aberrations are treated, (3.167) takes the very simple form

$$(k \cdot dr) = \frac{\tilde{\lambda}}{\lambda} (\tilde{k} \cdot d\tilde{r}) \ .$$

This is deduced by using also Gauss' relation (3.84) applied to the observer's point. Thus, in (3.167), the ratio $(\tilde{l}/\tilde{L}_R \tilde{p})/(l/L_R p)$ describes mainly the wavelength change $(\tilde{\lambda}/\lambda)$, whereas the two last terms in the same equations are correction terms, taking into account the oblicity and the change of the reference source position.

# 3.3 Particular Modifications at the Reconstruction

At the end of this chapter, we would like to treat two particular subjects of holography in which a modification of the image takes place at the reconstruction. First, we briefly review the basic concepts of so-called cylindrical composite holography, which is important for panorama white light reconstruction. Secondly, we need a preparation for fringes modification discussed in Chap. 5. For this purpose, we shall linearize here some equations of Sect. 3.1 and 3.2 in the case of small shifts and small wavelength changes of the reference source.

### 3.3.1 Basic Concepts of Cylindrical Composite Holograms

The purpose of cylindrical composite holograms, which are an assembly of holographic strips placed in the form of a cylinder, is to reconstruct an object whose image may be seen from all sides, and eventually reveal the interior of the considered body. In the literature of stereographic holography, one distinguishes between multiplex and composite holography. In multiplex holography, the ordinary pictures of an object taken from different directions are holographically recorded on the whole surface of a holographic plate. At the reconstruction, some kind of filtering must be done, so that the observer only sees one picture with each eye, thereby creating a three-dimensional stereoscopic impression [3.298–301]. In composite holography, each picture is recorded on a separate area of the holographic plate. In general, these small areas are vertical strips. Such strips dramatically simplify the recording process and allow e.g., the generation of a computer hologram, but make the vertical parallax vanish. At the reconstruction, each eye of the observer sees only one picture through one holographic strip, i.e., projected image of the object. Since the two eyes perceive different projections, the stereoscopic effect is thus reproduced [3.302–310]. If the hologram is not a plane but a cylinder, 360° viewing is possible [3.311–323]; if the original pictures were x-ray photographs, the three-dimensional reconstruction would show the interior of the object [3.299, 305, 324–331]. Here, we consider an installation similar to that used by *Honda* et al. in [3.313]. A white-light reconstruction brings an additional distortion due to the change of wavelength. For this short theoretical development, we restrict our discussion to laser-light reconstruction by introducing the basic relations needed for the description of the distortions due to the geometry of the system. Figure 3.21 illustrates the principle. A camera placed at some point R outside the region of the object takes an ordinary picture. This could also be an x-ray photo, covering then the three-dimensional domain formed by the neighborhood {P} of point P. Such a negative is later magnified so that the picture represents, in fact, the projection {$P_R$} of the points {P} from R onto the plane $\pi_R$ with unit normal $\hat{n}$ and passing through the axis, whose direction is given by the unit vector $\hat{n}$ (Fig. 3.21). This ordinary picture is thereafter recorded on a holographic strip, which constitutes an element along a generating line of a cylindrical surface of radius $a$ and of axis $\Delta$. The procedure is repeated for other locations of the camera on the circle of radius $b$ around $\Delta$. For instance, a neighboring point S of R gives the projection {$P_S$} of the set {P} onto the plane $\pi_S$, which is slightly tilted with respect to $\pi_R$. Both planes $\pi_S$ and $\pi_R$ intersect along axis $\Delta$. The neighboring holographic strip

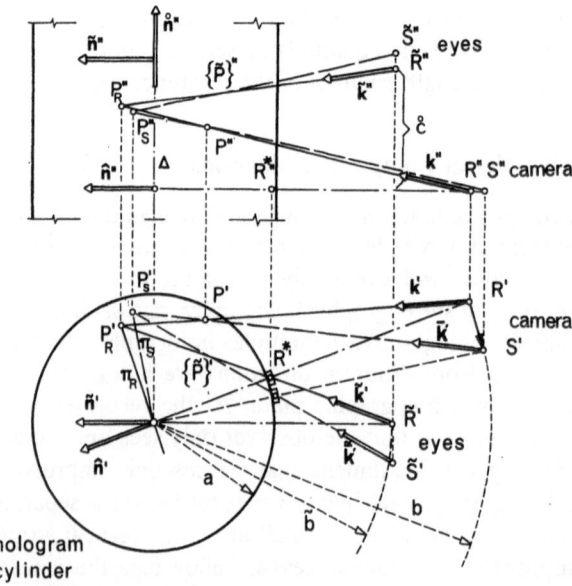

**Fig. 3.21.** Cylindrical composite hologram (' denotes quantities in the top-view, "denotes quantities in the side-view). R, S: center of the camera at two neighboring positions at the "recording"; photographs of ordinary images $P_R$, $P_S$ of P in the planes $\pi_R$, $\pi_S$. $\tilde{R}$, $\tilde{S}$: positions of the two eyes at the reconstruction, viewing through holographic strips corresponding to R, S. ($\hat{n}$, $\bar{n}$, $\hat{n}$, $k$, $\bar{k}$: unit vectors)

records the points $\{P_S\}$. Suppose now that an observer's eyes correspond to the points $\tilde{R}$ and $\tilde{S}$ at a distance $\tilde{b}$ from $\Delta$ at the reconstruction. The right eye looks through the strip associated with R, whereas the left one looks through the strip associated with S. Since $P_R$ and $P_S$ are the projections from the one and same point P, the observer will recognize the picture $\tilde{P}$ near the shortest distance between the slightly skewed rays $\tilde{R}P_R$ and $\tilde{S}P_S$.

In order to analyse this modified image $\{\tilde{P}\}$, one may refer to [3.313]. For our purposes, we introduce here the unit vectors $k$, $\tilde{k}$ relative to the rays RP and $\tilde{R}\tilde{P}$, respectively, as well as a horizontal unit vector $\tilde{n}$ whose support line connects point $\tilde{R}$ with the axis $\Delta$. We then first write the condition of oblique projection onto the plane $\pi_R$

$$\tilde{M}^{\mathrm{T}}(\tilde{b}\tilde{n} - \mathring{c}\hat{n}) = \hat{M}^{\mathrm{T}}(b\hat{n}) \ , \tag{3.168}$$

where $\mathring{c}$ marks the level of the eye $\tilde{R}$ above the plane $\pi_0$ in which the camera was situated. As previously, $\tilde{M} = I - \hat{n} \otimes \tilde{k}/\hat{n} \cdot \tilde{k}$ and $\hat{M} = I - \hat{n} \otimes k/\hat{n} \cdot k$ denote oblique projectors. As a second condition concerning the point R* of intersection with the strip, we may write:

$$\tilde{K}'(\tilde{b}\tilde{n} - a\hat{n}) = 0 \ , \tag{3.169}$$

where $\tilde{K}'$ is the normal projector along the horizontal direction $\mathring{N}\tilde{k}$ with the normal projector along $\Delta$, $\mathring{N} = I - \hat{n} \otimes \hat{n}$. Equations (3.168, 169), together with the auxiliary conditions

$$|k| = 1 \ , \quad |\hat{n}| = 1 \ ,$$

determine the two unit vectors $\hat{n}$ and $k$, if we assume, for example, $\tilde{k}$ to be a given direction, as was done in Sect. 3.2.2.

Equations (3.168, 169) read similarly for the other eye situated at point $\tilde{S}$: (and determine $\check{n}$, $\bar{k}$, i.e., with $k$ the point P)

$$\bar{\tilde{M}}^{\mathrm{T}}(\bar{b}\bar{\tilde{n}} - \bar{\tilde{c}}\check{n}) = \bar{\tilde{M}}^{\mathrm{T}}(b\check{n}) \ , \quad \bar{\tilde{K}}'(\bar{b}\bar{\tilde{n}} - a\check{n}) = 0 \ .$$

The two lines defined by the directions $\tilde{k}$ and $\bar{\tilde{k}}$ associated to the points $\tilde{R}$ and $\tilde{S}$ are, in general, skew. The reconstruction $\tilde{P}$ of point P is not perfect. The study concerning aberrations of this point may be done similarly to that in Sect. 3.1.2. Since the location of this point $\tilde{P}$ is different from that of point P, a deformation of the whole image is produced. Image deformation and point aberration are somehow connected, which may be best described along the same line of thought as that developed in Sect. 3.2.2.

## 3.3.2 Small Modifications

Let us now consider a special case in which the reconstruction arrangement is subjected to *small* optical modifications. We shall again restrict ourselves to the case where both the wavelength and position of the reconstruction source $\tilde{Q}$ differ *little* from those of the reference source Q, with the position of the hologram, for simplicity, remaining unchanged. The primary image wave field then forms a virtual image $\tilde{P}$ very *near* to the object point P. The image of a whole object, or rather, the vicinity $\overline{P}$ of each point P suffers small linear and angular dilatations. If $\tilde{Q}$ differs little from Q*, similar effects would be produced for the real image $\tilde{P}$*, formed by the conjugate image wave field and situated near P.

Such small modifications are of interest for several reasons. First, the equations for a large modification assume a linearized form in the case of small changes, so that they become easier to discuss. Secondly, all quantities may be assigned to a reference configuration, which we choose to be the recording arrangement. Thirdly, unintentionally produced small modifications are often encountered in practice as, for instance, in holographic interferometry, where the small optical shifts in the real-time technique are combined with the small mechanical displacements to be investigated.

In the present case (Fig. 3.22), the changes are primarily characterized by the displacement vector $d$ from Q to $\tilde{Q}$ and by the wavelength change $\Delta\lambda = \lambda - \tilde{\lambda}$. The assumption

$$\frac{d}{q} = O(\delta) \, , \qquad \delta \ll 1 \, , \qquad \frac{|\Delta\lambda|}{\lambda} = O\left(\frac{d}{q}\right) = O(\delta) \tag{3.170}$$

indicates that a common small parameter $\delta$ exists which introduces an asymptotic behavior of certain quantities and of some equations. This allows us to write the

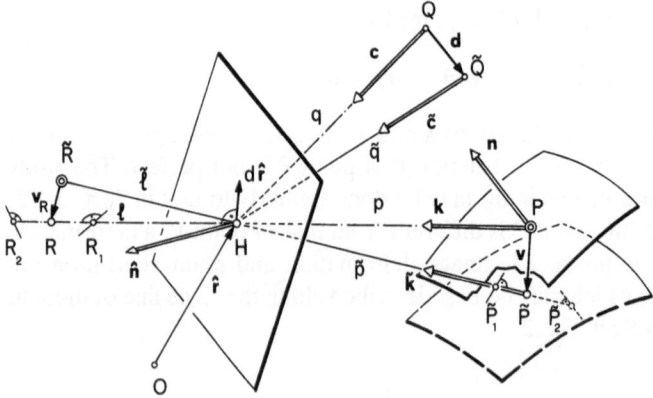

**Fig. 3.22.** Small optical modification. ($d$: shift of the reference source Q; $v$: optical displacement of an object point P at the recording to $\tilde{P}$ at the reconstruction; $v_R$: "dual displacement" of the center $\tilde{R}$ of the observing instrument at the reconstruction to R at the recording)

following relations:

$$\frac{1}{\tilde{\lambda}} = \frac{1}{\lambda - \Delta\lambda} \simeq \frac{1}{\lambda} + \frac{\Delta\lambda}{\lambda^2} \ ,$$

$$\tilde{q} = |qc - d| \simeq q - d \cdot c \ , \tag{3.171}$$

$$\tilde{c} = \frac{qc - d}{|qc - d|} \simeq c - \frac{1}{q} Cd \ ,$$

so that, with the stationarity condition $d\theta = 0$, (3.15) becomes

$$\hat{N}\left[\left(1 + \frac{\Delta\lambda}{\lambda}\right)\tilde{k} - k\right] = -\hat{N}\left(-\frac{\Delta\lambda}{\lambda}c + \frac{1}{q}Cd\right) \ . \tag{3.172}$$

As the right side of this equation is of the order of $\delta$ and $\tilde{k}$ is a unit vector like $k$, we may express it in the same manner as $c$:

$$\tilde{k} \simeq k - \frac{1}{p}Kv \simeq k - \frac{1}{l}Kv_R \ . \tag{3.173}$$

Here, $v$ is a small aberration of the object point P and $v_R$ is the dual aberration of the observer's point $\tilde{R}$; both of the order of $\delta$. Therefore, (3.172) assumes the following linearized form

$$\frac{1}{p}\hat{N}Kv = \frac{1}{l}\hat{N}Kv_R = \hat{N}\left[\frac{\Delta\lambda}{\lambda}(k - c) + \frac{1}{q}Cd\right] \ , \tag{3.174}$$

for the two lateral aberrations $Kv$ and $Kv_R$. By means of the oblique projector, $\hat{M} = I - \hat{n} \otimes k/\hat{n} \cdot k$, and the auxiliary conditions, $\hat{M}\hat{N} = \hat{M}$, $\hat{M}K = K$, we explicitely find for these lateral aberrations the result

$$\frac{1}{p}Kv = \frac{1}{l}Kv_R = \hat{M}\left[\frac{\Delta\lambda}{\lambda}(k - c) + \frac{1}{q}Cd\right] \ . \tag{3.175}$$

Now, to determine the longitudinal components $k \cdot v$ and $k \cdot v_R$, we have to consider the condition $d^2\theta = 0$ and develop the quantities which are contained in the tensor $\tilde{T}$, see (3.31). Thus, for example, the projector $\tilde{C}/\tilde{q}$ takes the linearized form:

$$\frac{1}{\tilde{q}}\tilde{C} \simeq \frac{1}{q\left(1 - \frac{1}{q}d \cdot c\right)}\left[I - \left(c - \frac{1}{q}Cd\right) \otimes \left(c - \frac{1}{q}Cd\right)\right]$$

$$\tag{3.176}$$

$$\simeq \frac{1}{q}C + \frac{1}{q^2}\left[c \otimes Cd + C(c \cdot d) + Cd \otimes c\right] = \frac{1}{q}C + \frac{1}{q^2}Cd$$

with the superprojector $C$. Moreover, by means of (3.173), the development of the oblique projector $\tilde{M}$ results in

$$\tilde{M} = I - \frac{\hat{n} \otimes \tilde{k}}{\hat{n} \cdot \tilde{k}} \simeq I - \frac{1}{\hat{n} \cdot k}\left(1 + \frac{\hat{n} \cdot Kv}{p\hat{n} \cdot k}\right)(\hat{n} \otimes k - \hat{n} \otimes \frac{1}{p}Kv) ,$$

$$\tilde{M} = \hat{M} + \frac{\hat{n} \otimes Kv}{p(\hat{n} \cdot k)}\, \hat{M} .$$

(3.177)

With these auxiliary relations and for the same tensor $\tilde{T}$ describing the astigmatism, we now get

$$\tilde{T} = \tilde{\lambda}\tilde{M}\left(\frac{1}{\lambda p}K - \frac{1}{\lambda q}C + \frac{1}{\tilde{\lambda}\tilde{q}}\tilde{C}\right)\tilde{M}^{\mathrm{T}}$$

$$\simeq (\lambda - \Delta\lambda)\left(\hat{M} + \frac{\hat{n} \otimes Kv}{p(\hat{n} \cdot k)}\hat{M}\right)\left[\frac{1}{\lambda p}K - \frac{1}{\lambda q}C\right.$$

$$\left. + \left(\frac{1}{\lambda} + \frac{\Delta\lambda}{\lambda^2}\right)\left(\frac{1}{q}C + \frac{1}{q^2}Cd\right)\right]\left(\hat{M}^{\mathrm{T}} + \hat{M}^{\mathrm{T}}\frac{Kv \otimes \hat{n}}{p(\hat{n} \cdot k)}\right)$$

$$= \frac{1}{p}K + \hat{M}\left[-\frac{\Delta\lambda}{\lambda}\left(\frac{1}{p}K - \frac{1}{q}C\right) + \frac{1}{q^2}Cd\right]\hat{M}^{\mathrm{T}} + \frac{\hat{n} \otimes Kv + Kv \otimes \hat{n}}{p^2\hat{n} \cdot k} .$$

Finally, using the decomposition

$$\frac{\hat{n}}{\hat{n} \cdot k} = \frac{K\hat{n}}{\hat{n} \cdot k} + k = -\hat{M}k + k ,$$

(3.178)

we obtain the sought after development of tensor $\tilde{T}$, now expressed in terms relative to $k$ but not to $\tilde{k}$

$$\tilde{T} = \frac{1}{p}K + \frac{1}{p^2}\tilde{U} + \frac{1}{p^2}(k \otimes Kv + Kv \otimes k) .$$

(3.179)

The tensor $\tilde{U}$, defined as

$$\tilde{U} = \hat{M}\left[-\frac{p^2\Delta\lambda}{\lambda}\left(\frac{1}{p}K - \frac{1}{q}C\right) + \frac{p^2}{q^2}Cd - (k \otimes Kv + Kv \otimes k)\right]\hat{M}^{\mathrm{T}} ,$$

(3.180)

is of the order of $\delta$ and characterizes a part of $\tilde{T}$ in the plane normal to $k$. The remaining out-of-plane component, $(k \otimes Kv + Kv \otimes k)/p^2$, is also of order $\delta$, whereas $K/p$ represents a constant finite tensor.

Since

$$\frac{1}{\tilde{p}}\tilde{K} = \frac{1}{p}K + \frac{1}{p^2}\left[Kv \otimes k + K(k \cdot v) + k \otimes Kv\right] ,$$

(3.181)

we see that the equation in the plane normal to $\tilde{k}$, determining the distance $\tilde{p}$ in case of *large* modifications, see (3.30),

$$\tilde{m} \cdot \left( \tilde{T} - \frac{1}{\tilde{p}} \tilde{K} \right) \tilde{m} = 0 \ , \tag{3.182}$$

has a linearized counterpart in the plane normal to $k$. Therefore, for the determination of the longitudinal shift $k \cdot v$, in the case of *small* modifications, we have

$$m \cdot [\tilde{U} - (k \cdot v)K]m = 0 \ , \tag{3.183}$$

where $\tilde{m}$, $m$ here are unit vectors perpendicular to $\tilde{k}$, $k$, respectively.

Both equations (3.180 and 183) were already deduced in [3.31] in another manner, which avoided the developments of (3.177 and 179) (Note that we use the letter $\tilde{U}$ and not $T$, since here $T$ denotes the dual of the tensor $\tilde{T}$ having the same structure.) The present investigation mainly makes apparent the relation (3.179) between $\tilde{T}$ and $\tilde{U}$.

Indeed, we may also linearize the dual tensor $T$ of $\tilde{T}$ and the dual equation of (3.182), $m \cdot (T - K/l) \, m = 0$. We begin with (3.109) for the tensor $T$, which, as for the other dual equations, we express relative to $k$. The linearized form of this tensor

$$T = \hat{M} \left( \frac{\lambda}{\tilde{\lambda}\tilde{l}} \tilde{K} + \frac{\lambda}{\tilde{\lambda}\tilde{q}} \tilde{C} - \frac{1}{q} C \right) \hat{M}^{T}$$

becomes

$$T = \hat{M} \left[ \frac{\Delta\lambda}{\lambda} \left( \frac{1}{l} \tilde{K} + \frac{1}{q} \tilde{C} \right) + \frac{1}{l} K + \frac{1}{l^2} K v_{R} + \frac{1}{q^2} Cd \right] \hat{M}^{T} \ ,$$

and may be written

$$T = \frac{1}{l} K + \frac{1}{l^2} U + \frac{1}{l^2} K(k \cdot v_{R}) \quad \text{with} \tag{3.184}$$

$$U = \hat{M} \left[ \frac{\Delta\lambda}{\lambda} l^2 \left( \frac{1}{l} K + \frac{1}{q} C \right) + \frac{l^2}{q^2} Cd + k \otimes K v_{R} + K v_{R} \otimes k \right] \hat{M}^{T} \ . \tag{3.185}$$

Thus, the counterpart of (3.183), always in its linearized form, becomes

$$m \cdot (U + (k \cdot v_{R})K)m = 0 \ . \tag{3.186}$$

As stated in (3.173), the reduced lateral displacements of $\tilde{R}$ and P are the same ($K v_{R}/l = K v/p$), whatever modification occurs. Starting from (3.183, 186), we may form the difference and the sum of reduced longitudinal parts

$$\frac{1}{l}\,(k \cdot v_R)\;-\frac{1}{p}\,(k \cdot v)$$

$$= -\frac{l+p}{q}\left[\frac{\Delta\lambda}{\lambda}\,m \cdot \hat{M}C\hat{M}^{\mathrm{T}}m + \frac{1}{q}\,m \cdot \hat{M}(Cd)\hat{M}^{\mathrm{T}}m\right]\,, \tag{3.187}$$

and

$$\frac{1}{l^2}\,(k \cdot v_R)\;+\frac{1}{p^2}\,(k \cdot v)$$

$$= -\left(\frac{1}{p}+\frac{1}{l}\right)\left[\frac{\Delta\lambda}{\lambda}+\frac{2}{p}\,(m \cdot \hat{M}k)(m \cdot Kv)\right]\,. \tag{3.188}$$

Equation (3.188) could also be found by linearizing (3.118).

In order to gain an insight into some applications of these two equations, let us now consider some special cases.

First, if the modification is produced by a wavelength change, $\Delta\lambda \neq 0$, with a collimated reference beam ($q \to \infty$), we get directly from (3.187)

$$\frac{1}{l}\,(k \cdot v_R) = \frac{1}{p}\,(k \cdot v)\,. \tag{3.189}$$

This also holds true for finite changes in direction of the reference beam, $d/q \neq 0$, ($d/q^2 \to 0$), and for oblique observation directions, $\hat{M}k \neq 0$. In this case, the displacement $v_R = \tilde{R}R$ is parallel to the displacement $v = P\tilde{P}$ and the ratio of their respective lengths, $|v_R|/|v|$ is just equal to the ratio $l/p$, see (3.175, 189). In the special case considered in Sect. 3.2.1, when studying the Seidel aberrations, we have further considered an observation direction normal to the hologram plate, $\hat{M}k = 0$, and a direction change of the reference beam in such a way as to get $\tilde{k} = k$ or, according to (3.173), $Kv/p = Kv_R/l = 0$. From (3.188, 189), we then obtain

$$\frac{v_R}{l} = \frac{v}{p} = -\frac{\Delta\lambda}{\lambda}\,,$$

which could also be directly deduced from linearization of the Gaussian equation (3.86).

Secondly, we may consider a geometrical modification of the reference source, $d \neq 0$, $q \neq \infty$, without a wavelength change ($\Delta\lambda = 0$). For an observation direction normal to the hologram, $\hat{M}k = 0$, (3.188) leads to

$$\frac{1}{l^2}\,(k \cdot v_R) = -\frac{1}{p^2}\,(k \cdot v)\,. \tag{3.190}$$

The longitudinal parts of $v_R$ and $v$ are opposite in sign.

Finally, a combined example of these two special cases, ($q \to \infty$, $\Delta\lambda = 0$ and $\hat{M}k = 0$) must simultaneously satisfy (3.189 and 190), which is only possible when $k \cdot v_R = k \cdot v = 0$.

In Chap. 5, we shall again make use of the relationships between $v$ and $v_R$.

# 4. Holographic Interferometry or Fringe Interpretation

As seen in Chap. 3, the principal property of holograms is the storage of wave fields which may be reconstructed at any later time. As was already mentioned in the introduction, it is for this reason that two slightly different wave fields, which do not exist simultaneously, may be brought to interfere. Analysis of the resulting interference fringes (Fig. 4.1) constitutes the principal problem in holographic interferometry, for which the first experiments were performed around 1965 [4.1–14]. In particular, if both wave fields are recorded successively on the same hologram (with the same reference source) and later reconstructed together, we obtain the well-known *double exposure* or frozen fringe effect. Using this technique, measurements are almost completely insensitive to small modifications of the optical set-up at the reconstruction, since both wave fields undergo the same small changes. Thus, the fringes disclose information about the *difference* of relative positions of the two (stationary) configurations from which these wave fields emanate. In this chapter, we shall investigate two cases of special interest in engineering. First, we present the elements concerning holographic determination of the *deformation* of an opaque body with a diffusely reflecting surface. Here, the concepts of Sect. 2.3 will be combined with those outlined in Sect. 3.1.1. Secondly, we intend to study holographic interferometry of phase objects, i.e., the determination of the spatial *variation* of the *index of refraction* of a transparent medium. This topic relates Sects. 2.2 to 3.1.1.

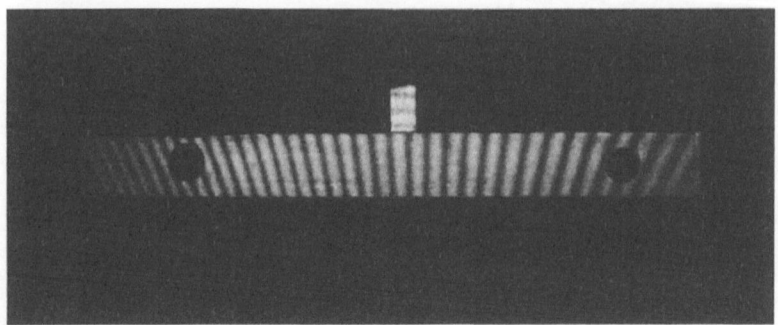

**Fig. 4.1.** Example of a fringe pattern: Bending of a beam by a concentrated vertical load

Concerning the most commonly used techniques, we note that in a nonstationary dynamic process, in addition to the possibility of stroboscopic double exposure, in each of these two fields there exists the familiar *time average* method. In this case, a continuously varying configuration is recorded over some finite time interval (i.e., the period of oscillation). Finally, the so-called *real-time* or live-fringe technique, where a reconstructed wave field interferes with an actual existing field, will be treated later in Sect. 5.1, since the respective measurements are sensitive to optical modifications.

## 4.1 Basic Relations for Deformed Opaque Bodies and Isotropic Nonhomogeneous Transparent Media

The principal quantity used in holographic interferometry is the difference of the optical paths to an arbitrary point in the two exposures at the recording. The basic relation assigns this quantity either to the displacement (in case of the deformation of an opaque object) or to an integral of the refractive index change (in case of a transparent medium). Moreover, we shall see that derivatives of the optical path difference will come into play for several reasons.

### 4.1.1 Optical Path Difference and Its Derivative for a Deformed Opaque Object

The determination of the deformation of an opaque body by holographic interferometry is the most important engineering application in this area, since it pertains to methods of non-destructive testing. Nevertheless, this subject will be outlined here as briefly as possible in order both to avoid unnecessary duplication of other recently published books [4.15–23] and to prevent repetition of its exhaustive treatment in [Ref. 4.24, Sect. 4.1].

A simple preliminary explanation of the basic principle may now be introduced with Fig. 4.2, which depicts the surfaces of the undeformed and deformed bodies in question. The reconstruction arrangement is supposed to be ideal (no modification), so that this figure may be used for an elementary illustration of the interference formation. Point S is the object source (at the recording), P and P' are the two positions of a material point, separated by the *displacement vector u*, and K denotes an arbitrary point in space where the fringes will be considered at the reconstruction. Let us start with the reconstruction process explained in Sect. 3.1. In the case of the double exposure technique, instead of (3.6) for the single exposure, for the transmitted wave we now have, at any point of the hologram (in particular at point H on KP or at H' on KP') the equation

$$\tilde{V}T = \tilde{V}T_0 - \frac{\beta\tau}{4}\left[(U + U')V^*\tilde{V} + (U^* + U'^*)V\tilde{V}\right], \tag{4.1}$$

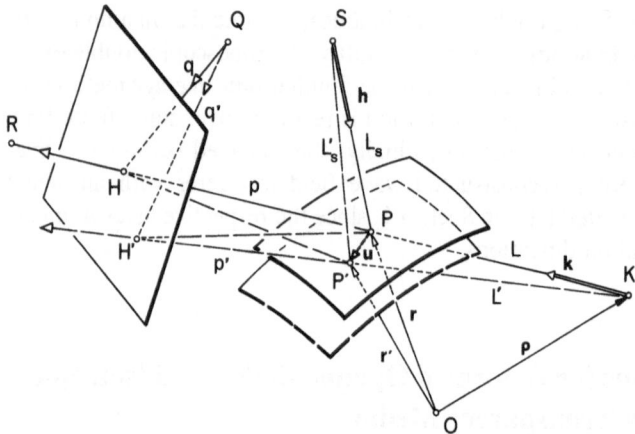

**Fig. 4.2.** Principle of the double exposure technique. Simplified model explaining the formation of fringes in space. (K: point where the interference is considered, S: object source; Q: reference source; P, P': object point before and after deformation; **h**, **k**: unit vectors)

where $U$, $U'$ are the complex amplitudes of the two object waves for the unde-formed and the deformed state, respectively, assuming equal exposure times $\tau/2$. Excluding any optical modifications, the reconstructing wave is supposed to be identical with the reference wave ($\tilde{V} \equiv V$). The primary wave field is then $-(\beta\tau/4)|V|^2(U + U')$ and therefore reconstructs the superposition of $U$ and $U'$, which may interfere. If the points P, P' were present *alone*, (without considering any neighboring region on the surface), for the space point K, we could now write other amplitudes $U = (A_P/L)\exp(-2\pi i\,L/\lambda)$, $U' = (A'_P/L')\exp(-2\pi i\,L'/\lambda)$, where the distances $L = \overline{KP}$ and $L' = \overline{KP'}$ intervene. When K lies behind the object surface, the distances $L$ and $L'$ are positive. At this point K, we have therefore the intensity (corresponding to the amplitude $U + U'$):

$$J = \frac{1}{2}(U + U')(U + U')^* = \frac{1}{2}\left\{|U|^2 + |U'|^2\right.$$

$$\left. + \frac{A_P^* A'_P}{LL'}\exp\left[\frac{2\pi i}{\lambda}(L-L')\right] + \frac{A_P A'^*_P}{LL'}\exp\left[-\frac{2\pi i}{\lambda}(L-L')\right]\right\}.$$

In addition to the difference of phases contained in $A_P A'^*_P$, the difference of paths $L - L'$ appears.

The following development must be independent of the type of observation system. That is why the intensity $J$ is considered at the (virtual) point K and not at the corresponding image point K* on the image plane (retina of the eye, film of the camera, plane of the photodetector, etc.). Due to Malus' theorem, the optical paths between point K on the fringe in space and the fringe point K* lying on the image plane and passing through points P and P', respectively, are the same. For the two object positions P and P', illuminated by the same source S, the phase difference contained in $A_P A_P'^*$ may be expressed by $2\pi i \, (L_s - L_s')/\lambda$, where $L_s$ denotes the distance $\overline{SP}$ and $L_s'$ the distance $\overline{SP'}$. The optical path difference at point K on the fringe is thus expressed as

$$D = \lambda(\nu - \tfrac{1}{2}) = (L_s - L) - (L_s' - L') , \qquad (4.2)$$

where $\nu$ is the fringe order. Since we assume here the double exposure technique, a dark fringe corresponds to $\nu$ being an integer. We postulate that the displacement is small compared with the distances of S and K from P ($|u| \ll L_s, L$) so that, with the unit direction vector $k$, we may write

$$L' = |Lk + u| = L\sqrt{1 + 2\,\frac{u \cdot k}{L} + \frac{|u|^2}{L^2}}$$

$$= L + u \cdot k + \frac{1}{2L}\, u \cdot Ku + \dots , \qquad (4.3)$$

where $K = I - k \otimes k$ is a projector, see (2.25). A similar expansion holds true for $L_s'$ with unit vector $h$. Neglecting the second-order terms, we then find the basic equation

$$D = \lambda(\nu - \tfrac{1}{2}) = u \cdot (k - h) = u \cdot g , \qquad (4.4)$$

which relates, in a simple way, the optical path difference $D$ (or fringe order $\nu$) to the displacement vector $u$ and to the so-called *sensitivity vector* $g = k - h$.

As $h(r)$ and $u(r)$ are vector functions of the position $r$, the scalar function $D = D(r, k)$ depends, in fact, on two independent variables $r$ and $k$; however, the linearized Eq. (4.4) does not depend on the distances $L_s$, $L$, separating the object surface from points S and K. Consequently, the exact location of point K on the line of sight is not important for determining $u$. If we now observe (as if through a window) a field of fringes from a fixed point R through the hologram, we obtain an additional relation, $r = r(k)$, which performs a mapping of the function $D = D_R(k)$ for all points P. This elimination of $r$ makes this function dependent on $k$ alone. On the other hand, if we always look at the same point P ($r = $ const) from different positions R, we obtain another function, $D = D_P(k)$. The determination of all three components of the displacement vector $u$ at a specific point P must be performed by this latter function with at least three sufficiently different observation directions. We do not want to expand on the technical details here nor present any special applications. The large technical bibliography given by *Vest* [4.25], shows that holographic interferometry has been used for a great number of different applications. Recent works not cited in [4.25] may be found in [4.26–4.68]. Much effort has been spent in deriving automatic evaluations (in particular, see e.g. [4.69–90]). Let us only mention the acknowledged classification of the absolute and relative fringe-order methods.

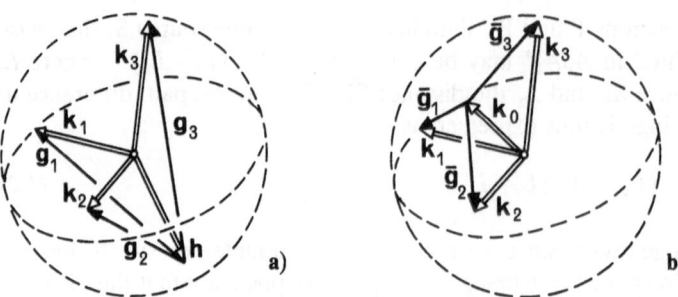

**Fig. 4.3.** (a) Complete base formed by three sensitivity vectors $g_1$, $g_2$, $g_3$ (absolute fringe-order method). (b) Complete base formed by three vectors $\bar{g}_1$, $\bar{g}_2$, $\bar{g}_3$ (relative fringe-order method)

The *absolute fringe-order method*, introduced by *Ennos* [4.91], (see also [4.92–123]) implies knowledge of the location of a point without displacement ($D = 0$), so that $v = 1/2$ (zero-order bright fringe). We just have to count the number of dark fringes from that point to point P

$$D_i = \lambda(v_i - \tfrac{1}{2}) = u \cdot g_i \, , \tag{4.5}$$

where $D_i = D_P(k_i)$ (on the absolute fringe orders $v_i$) may be interpreted as covariant components of $u$ with respect to the complete base, $g_i = k_i - h$, $i = 1, 2 \ldots$ (Fig. 4.3).

By contrast, the *relative fringe-order method*, given by *Aleksandrov-Bonch-Bruevich* [4.124] (see also [4.117–123, 125–135]) relates to the system

$$\Delta D_i = \lambda \Delta v_i = D_i - D_{i0} = u \cdot (k_i - k_{i0}) = u \cdot \bar{g}_i \, , \tag{4.6}$$

with another base, $\bar{g}_i = k_i - k_{i0}$, and relative fringe orders $\Delta v_i$. The determination of the zero-order fringe is replaced here by supplementary observation directions $k_{i0}$, from which one must be able to pass continuously to $k_i$ (Fig. 4.3).

The simplest way to find the three components of $u$ would be to directly solve the three equations (4.6) or (4.5) (for $i = 1, 2, 3$) obtained by "inversion" using the contravariant base vectors $\bar{g}^j$, defined by the scalar product $\bar{g}_i \cdot \bar{g}^j = \delta_i^j$. Formally, one would get

$$u = \sum_{j=1}^{3} \Delta D_j \bar{g}^j \, . \tag{4.7}$$

However, for each practical purposes, the vectors $\bar{g}_i$ differ little from one another, so that we may obtain sufficient precision by applying the least-squares method after making a certain number of measurements. Indeed, let us assume a number $n$ of direction measurements $\bar{g}_i$, $n > 3$, $i = 1, \ldots, n$, with corresponding measured values $\Delta v_i$ of the relative fringe order differing slightly from the expected values $\overline{\Delta v_i}$ [4.126, 128, 136–145]. We then have the condition

$$\sum_{i=1}^{n} (\Delta v_i - \overline{\Delta v_i})^2 = \text{Min} \, , \tag{4.8}$$

which, after differentiation, leads to the following necessary condition (for $\partial_u$ refer to the end of Sect. 2.1.2)

$$2 \sum_{i=1}^{n} (\Delta v_i - \overline{\Delta v_i}) \partial_u (\Delta v_i - \overline{\Delta v_i}) = \frac{2}{\lambda} \Sigma (\overline{\Delta v_i} - \Delta v_i) \bar{g}_i = 0 .$$

Considering (4.6) once more, we deduce the linear system

$$u \left( \sum_{i=1}^{n} \bar{g}_i \otimes \bar{g}_i \right) = \lambda \sum_{i=1}^{n} \Delta v_i \bar{g}_i , \tag{4.9}$$

for the components of $u$. Equation (4.9) then replaces (4.6). Alternatively, by known displacements, one could use (4.5) to determine the sensitivity vectors $g_i$, as well as the location of the whole object [4.111, 146–148].

Usually, strain and rotation components are subsequently found by means of finite differences [4.149–158]. However, it is more advantageous to study the direct use of the *derivative* of $D$. Thus, let us now consider the differential $dD$ of the optical path difference, first taken from the general form

$$D(r, k) = u(r) \cdot [k - h(r)] .$$

The vector $r$ varies on the object surface of unit normal $n$, whereas the vector $k$ varies within a unit sphere. Therefore, referring to (2.59), we obtain for the total differential or increment of $D$:

$$dD = dr \cdot \partial_r D + dk \cdot \partial_k D$$
$$= dr \cdot [(\nabla_n \otimes u)g - (\nabla_n \otimes h)u] + dk \cdot Ku . \tag{4.10}$$

We then successively write the detailed derivatives. First, from (2.24) (Sect. 2.1), we get

$$\nabla_n \otimes h = N\nabla \otimes h = \frac{1}{L_s} NH , \tag{4.11}$$

where $H = I - h \otimes h$ is a normal projector as is $N = I - n \otimes n$.

Secondly, the decomposition of the surface deformation gradient (2.143) (Sect. 2.3) reads

$$\nabla_n \otimes u = \gamma + \Omega E + \omega \otimes n , \tag{4.12}$$

with the symmetric inplane strain tensor $\gamma$, the pivot rotation scalar $\Omega$, the antimetric permutation tensor $E$, and the inclination vector $\omega$ of the surface element at point P. In the case of moderate rotations, the decomposition of the projected deformation gradient may be deduced from (2.152 and 150).

For convenience, we now introduce the abbreviation

$$Nw = (\gamma + \Omega E + \omega \otimes n)g - \frac{1}{L_s} NHu , \tag{4.13}$$

which reduces the increment of $D$ to the form

$$dD = dr \cdot Nw + dk \cdot Ku \ . \tag{4.14}$$

It should be noted that (4.13) defines only the projection $Nw$ of $w$. For a univocal definition, the vector $w$ could be written with the complete deformation gradient

$$w = (\nabla \otimes u)g - \frac{1}{L_s}Hu \tag{4.15}$$

but, as yet, this expression is purely formal as far as the normal component $n(n \cdot w)$ is concerned, since a hologram is not able to store what lies "behind" the object surface.

Inasmuch as the increments $dr$ and $dk$ are independent, the ray through the neighboring point $\overline{P}$ of P may be arbitrary, e.g., it can be slightly skew with respect to the one passing through P. However, we can now add the supplementary condition that all the considered rays intersect at a fixed point, e.g., at the center R of the observing instrument or the center of the eye (Fig. 4.4). This is useful when we want to investigate the fringe direction and fringe interspace. The condition of collineation then reads

$$r = \varrho_R - L_R k \ , \tag{4.16}$$

where $L_R$ is the distance of R (position vector $\varrho_R$) from P. The differentials $dr$ and $dk$ become related by the affine connection, see (2.47),

$$dr = -L_R M^T dk \ , \tag{4.17}$$

containing the oblique projector $M = I - n \otimes k/n \cdot k$.

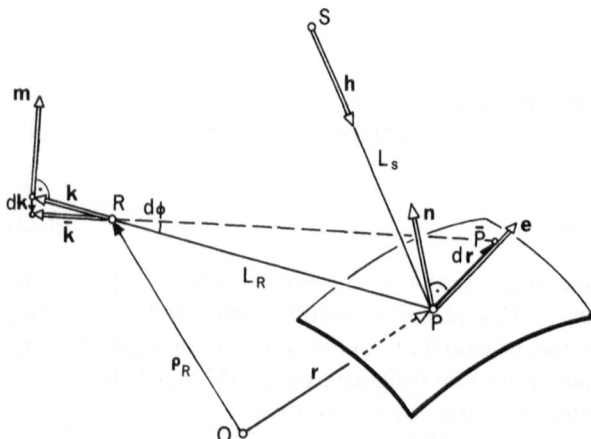

Fig. 4.4. Collineation between the increments $dr$ on the object surface, and $dk$ on the unit sphere around the observation center R

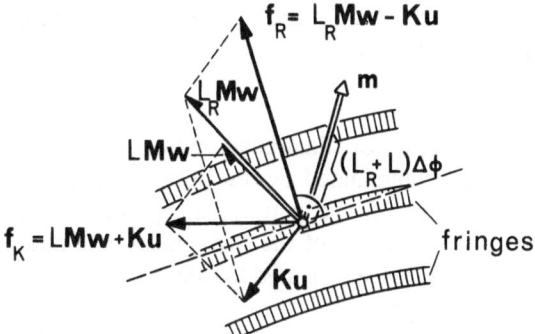

Fig. 4.5. Fringe interspace and direction. ($f_R$: fringe vector; $f_K$: visibility vector; $Ku$: normal projection of the displacement $u$; $Mw = M[(\nabla \otimes u)g - Hu/L_s]$: vector containing the deformation gradient)

Thus, with $dk = -m\, d\phi$, where $m$ is a unit vector and $d\phi$ an angular increment, and with the auxiliary relations of successive projectors $MN = M$, $MK = K$, see (2.52, 61), from (4.14) we find the derivative of $D$ with respect to $\phi$ and relative to the fixed point R in the form

$$\frac{dD_R}{d\phi} = m \cdot f_R , \quad \text{where} \tag{4.18}$$

$$f_R = L_R M \nabla_n D_R = L_R Mw - Ku . \tag{4.19}$$

The derivative $dD_R/d\phi$ vanishes along the fringes. For $m_t$ parallel to the projected fringes onto the normal plane to $k$, we have:

$$m_t \cdot f_R = 0 . \tag{4.20}$$

If $f_R$ does not disappear (i.e., in P there should be no singularity of the fringe field), this so-called *fringe vector* $f_R = L_R Mw - Ku$, which was introduced by Stetson [4.159–166], is normal to the fringes (Fig. 4.5). Taking now a unit vector $m_n$ in this normal direction for a finite increment $\Delta v_R = 1$, (i.e., $\Delta D_R = \lambda$), we find the apparent *angular fringe spacing*

$$|\Delta\phi|_{min} = \frac{\lambda}{|f_R|} . \tag{4.21}$$

The relations (4.20 and 21) are useful when strain and rotation components encountered in (4.13) should be determined from the fringe spacing and fringe direction measurements [4.161, 162, 167–171]. In the particular case of rigid body motion, we have constant values for the rotation components over the whole object. Assuming further a nearly collimated object beam ($L_s \to \infty$) and the observer placed far away from the object, the term $Ku/L_R$ becomes negligible and $k \simeq$ const. According to (4.13, 18), we then get a fringe vector which is nearly constant over the whole fringe pattern. Therefore, the fringes are nearly straight, parallel, and equally spaced. When changing the observer's position, these fringes will change their spacing and direction according to the change of the sensitivity vector $g$, but will remain nearly straight and parallel. For such a rigid body motion, the constant rotation vector may be alternatively determined by a finite difference of the displacement vector $\bar{u} - u = \omega_r \times (\bar{r} - r)$ (formula of elementary kinematics) [4.172–175].

In the holographic Moiré technique or dual-beam holographic interferometry, the object is simultaneously illuminated by two beams (illumination direction $h_1$, $h_2$), so that two interference fringe patterns are produced. The moiré fringes correspond then to the difference of the two optical path differences and thus, the derivative of this difference is related to a difference of fringe vectors which may be called "Moiré vector":

$$\Delta f_R = -L_R M \left[ (\nabla_n \otimes u)\Delta h - \frac{1}{L_{s_1}} NH_1 u + \frac{1}{L_{s_2}} NH_2 u \right] ,$$

where $\Delta h = h_2 - h_1$ [4.176–185]. When the two object beams are collimated $(L_{s_1} \to \infty, L_{s_2} \to \infty)$, one gets a direct relationship between $\Delta f_R$ and $\nabla_n \otimes u$, without any displacement term.

Instead of taking R as a collineation center, we provisionally fix point K of the fringe locus in space. The whole bunch of rays passing through the aperture which "intersecting" there, produces, by accumulation, the effect which ultimately allows the appearence of an interference fringe. Equation (4.17) must be replaced here by

$$dr = LM^T dk , \qquad (4.22)$$

which, instead of (4.18), leads to the relation

$$dD_K = dk \cdot f_K , \qquad (4.23)$$

with $f_K = LMw + Ku$.

This equation plays a role in fringe contrast considerations (Sect. 4.2). Consequently, this vector, $f_K = LMw + Ku$, may be called the *visibility vector*.

## 4.1.2 Optical Path Difference and Its Derivative for a Transparent Isotropic Medium with a Varying Index of Refraction. The Analogy

We now turn to the other field of holographic interferometry: the analysis of interference fringes caused by changing the index of refraction in an isotropic transparent medium [4.186–189]. Its application is of special interest in subjects such as aerodynamic flow studies [4.190–209], heat transfer problems [4.210–226], mass transfer examinations [4.227–234], the determination of liquid properties [4.235–247], the observation of combustion process [4.248–250], plasma diagnostics [4.251–268] and the determination of transparent solid properties [4.269–271] (holographic photoelasticity is excluded here). We may first assume that, by two very short exposures, we record two quasi-stationary nonhomogeneous spatial distributions $n(x)$ and $n'(x)$ of the index of refraction, $x$ denoting the position vector of any interior point X, belonging to the domain $\mathbb{G}$ occupied by the gas building the nonhomogeneous medium. The light coming from the source S is diffusely reflected by a rough opaque surface $A$. As far as reconstruction with the recorded hologram is concerned, reasoning similar to that of (4.1) holds. Usually it is supposed that the spatial change of the index of refraction is so small that the curvature of the light rays inside $\mathbb{G}$ may be ignored. The object delimitting domain $\mathbb{G}$ is then called a *phase object*. However, we do not want to exclude the possibility of generalization and also intend to consider the optical path difference and its derivative in a way similar to that for the case of deformed opaque objects; finally, this should introduce the developments of the homology and visibility concepts. It is therefore advisable to refer provisionally to the two different slightly curved rays travelling from an arbitrary point P on the surface $A$ to points H and H' on the hologram, where they interfere with the reference wave coming from Q. At reconstruction, these two virtual and straight rays are assumed to intersect at point K. The distances $L_1^s$, $L_2^s$, $L_1^{s'}$, $L_2^{s'}$ denote the straight lengths of the rays outside $\mathbb{G}$ ($n = n_0$), whereas $s_{12}$, $s'_{12}$ are the arc-lengths inside $\mathbb{G}$ ($n \neq n_0$). We consider the lengths $L_2^s$, $L_2^{s'}$ positive when K lies in front of $\mathbb{G}$, but negative when point K lies inside $\mathbb{G}$ or behind it (Fig. 4.6). Note that point K is defined at reconstruction as if the domain $\mathbb{G}$ defined earlier is replaced by an air medium of uniform refractive index. Therefore, the rays $X_2K$, $X_2'K$ are the straight-line backwards prolongation of the rays diffracted at H and H'. It must be noted that, in practice, such a gas could be separated from the surrounding medium by a window that causes an additional refraction effect. For simplicity we disregard this refraction here so that the index of refraction changes

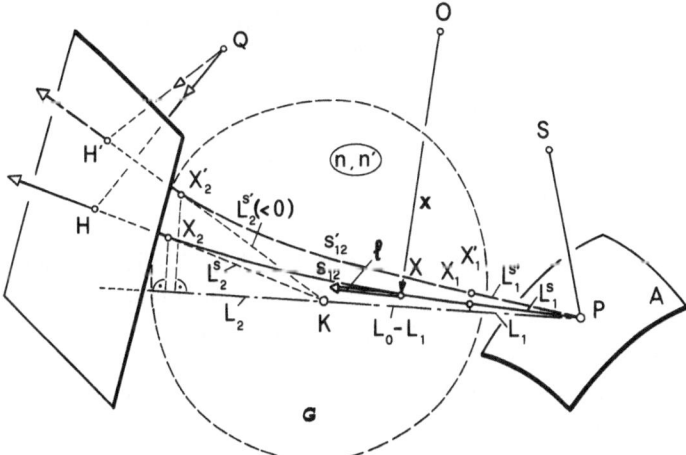

**Fig. 4.6.** Double exposure through a medium combined by a domain $\mathbb{G}$ with a variable index of refraction $n$, $n'$. (A: diffusing surface; $l$: unit vector along the curved ray. $\overline{PX_1} = L_1^s$, $\overline{PX_1'} = L_1^{s'}$, $\widehat{X_1X_2} = s_{12}$, $\widehat{X_1'X_2'} = s'_{12}$, $\overline{X_2K} = -L_2^s$, $\overline{X_2'K} = -L_2^{s'}$)

continuously from $n_0$ to $n(x)$ and back to $n_0$ while the ray goes from P to H. At point K we then have the optical path-length difference in the form

$$D = n_0(L_1^s - L_1^{s'}) + \int_0^{s_{12}} n \, ds - \int_0^{s_{12}} n' \, ds' + n_0(L_2^s - L_2^{s'}) , \tag{4.24}$$

where the arc-lengths $s$, $s'$ are measured from the entrance points $X_1$, $X_1'$. Applying an integration by parts, we may write the same function in the alternative form

$$D = n_0(L_0^s - L_0^{s'}) - \int_0^{s_{12}} \frac{\partial n}{\partial s} s \, ds + \int_0^{s_{12}} \frac{\partial n'}{\partial s'} s' \, ds' . \tag{4.25}$$

The quantities

$$L_0^s = L_1^s + s_{12} + L_2^s , \qquad L_0^{s'} = L_1^{s'} + s_{12}' + L_2^{s'} \tag{4.26}$$

are the total path-lengths of the curved indirect ray between P and K.

We now assume the spatial change of the index of refraction to be small compared to the inverse of some characteristic lengths, for instance, the distance $\overline{PK} = L_0$ ($L_0 > 0$ when K is in front of the surface $A$):

$$|L_0| \, |\nabla n|_{\max} = \varepsilon \ll 1 . \tag{4.27}$$

We also recall the ray equation (2.85) (Fig. 2.13)

$$\frac{dl}{ds} = \nabla_1(\log n) = \frac{m_0}{\varrho_s} , \tag{4.28}$$

where $l$ is the tangential unit vector, $\varrho_s$ the radius of curvature, $m_0$ the principal unit normal and $\nabla_1 = (I - l \otimes l)\nabla$ the lateral derivative operator. Since the ratio $L_0/\varrho_s$ is also of the order of $\varepsilon$, we conclude that

$$L_0^s = L_0 + O(\varepsilon^2) , \qquad L_0^{s'} = L_0 + O(\varepsilon^2) ,$$

$$\int \frac{\partial n}{\partial s} s \, ds = O(\varepsilon) , \qquad \int \frac{\partial n'}{\partial s'} s' \, ds' = O(\varepsilon) . \tag{4.29}$$

Such a discussion could be extended to the case where the lateral and longitudinal parts of $\nabla n$ or of $\nabla(n - n')$ are of different orders of magnitude. Likewise, $s_{12}$ and $L_1^s$, $L_2^s$ could be of different orders of magnitude. The conclusions would then be different from those deduced here. In the present "quasi-uniform" case, we see that the optical path difference is mainly of order $\varepsilon$. When considering the terms of the same order of magnitude in (4.25), the term $n_0(L_0^s - L_0^{s'})$ may be neglected. Further, with the approximations

$$\frac{\partial n}{\partial s} ds = dn = \frac{\partial n}{\partial \zeta} d\zeta + O(\varepsilon^2) , \quad s = \zeta + O(\varepsilon^2) ,$$

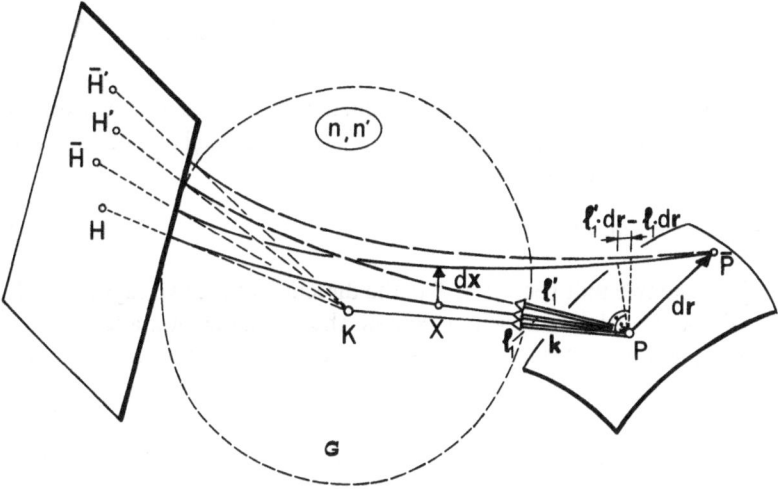

**Fig. 4.7.** Neighboring curved rays passing through a medium combined by a domain $\mathbb{G}$ of variable refractive index $n$, $n'$. ($l_1$, $l_1'$: unit vectors at $P \in A$ in the direction of the true curved rays. $k$: unit vector on the straight reference line through K; $dr$: increment relating the neighboring point $\bar{P}$ on the diffusing surface $A$)

we may connect $D$ to the abscissa $\zeta$ on the direct straight line $\overline{PK}$ instead of to the arclengths $s$, $s'$. Hence (4.25) is approximated by ($L_1^s$, $L_1^{s'}$ are now replaced by $L_1$ on $\overline{PK}$):

$$D = - \int_0^{\zeta_{12}} \left( \frac{\partial n}{\partial \zeta} - \frac{\partial n'}{\partial \zeta} \right) \zeta \, d\zeta = - \int_0^{\zeta_{12}} (L_1 + \zeta) \left( \frac{\partial n}{\partial \zeta} - \frac{\partial n'}{\partial \zeta} \right) d\zeta . \quad (4.30)$$

With a second integration by parts, we find alternatively the expression

$$D = \int_0^{\zeta_{12}} (n - n') \, d\zeta \quad (4.31)$$

which is the commonly used relation [4.272–277]. However, for further development, we shall in this instance use the first expression (4.30), where small quantities appear directly and not as a small difference of two finite quantities as in (4.31).

Since the value $(n - n')$ is a function to be integrated, its determination would need a great number of measurements of $D$ (Sect. 4.1.3). Before discussing this important problem of inversion, let us first pass to the *derivative* of $D$ [4.278]. In a preliminary step again, we consider the general case of slightly curved rays passing through the point K where the fringe is considered fixed. In order to determine the increment $dD_K$ of $D$, we look at the rays coming from a neighboring point $\bar{P}$ of P on the diffusing surface. This point is separated from $P$ by an increment $dr$, to which corresponds another increment $dx$ at point X, appertaining to the domain $\mathbb{G}$ and perpendicular to the ray through P (Fig. 4.7). Since

$$\tilde{n} = n[1 + dx \cdot \nabla_1(\log n)] \quad \text{and} \quad d\tilde{s} = ds \left( 1 - \frac{1}{\varrho_s} dx \cdot m_0 \right) ,$$

we have, of course, with (4.28) the acknowledged property of normal wavefronts $\tilde{n} \, d\tilde{s} = n \, ds$. Therefore, in the difference $\overline{D} - D$, only the small part of the longer path near P remains, namely,

$$dD_K = \overline{D} - D = n_0 dr \cdot (l_1' - l_1) , \tag{4.32}$$

where $l_1$, $l_1'$ are the two unit vectors at $P$ in the direction of the two rays $PX_1$, $PX_1'$ (Fig. 4.7). These directions implicitly contain the variations of $n$ and $n'$.

We may express this interdependence explicitly by using the geometric condition of path connection (Fig. 4.6)

$$L_1^s l_1 + \int_0^{s_{12}} l \, ds + L_2^s l_2 = L_1^{s'} l_1' + \int_0^{s_{12}'} l' \, ds' + L_2^{s'} l_2' = L_0 k \tag{4.33}$$

where $l_2$, $l_2'$ are the two direction unit vectors of the straight parts $X_2 K$, $X_2' K$. Using the ray equation (4.28) and again integrating by parts, we successively have

$$l_2 = l_1 + \int dl = l_1 + \int_0^{s_{12}} \nabla_1(\log n) ds ,$$

$$\int_0^{s_{12}} l \, ds = sl \Big|_0^{s_{12}} - \int_0^{s_{12}} s \frac{dl}{ds} ds = s_{12} l_2 - \int_0^{s_{12}} s \nabla_1(\log n) ds$$

$$= s_{12} l_1 + \int_0^{s_{12}} (s_{12} - s) \, \nabla_1(\log n) ds, \dots .$$

Therefore, with the definitions (4.26), expression (4.32) becomes

$$dD_K = n_0 dr \cdot \left\{ \left( \frac{L_0}{L_0^{s'}} - \frac{L_0}{L_0^s} \right) k + \frac{1}{L_0^s} \int_0^{s_{12}} [L_0^s - (L_1^s + s)] \nabla_1(\log n) ds \right.$$

$$\left. - \frac{1}{L_0^{s'}} \int_0^{s_{12}'} [L_0^{s'} - (L_1^{s'} + s')] \nabla_1'(\log n') ds' \right\} \tag{4.34}$$

representing a general formula for the differential of the optical path difference.

Again, for a small gradient of the index of refraction, see (4.27), this equation simplifies to

$$dD_K = \frac{1}{L_0} dr \cdot \int_0^{\zeta_{12}} [L_0 - (L_1 + \zeta)] \nabla_k(n - n') d\zeta , \tag{4.35}$$

where the lateral derivative operator $\nabla_k = (I - k \otimes k) \nabla$ at point X is now referred to the unit direction vector $k$ along $\overline{PK}$. Equations (4.30, 35) have a common structure. They show that the fringe order and its differential let the longitudinal and the lateral derivative of the change of refractive index appear separately.

As in Sect. 4.1.1, we can consider the observer's center R as a fixed point instead of K. With the affine connection $K \, dr = L_R m \, d\phi$, see also (4.17), we then get [Ref. 4.279, Eq. (7)]:

$$\frac{dD_R}{d\phi} = m \cdot \int_0^{\zeta_{12}} [L_R - (L_1 + \zeta)] \nabla_k (n - n') d\zeta \ . \tag{4.36}$$

Due to the assumption of quasi-straight rays, this formula may also be directly obtained from

$$dD_R = \int_0^{\zeta_{12}} dx \cdot \nabla_k (n - n') d\zeta$$

by means of the affine connection (Fig. 4.8)

$$L_R \, dx = [L_R - (L_1 + \zeta)] K \, dr \ .$$

Equation (4.36) will be used practically when fringe spacing and fringe direction measurement come into account. In particular, if $m_t$ is the unit vector parallel to the projected fringe, we have

$$m_t \cdot \int_0^{\zeta_{12}} [L_R - (L_1 + \zeta)] \nabla_k (n - n') d\zeta = 0 \ . \tag{4.37}$$

The vector

$$\int_0^{\zeta_{12}} [L_R - (L_1 + \zeta)] \nabla_k (n - n') d\zeta$$

is therefore called the *fringe vector*. As far as fringe spacing is concerned, a similar equation to (4.21) may be given for $\Delta v_R = 1$, $\Delta D_R = \lambda$:

$$|\Delta\phi|_{min} = \frac{\lambda}{\left| \int_0^{\zeta_{12}} [L_R - (L_1 + \zeta)] \nabla_k (n - n') d\zeta \right|} \ . \tag{4.38}$$

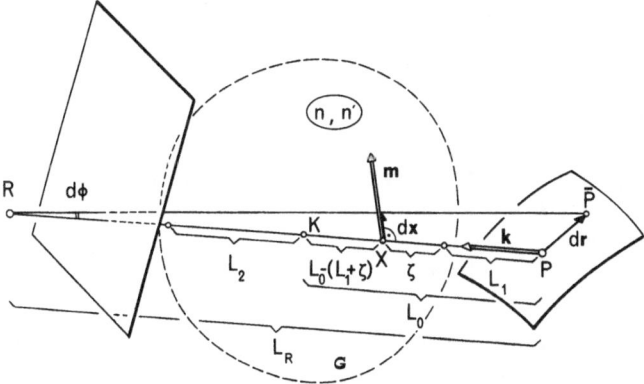

**Fig. 4.8.** Affine connection between an interior increment $dx \in G$ and the surface increment $dr$ on $A$. ($d\phi$: angular increment at the observation center R)

Generally speaking, it turns out that there exists an *analogy* between some quantities encountered in Sect. 4.1.1 (the displacement $u$ and its gradient) and the corresponding quantities in the present paragraph, namely ($L$ corresponds to $-L_0$):

|  | Deformation of an opaque body | Change of the refractive index in a transparent medium |
|---|---|---|
| Fringe order: $(v - 1/2) = D/\lambda$ longitudinal components | $D = u \cdot k - u \cdot h$ | $D = -\int_0^{\zeta_{12}} (L_1 + \zeta) \dfrac{\partial(n - n')}{\partial \zeta} d\zeta$ |
| Derivative of fringe order: proper gradients | $Mw = M[(\nabla \otimes u)g - Hu/L_s]$ | $\int_0^{\zeta_{12}} \nabla_k(n - n') d\zeta$ |
| lateral components | $Ku$ | $\int_0^{\zeta_{12}} (L_1 + \zeta) \nabla_k(n - n') d\zeta$ |
| fringe vectors $f_R$ | $L_R Mw - Ku$ | $\int_0^{\zeta_{12}} [L_R - (L_1 + \zeta)] \nabla_k(n - n') d\zeta$ |
| visibility vectors $f_K$ | $LMw + Ku$ | $-\int_0^{\zeta_{12}} [L_0 - (L_1 + \zeta)] \nabla_k(n - n') d\zeta$ |

This analogy will be useful in the following considerations. For instance, we can restrict the discussion on fringe visibility to opaque or transparent objects.

Another application of the analogy will occur in the case of a superposition. For example, a gas surrounding an opaque body could change its temperature simultaneously with the deformation of the opaque body being considered [4.280]. We then have at recording two different diffusing surfaces $A, A'$, namely the two surfaces of the body in its undeformed and its deformed state. In order to envisage a realistic situation, we may consider the set-up depicted in Fig. 4.9, where the transparent medium that changes its refractive index is supposed to be in contact with the surface of the deformed body, but is separated by a window i from the free air (index $n_0$). If the rays are inclined with respect to the window, we must take into account (and this especially for the derivative) the refractive effects at this interface.

The optical path difference for the two corresponding rays, going from the source $S_0$ to the object points P, P' and then interfering (after a second refraction on i) virtually at some point $K_0$, is primarily given by

$$D = n_0(l_{0s} - l'_{0s}) + \int_0^{\sigma_p} n \, d\sigma - \int_0^{\sigma'_p} n' d\sigma'$$

$$+ n_0(l_0 - l'_0) + \int_0^{s_p} n \, ds - \int_0^{s'_p} n' ds' \,, \tag{4.39}$$

where $l_{0s}$, $l'_{0s}$ are the paths from point $S_0$ to points J, J', situated on the surface $i$, and where $l_0$, $l'_0$ are those from I, I' to $K_0$, and where finally $\sigma, s, \sigma', s'$ denote the

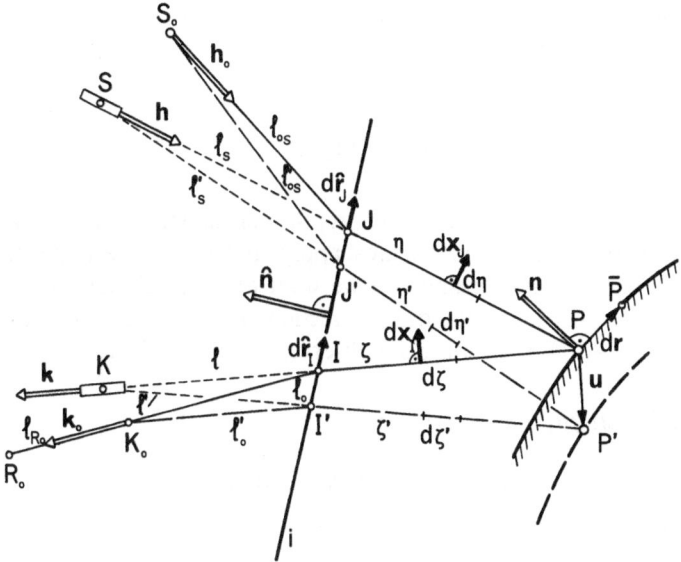

**Fig. 4.9.** Deformation of an object immersed in a medium with a varying index of refraction and separated from outside air by a window i. ($S_0$: object source; $R_0$: observer)

arc-lengths inside the medium between i and P, P', the origins being J, I and J', I', respectively. The terms denoting the paths outside the window may be transformed by the law of refraction; however, let us provisionally introduce the distances $l_s$, $l_s'$, $l$, $l'$ joining the points representing the corresponding astigmatic images {S} {K} of $S_0$, $K_0$ (Fig. 4.9). As before, we now assume that $n'$ differs little from $n$ and that we have small gradients; nevertheless neither $n$ nor $n'$ equals $n_0$. Hence, for example

$$n_0(l_0 - l_0') = n_I(l - l')$$

and therefore, with integrations by parts both forwards and backwards, we get

$$n_0(l_0 - l_0') + \int_0^{\zeta_p} n \, d\zeta - \int_0^{\zeta_p'} n' d\zeta' = n_I(l - l') + [n(l + \zeta)]_0^{\zeta_p}$$

$$- \int_0^{\zeta_p} (l + \zeta) \frac{\partial n}{\partial \zeta} d\zeta - [n'(l' + \zeta')]_0^{\zeta_p'} + \int_0^{\zeta_p'} (l' + \zeta') \frac{\partial n'}{\partial \zeta'} d\zeta'$$

$$\cong n_p \boldsymbol{u} \cdot \boldsymbol{k} - (n_I - n_I')l + (n_p - n_p')(l + \zeta_p) - \int_0^{\zeta_p} \frac{\partial(n - n')}{\partial \zeta} (l + \zeta) d\zeta$$

$$= n_p \boldsymbol{u} \cdot \boldsymbol{k} + \int_0^{\zeta_p} (n - n') d\zeta , \qquad (4.40)$$

because $n_p(l + \zeta_p - l' - \zeta_p') \cong n_p \boldsymbol{u} \cdot \boldsymbol{k}$. As before, the vector $\boldsymbol{u}$ denotes the displacement vector of P, whereas $\boldsymbol{k}$ is now the unit direction vector along the ray

PK, which should not be confused with the unit vector $k_0$ along $IK_0$. With another unit vector $h$ (not to be confused with $h_0$), and a reference variable $\eta$ replacing $\sigma$, $\sigma'$, we find the expected formula ($g = k - h$)

$$D = n_p u \cdot g + \int_0^{\zeta_p} (n - n')d\zeta + \int_0^{\eta_p} (n-n')d\eta . \tag{4.41}$$

Whereas this result is arrived at rather simply, the calculation of the derivative becomes rather cumbersome, since we must apply the affine connection with the interface as in Sect. 2.2. For example, if we want to get a relation for the fringe spacing $\Delta\phi_0$, like (4.21), and the fringe direction $m_{0t}$, like (4.20), we must form the increment of (4.41) by keeping the observer's point $R_0$ fixed:

$$\begin{aligned} dD_R &= dr \cdot [(\nabla_n \otimes u) g n_p + (u \cdot g) \nabla_n n_p] \\ &+ d\bar{k} \cdot un_p - d\bar{h} \cdot un_p + \int_0^{\zeta_p} dx_I \cdot \nabla_k (n - n')d\zeta \\ &+ \int_0^{\eta_p} dx_J \cdot \nabla_h (n-n')d\eta \\ &- dr \cdot k(n_p - n'_p) + d\hat{r}_I \cdot k(n_I - n'_I) + dr \cdot h(n_p - n'_p) - d\hat{r}_J \cdot h(n_J - n'_J) . \end{aligned} \tag{4.42}$$

The different increments $dr$, $d\bar{k}$, $d\bar{h}$, $dx_I$ and $dx_J$ must then be related to $-m_0 d\phi_0 = dk_0$. First, for the increment $dr$ on the object surface, we directly may apply (2.123)

$$dr = - l_{R_0} M^T \hat{M} \left( \frac{n_0 \zeta_p}{n l_{R_0}} K_0 + K \right) M_0^T dk_0 . \tag{4.43}$$

Secondly, for the increment $d\bar{k}$ between the neighboring reflecting *skewed* rays through P, $\bar{P}$, we have, see (2.121),

$$d\bar{k} = - \frac{1}{\zeta_p} K(dr - d\hat{r}_I) , \tag{4.44}$$

where the increment $d\hat{r}$ of the interface at I is

$$d\hat{r}_I = - l_0 M_0^T dk_0 .$$

A similar relation is obtained for $d\bar{h}$, except that the point $S_0$ is now the collineation center, so that ($H = I - h \otimes h$)

$$d\bar{h} = \frac{1}{\eta_p} H(dr - d\hat{r}_J) \tag{4.45}$$

with

$$d\hat{r}_J = l_{0s} M_{0s}^T dh_0 .$$

Also, the vector $dh_0$ must be expressed by $dk_0$ by means of two affine connections of type (4.43) with $dr$ as a bridge term. Thirdly, we could calculate the interior increment $dx_I$; here, we can apply a similar affine connection to (4.43) without $M^T$, namely,

$$dx_I = -l_{R_0}\hat{M}\left(\frac{n_0\zeta}{nl_{R_0}}K_0 + K\right)M_0^T dk_0 . \tag{4.46}$$

Finally, $dx_J$ could be found in a similar manner, but with the additional difficulty of transitivity, which we encountered for $d\bar{h}$.

### 4.1.3 Problem of Inversion

As already mentioned, the practical application of the basic formula (4.31) for transparent media cannot be made directly, because the measured quantity is usually the path difference $D$, whereas the change of refractive index $n - n' = f$ (which we are seeking) is a quantity to be integrated. This problem has been investigated by *Vest* and his collaborators in holographic interferometry [4.17, 281–285], and by others [4.286–299]. Here, let us briefly repeat some main principles in order to relate the remainder of this chapter to these important publications as well as to indicate what is necessary for practical applications in this field.

Suppose that the observer is situated very far away in a fixed direction $k$, so that we may rewrite (4.31) in the form

$$D(\xi) = \int_{-\infty}^{\infty} f(x)d\zeta . \tag{4.47}$$

The integration limits $-\infty, +\infty$ are convenient, but, of course, the function $f(x)$ is zero outside the domain considered. The position vector $\xi = Kx$ is the projection of $x$ (with components $x$, $y$, $z$ in a given system) onto the plane normal to $k$ through the origin O. The components of $\xi$ relative to another system in the plane are, for instance, $\xi$, $\eta$ (the component $\xi$ should not be confused with the vector $\xi$). The function $D(\xi)$ now represents a two-dimensional scalar field, whereas $f(x)$ is a three-dimensional scalar field. We may tentatively distinguish two cases:

a) There exists beforehand an additional condition limiting the dependence of $f(x)$ on two variables only, so that this function may be determined from one fringe pattern corresponding to a single direction $k$. Examples of this case are uniform distributions along a direction $k$ or radially symmetric distributions around an axis perpendicular to $k$. Or,
b) the above conditions do not exist; the field is general and asymmetric. Hence, several fringe fields corresponding to different directions $k$ (multi-directional measures) must then be used for the determination of $f(x)$. This constitutes the true problem of inversion, which is also known from other fields in physics.

Let us first look at case (a). Suppose there is a family of curves $\boldsymbol{\xi} = \boldsymbol{\xi}(s, c)$ in the plane normal to $\boldsymbol{k}$, where $s$ is a parameter along each curve (e.g., the arc) and where $c$ is a parameter of the family. Then, the vector

$$\boldsymbol{x}(s, c, \zeta) = \boldsymbol{\xi}(s, c) + \zeta\boldsymbol{k}$$

is an ensemble of cylindrical surfaces with the curves just described as base and generators parallel to $\boldsymbol{k}$. If we further set $\zeta = \zeta(s, r, c)$, we get curves with another family parameter $r$ on each cylindrical surface. Suppose then that $f$ is known to be constant on each of these curves, i.e. $f = f(r, c)$. The two independent variables for the two-dimensional field of $f$ are thus now $r$ and $c$. If these curves possess a property of symmetry and if $r \to +\infty$ when $\zeta \to \pm\infty$, then (4.47) takes the form

$$D(s, c) = 2 \int_{r_0(s)}^{\infty} f(r, c) \frac{\partial\zeta}{\partial r} dr \, , \tag{4.48}$$

which is an integral equation for $f(r, c)$. The radially symmetric distribution forms a special case of (4.48), for which the cylindrical surfaces become plane and the curves on them are circles. From Fig. 4.10, we remark that

$$\zeta = \sqrt{r^2 - s^2} \, , \qquad \frac{\partial\zeta}{\partial r} = \frac{r}{\sqrt{r^2 - s^2}} \, ,$$

so that

$$D(s, c) = 2 \int_{s}^{\infty} \frac{f(r, c)r \, dr}{\sqrt{r^2 - s^2}} \tag{4.49}$$

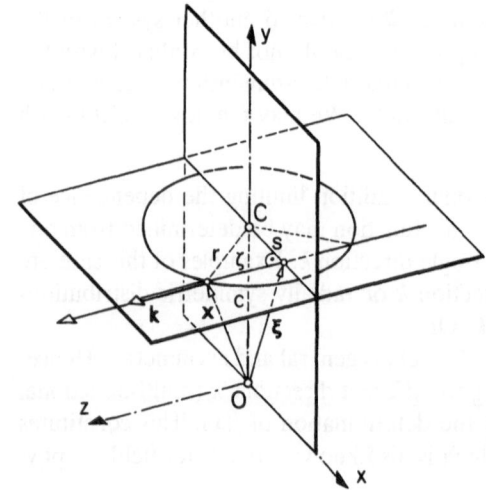

Fig. 4.10. Radial symmetric distribution of the refractive index variation in a transparent medium (phase object). ($\boldsymbol{k}$: unit direction vector on the ray where the integrated effect is considered)

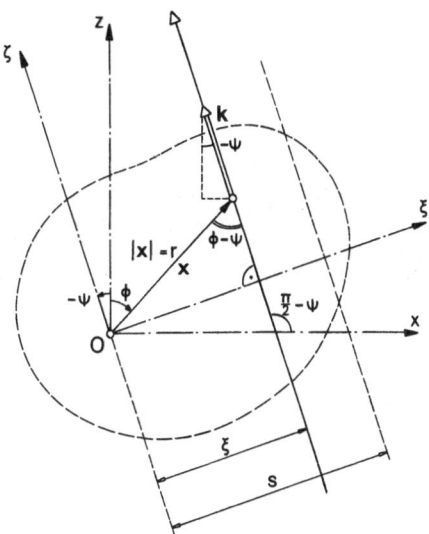

**Fig. 4.11.** Asymmetric and general distribution of the variation of refractive index in a transparent medium. (**k**: unit vector on the ray where the integrated effect is considered)

is the *Abel* transform of $f(r, c)$. The inverse of (4.49) is well known and reads [4.283, 284, 288, 290, 292, 294]

$$f(r, c) = -\frac{1}{\pi} \int_r^\infty \frac{\dfrac{\partial D(s, c)}{\partial s} ds}{\sqrt{s^2 - r^2}} \qquad (4.50)$$

This result shows that, with measurements of the *interspace* and *direction* of the fringes (leading to $\partial D/\partial s$), we may determine a radially symmetric distribution of the change of refractive index.

Let us next turn to the asymmetric fields in case (b), where one must perform multi-directional measurements. First, we assume that $k$ varies only in a plane $y = c$, the components of $k$ are, for instance, $(\cos \psi, \sin \psi)$, referring to an azimuth $\psi$. If $\zeta$, $\xi$ are cartesian coordinates parallel and perpendicular to the direction of the rays now in the plane $y = \eta = c$ (Fig. 4.11), and if $\xi = s$ is the considered ray where we express the path difference $D$, we may write (4.47) in the form

$$D(s, \psi, c) = \int_{-\infty}^\infty d\zeta \int_{-\infty}^\infty f(\xi, \zeta, c)\delta(s - \xi)d\xi , \qquad (4.51)$$

since the integral of the Dirac-function $\delta(s - \xi)$ yields unity: $\int_{-\infty}^\infty \delta(s - \xi)d\xi = 1$, and since $f(\xi, \zeta, c)$ plays no role in this integration. Using polar coordinates $r, \phi$ instead of $\zeta, \xi$, (4.51) becomes, with the surface element $d\zeta \, d\xi = r \, dr \, d\phi$,

$$D(s, \psi, c) = \int_0^{2\pi} d\phi \int_0^\infty f(r, \phi, c)\delta(s - r\sin(\phi - \psi))r \, dr$$

which is the two-dimensional *Radon* transform [4.17, 283, 287, 288, 300]. The inversion formula is given by (see also [4.17, 283, 288])

$$f(r, \phi, c) = \frac{1}{2\pi^2} \int\limits_{-\pi/2}^{\pi/2} d\psi \int\limits_{-\infty}^{\infty} \frac{\dfrac{\partial D(s, \psi, c)}{\partial s} \, ds}{r\sin(\phi - \psi) - s} \ . \tag{4.52}$$

The practical application of (4.52) is only possible if multi-directional measurements of fringe interspace can be made for the complete half circle $-\pi/2 < \psi < \pi/2$. For a radially symmetric distribution, $D$ is no longer a function of $\psi$. Further, with $\phi = 0$ (choice of the reference line) and with $s + r\sin\psi = t$, we get

$$\int\limits_{-\pi/2}^{\pi/2} \frac{d\psi}{s + r\sin\psi} = \int\limits_{s-r}^{s+r} \frac{dt}{t\sqrt{r^2 - (t-s)^2}}$$

$$= \frac{1}{\sqrt{s^2 - r^2}} \arccos\left(\frac{s^2 - r^2 - st}{tr}\right)\Bigg|_{s-r}^{s+r} = \frac{\pi}{\sqrt{s^2 - r^2}} \ ,$$

so that (4.52) corresponds to (4.50).

Another possible approach for the inversion in the asymmetric case would be made by means of a *Fourier transform* [4.17, 281, 282, 289]. Consider, besides $x$, another vector $u$ with components $u$, $v$, $w$ in the given system. The projection $\mu = Ku$ has the components $\mu$, $v$ in the system bound to the plane normal to $k$. With (4.47) we then obtain

$$\Delta(\mu) = \int\limits_{-\infty}^{\infty} \int\limits_{-\infty}^{\infty} D(\xi) \exp(-2\pi i \mu \cdot \xi) d\xi \, d\eta = \mathscr{F}_2[D]$$

$$= \int\limits_{-\infty}^{\infty} \int\limits_{-\infty}^{\infty} \int\limits_{-\infty}^{\infty} f(x) \exp(-2\pi i u \cdot x) d\xi \, d\eta \, d\zeta \big|_{k \cdot u = 0}$$

$$= \mathscr{F}_3[f]\big|_{k \cdot u = 0} = F(u)\big|_{k \cdot u = 0} \tag{4.53}$$

which can be interpreted either as the two-dimensional Fourier transform of the integral $D = \int_{-\infty}^{\infty} f \, d\xi$ (along the straight line of direction $k$) or as the three-dimensional Fourier transform of $f$ taken in the plane $k \cdot u = 0$ normal to $k$.

Proceeding in the same way for other directions $k$, we actually obtain $F$ on the group of planes through the origin O in the $\mathscr{F}_3$-space. The coordinates $\mu$, $v$ in each plane must therefore be transformed to the fixed system $u$, $v$, $w$ by a relation

$$\begin{Bmatrix} u \\ v \\ w \end{Bmatrix} = [C] \begin{Bmatrix} \mu \\ v \\ 0 \end{Bmatrix} \tag{4.54}$$

where $[C(k)]$ is an orthogonal transformation matrix. When the data of $F$ are found numerically at discrete points in each plane, an interpolation leading to values at points on a rectangular grid in the $u$, $v$, $w$-space is necessary. There one could apply the inversion formula

$$f(x) = \int\limits_{-\infty}^{\infty} \int\limits_{-\infty}^{\infty} \int\limits_{-\infty}^{\infty} F(u) \exp(2\pi i u \cdot x) du \, dv \, dw \ . \tag{4.55}$$

If the experimental data of $D$ cannot be obtained for a complete half sphere, but only for a limited angle, then the values of $F$ remain undetermined within a certain cone in the $\mathscr{F}_3$-space. If this cone is not too large, we could think of an extrapolation from the complementary part into it. Let us finally add that, instead of $\mathscr{F}_2$ and $\mathscr{F}_3$, we can also work, with $\mathscr{F}_2$ and $\mathscr{F}_1$ in a plane where $y$ is constant, which is the usual procedure.

## 4.2 Fringe Visibility and Localization

In the elementary explanation of the basic relation in Sect. 4.1, we have regarded the fringes as being formed by only two light rays for any given observation direction. In the present section, we want to investigate a refined theory describing how the neighboring rays of each of these rays contribute to the interference phenomena. In fact, this contribution results in fringes which are more or less visible, depending on the region in space where they are examined. In particular, to find the places of high contrast, one may apply the theory of homology which will be outlined here at length and which will reveal the spatial change of the optical path difference $D$. Secondly, we shall study the visibility and the perturbation of the fringes anywhere as a function of this spatial change of $D$ and as a function of the aperture of the observing system. Finally, when a non-stationary configuration is recorded during two small but finite time intervals, one must investigate the influence of the temporal change of the path difference (besides the spatial change) upon the visibility and the apparent fringe position.

### 4.2.1 Theory of Homologous Rays

The problem of localization and visibility of fringes has been studied by many authors [4.301–337], though primarily in reference to deformation of opaque bodies. Concepts of diffraction and coherence as well as purely geometrical considerations must be applied when studying the subject. In the field of transparent isotropic media with a varying index of refraction, several approaches to the problem of localization are given in [4.17, 338–342]. Here, we intend to present a complementary analysis. Due to the analogy introduced in Sect. 4.1.2, this theory will appear very similar to what was originally found by *Walles* in the case of deformed opaque objects [4.301, 302]. As far as the instrinsic notation is concerned, we shall deal with the problem concerning transparent isotropic media along a line of thought similar to that employed with deformed opaque bodies [Ref. 4.24, p. 97].

For simplicity, we refer to the first case considered in Sect. 4.1.2 where no window is present and where the diffusing surface $A$ is not in contact with the domain $\mathbb{G}$ (Fig. 4.12). Moreover, in order to keep the formulas as simple as possible while maintaining generality, we also assume that $n = n_0$, i.e., a medium homogeneous everywhere during the first exposure, so that the effect due to the

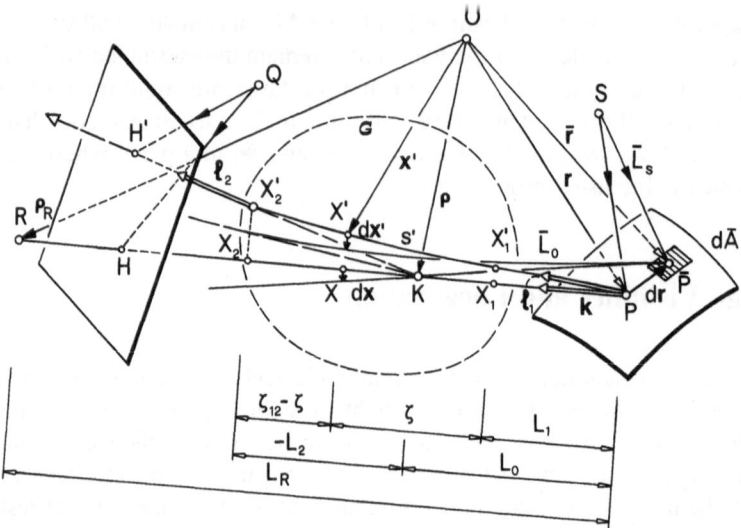

**Fig. 4.12.** Illustration of the rays producing the total complex amplitudes at a point K in the case of a phase object. At the variable point $X'$, the index of refraction changes from the constant value $n_0$ to the function $n'(x')$. (PK: reference line; $\overline{P}K$: neighboring ray of PK on the first exposure; $PX_1'X_2'K$: curved ray corresponding to PK on the second exposure; $d\overline{A}$: element on the diffusing surface)

variation of the index of refraction, e.g., $n'(x) \neq n_0$, only appears at the second exposure.

As in Sect. 4.1, we consider the interference that occurs at reconstruction at a point K of distance $\overline{L}_0$ from $\overline{P}$. Point K of position $\varrho$ could lie, for instance, inside $\mathbb{G}$, as in Fig. 4.12, so that the virtual path $\overline{L}_2^{s'} = \overline{X_2'K}$ is then considered to be negative.

In order to describe the diffuse reflection of the light coming from the source S, we use the following model [4.301, 305]. We assume a smooth, mean surface $A$ on which the points $\overline{P}$ with position vectors $\bar{r}$ are defined. Then there exists the actual surface $\tilde{A}$ with points $\tilde{P}$, having a microprofile that varies in a random way around $A$ (Fig. 4.13). Since this surface is not exactly known, we envisage an ensemble of surfaces $\{\tilde{A}\}$ all having the same mean surface $A$. The reflection is

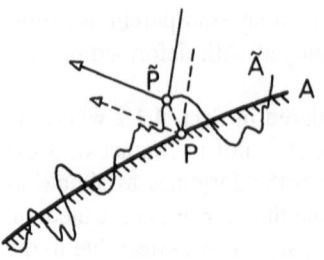

**Fig. 4.13.** Model of a rough diffusing surface $\tilde{A}$ and its corresponding smooth mean surface $A$

now described by means of a random complex reflection function $G(\bar{r})$ [4.343–346], defined as the ratio of the reflected amplitude to incident amplitude at $\bar{P}$. We also assume independent real and imaginary parts of $G(\bar{r})$, both having zero mean values which may be expressed by the disappearence of the expected value or ensemble average (for $\forall \bar{r}$ of the illuminated part of $A$)

$$E[G(\bar{r})] = 0 \ . \tag{4.56}$$

In the following paragraph, we shall use the fact that the auto-correlation of $G(\bar{r})$ may be approximated by the Dirac function $\delta$ [4.305, 344], namely

$$E[G(\bar{r})G^*(\bar{\bar{r}})] = C\delta(\bar{r} - \bar{\bar{r}}) \ , \tag{4.57}$$

where $C$ is a constant. This relation indicates that each point $\bar{P}$ acts as an independent scatterer. For opaque bodies, this last equation could not be fulfilled, for instance, when determining plastic surface deformations, or when studying corrosion processes [4.347–348] or rather smooth surfaces [4.349]. Now at the point $\bar{P}$, the light emitted by S has a complex amplitude $(S/\bar{L}_s) \exp(2\pi i \bar{L}_s/\lambda)$, where $S$ is a complex constant and $\bar{L}_s$ denotes the distance $\overline{SP}$.

We do not take into account the influence of the direction of illumination and of reflection by a random function, but simply multiply $G(\bar{r})$ by a deterministic function $K(\bar{n} \cdot \bar{h}, \bar{n} \cdot \bar{k})$. Here the vector $\bar{n}$ is the unit normal of $A$, $\bar{h}$ and $\bar{k}$ are direction unit vectors of incidence and reflection relative to point $\bar{P}$.

Therefore, according to these assumptions, the total complex amplitude at point K, after the first exposure, reads for the rays within the aperture

$$U(\varrho) = \iint\limits_A \frac{SKG}{\bar{L}_s |\bar{L}_0|} \exp\left[\frac{2\pi i}{\lambda_0} n_0(\bar{L}_s + \bar{L}_0)\right] d\bar{A} \ , \tag{4.58}$$

and corresponds to (2.77) of the Fresnel-Kirchhoff diffraction; $dA = |a_1 \times a_2| d\theta^1 d\theta^2$ being the surface element of the diffusing surface (with base $a_\alpha$ and curvilinear coordinates $\theta^\alpha$).

During the second exposure, when $n' \neq n_0$, the optical path also contains the difference, see (4.24),

$$\bar{D} = n_0(\bar{L}_0 - \bar{L}_1^s - \bar{L}_2^s) - \int_0^{\bar{s}_{12}} \bar{n}' d\bar{s}' \tag{4.59}$$

intervening in the complex amplitude at point K ($A' \sim A$ assumed)

$$U'(\varrho) = \iint\limits_A \frac{SKG}{\bar{L}_s |\bar{L}_0|} \exp\left\{\frac{2\pi i}{\lambda_0} [n_0(\bar{L}_s + \bar{L}_0) - \bar{D}]\right\} d\bar{A} \tag{4.60}$$

in which the influence of $\bar{D}$ has to be taken into account only in the phase factor.

Furthermore, since the aperture $\AA$ of the observing system and consequently the corresponding illuminated area $A$ on the diffusing surface are small compared to the distance $\bar{L}_0$, we may develop the optical path difference with respect to a constant reference ray through a point P. Thus, with point K remaining fixed, we may write:

$$\bar{D} = D + \Delta D_K \cong D + dD_K + \tfrac{1}{2}d^2 D_K \ , \tag{4.61}$$

where, with $f_K$ denoting the visibility vector, see (4.30, 35),

$$D = n_0(L_0 - L_1^s - L_2^s) - \int_0^{s_{12}} n'ds' \cong \int_0^{\zeta_{12}} (L_1 + \zeta) \frac{\partial n'}{\partial \zeta} d\zeta \tag{4.62}$$

$$dD_K = d\mathbf{k} \cdot \int_0^{\zeta_{12}} [L_0 - (L_1 + \zeta)] \nabla_k n' d\zeta = d\mathbf{k} \cdot f_K \ . \tag{4.63}$$

We will leave out the term $d^2 D_K$ for the moment, since the homology being sought refers to the first order approximation only. With $d\mathbf{k} \cong \bar{\mathbf{k}} - \mathbf{k}$, $\mathbf{k} \cdot \nabla_k = 0$ and $\bar{\mathbf{k}} \cdot \mathbf{k} \cong 1$, we obtain, instead of (4.61),

$$\bar{D} = \bar{\mathbf{k}} \cdot \left[ L_0 \int_0^{\zeta_{12}} \nabla_k n' d\zeta - \int_0^{\zeta_{12}} (L_1 + \zeta) \left( \nabla_k n' - \mathbf{k} \frac{\partial n'}{\partial \zeta} \right) d\zeta \right] \ . \tag{4.64}$$

We then replace $\bar{L}_0$ by $\bar{\mathbf{k}} \cdot (\mathbf{\varrho} - \bar{\mathbf{r}})$ in both expressions of the complex amplitudes (4.58 and 60) in order to reveal the argument $\mathbf{\varrho}$ of these functions in their phase terms; by comparison, we find that

$$U'(\mathbf{\varrho}) = U\left( \mathbf{\varrho} - \frac{L_0}{n_0} \int_0^{\zeta_{12}} \nabla_k n' d\zeta + \int_0^{\zeta_{12}} \frac{L_1 + \zeta}{n_0} \left[ \nabla_k n' - \mathbf{k} \frac{\partial n'}{\partial \zeta} \right] d\zeta \right)$$

or also that

$$U'\left( \mathbf{\varrho} - \int_0^{\zeta_{12}} \frac{L_1 + \zeta}{n_0} \left[ \nabla_k n' - \mathbf{k} \frac{\partial n'}{\partial \zeta} \right] d\zeta + \frac{L_0}{n_0} \int_0^{\zeta_{12}} \nabla_k n' d\zeta \right) = U(\mathbf{\varrho}) \ . \tag{4.65}$$

The following interpretation may be applied to this result: the change of the refractive index in $\mathbb{G}$ produces a modification of the wavefront, such that the complex amplitude at K with position vector $\mathbf{\varrho}$ (during the first exposure for $n = n_0$) is the same as the complex amplitude at another point K' with position vector

$$\mathbf{\varrho}' = \mathbf{\varrho} - \int_0^{\zeta_{12}} \frac{L_1 + \zeta}{n_0} \left( \nabla_k n' - \mathbf{k} \frac{\partial n'}{\partial \zeta} \right) d\zeta + \frac{L_0}{n_0} \int_0^{\zeta_{12}} \nabla_k n' d\zeta \tag{4.66}$$

during the second exposure, when $n' \neq n_0$.

If we vary the parameter $L_0$, the vector functions $\mathbf{\varrho}(L_0)$ and $\mathbf{\varrho}'(L_0)$ describe two generally *skew* straight lines PKH and P'K'H'' which are called *homologous* "rays" (Fig. 4.14) [4.303–306, 350]. The distance $t$ between them, measured at K perpendicularly to $\mathbf{k}$ is

$$t = \left| \int_0^{\zeta_{12}} \frac{L_0 - (L_1 + \zeta)}{n_0} \nabla_k n' d\zeta \right| = \frac{|f_K|}{n_0} \ . \tag{4.67}$$

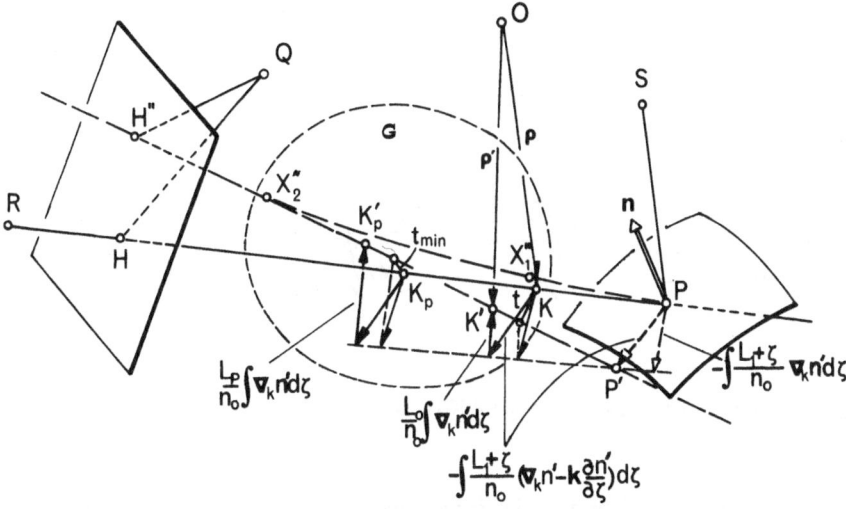

**Fig. 4.14.** Homologous "rays" PKH and P'K'H" in case of a phase object at reconstruction. (PX"₁X"₂H": true curved ray during the second exposure at the recording when $n' = n'(x)$

Several remarks should be added. First, in (4.67) only the distance $L_0 - (L_1 + \zeta)$ from X to K is involved, therefore the homologous rays do not depend upon the position of the diffusing surface if this surface is not deformed as was the case in the last example of superposition in Sect. 4.1.2. Secondly, the distance $t$ between the homologous rays depends only on the lateral part $\nabla_k n'$ of the gradient $\nabla n'$. In contrast the homologous point $K'$ of K, see (4.66), also depends on the longitudinal part $\partial n'/\partial \zeta$, but the reader should note the minus sign in (4.66). Thirdly, it must be noted that the virtual ray P'K'H" has to be considered at reconstruction, whereas the corresponding ray at recording is the partially curved path PX"₁X"₂H" which, in Fig. 4.14, has only the part X"₂H" in common with the former and which is not exactly identical with the path PX'₁X'₂H' of Fig. 4.12. The index variation has an effect as if P had been displaced to P'. Fourthly, it can happen that the homologous rays intersect, i.e. that there exists a distance $L_0 = L_{0c}$, for which

$$L_{0c} \int_0^{\zeta_{12}} \nabla_k n' \, d\zeta \; - \int_0^{\zeta_{12}} (L_1 + \zeta) \, \nabla_k n' \, d\zeta = f_K = 0 \; . \tag{4.68}$$

In this case, the optical path difference is stationary, since $dD_K$ expressed in (4.35) vanishes. The neighboring rays then contribute best to the interference phenomenon and, as will be later confirmed in detail, we thus expect the best contrast or black fringes. For the case of complete localization just described, with (4.68) and after cancelling the factor $L_R - L_{0c}$, (4.37) becomes

$$\boldsymbol{m}_t \cdot \int_0^{\zeta_{12}} \nabla_k n' \, d\zeta = 0 \; . \tag{4.69}$$

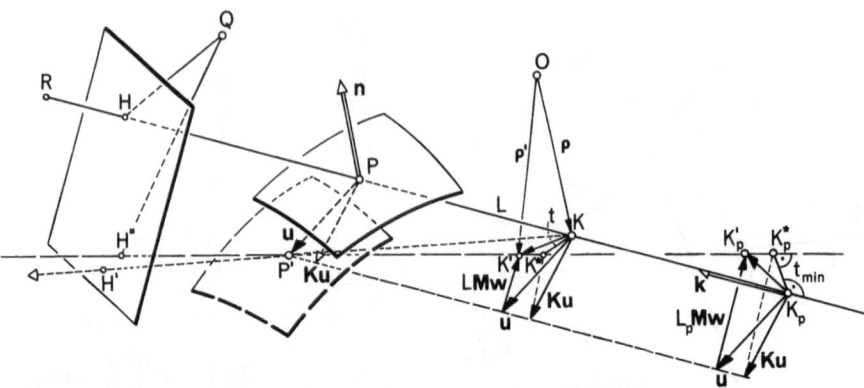

**Fig. 4.15.** Homologous "rays" PKH and P'K'H" in the case of a deformed opaque body

In other words: if and only if the fringes have the best contrast, then they are apparently perpendicular to the integral of the lateral gradient of the index of refraction (normality theorem) [4.159, 322, 326–328]. Finally, let us call attention to the analogy outlined in Sect. 4.1.2. In the case of a deformed opaque body (Fig. 4.15), (4.66, 67–69) find their correspondent relations, respectively, as

$$\varrho' = \varrho + u + LMw \; , \qquad \text{(homologous points)} \qquad\qquad (4.70)$$

$$t \;= |Ku + LMw| = |f_K| \; , \qquad \begin{array}{l}\text{("distance" at K between}\\ \text{homologous rays)}\end{array} \qquad (4.71)$$

$$L_c Mw = - Ku \; , \qquad \text{(condition of complete localization)} \quad (4.72)$$

$$m_t \cdot Ku = 0 \; , \qquad \text{(normality theorem)} \qquad\qquad (4.73)$$

where it must be noted that, in the last case, we have chosen to maintain $Ku$ which is not the analogue of $\int_0^{\zeta_{12}} \nabla_k n' \, d\zeta$ (instead of $Mw$).

## 4.2.2 Visibility and Shift of Fringes for a Finite Aperture

Having calculated the complex amplitudes $U$ and $U'$, (4.58, 60), for the homogeneous and the non-homogeneous distribution of the refractive index $n$ in $\mathbb{G}$ in the previous paragraph, we may now form the total amplitude $U_t(\varrho) = U(\varrho) + U'(\varrho)$ at K. We notice again that these functions are of random type with vanishing mean values. For the intensity $J$, we must therefore work with expected values [4.301] and write

$$J(\varrho) = \tfrac{1}{2}\{E[UU^*] + E[U'U'^*] + E[U'U^*] + E[UU'^*]\} \; . \qquad (4.74)$$

Using (4.57) and (4.61), these expected values are explicitly, on the one hand, the *autocorrelation* with equal argument $(A' \sim A)$

$$E[UU^*] = E\left[ \iint_A \iint_A \frac{SKG}{L_s|\bar{L}_0|} \exp\left(\frac{2\pi i}{\lambda_0} n_0(\bar{L}_s + \bar{L}_0)\right) \frac{S^*K^*G^*}{\bar{L}_s|\bar{L}_0|} \right.$$

$$\left. \times \exp\left(-\frac{2\pi i}{\lambda_0} n_0(\bar{L}_s + \bar{L}_0)\right) d\bar{A} \ d\bar{A} \right]$$

$$= \iint_A \frac{|S|^2|K|^2C}{L_s^2 L_0^2} d\bar{A} = 2I \quad (\simeq E[U'U'^*]) , \tag{4.75}$$

and, on the other hand, the *cross-correlation* with equal argument

$$E[U'U^*] = E\left[ \iint_A \iint_A \frac{SKG}{L_s|\bar{L}_0|} \exp\left(\frac{2\pi i}{\lambda_0} [n_0(\bar{L}_s + \bar{L}_0) - \bar{D}]\right) \frac{S^*K^*G^*}{\bar{L}_s|\bar{L}_0|} \right.$$

$$\left. \times \exp\left(\frac{-2\pi i}{\lambda_0} n_0(\bar{L}_0 + \bar{L}_s)\right) d\bar{A} \ d\bar{A} \right]$$

$$= \exp\left(\frac{-2\pi i}{\lambda_0} D\right) \iint_A \frac{|S|^2|K|^2C}{\bar{L}_s^2 L_0^2} \exp\left(-\frac{2\pi i}{\lambda_0} \Delta D_K\right) d\bar{A}$$

$$= 2\Gamma \quad (\simeq E[UU'^*]^*) . \tag{4.76}$$

Furthermore, we can separate $\Gamma$ into an amplitude and a phase factor

$$\Gamma = |\Gamma| \exp\left[-\frac{2\pi i}{\lambda_0}(D + \delta)\right] , \tag{4.77}$$

where

$$\delta = -\frac{\lambda_0}{2\pi} \arg\left[ \iint_A \frac{|S|^2|K|^2C}{\bar{L}_s^2 L_0^2} \exp\left(-\frac{2\pi i}{\lambda_0} \Delta D_K\right) d\bar{A} \right] \tag{4.78}$$

summarizes the influence of the spatial increment $\Delta D_K$ of $D$ upon the total phase and where

$$|\Gamma| = \frac{1}{2} \left| \iint_A \frac{|S|^2|K|^2C}{\bar{L}_s^2 L_0^2} \exp\left(-\frac{2\pi i}{\lambda_0} \Delta D_K\right) d\bar{A} \right| , \tag{4.79}$$

which is independant of $D$, shows the influence of $\Delta D_K$ upon the amplitude. Therefore, we can rewrite (4.74) in the form

$$J(\varrho) = 2I\left\{ 1 + \frac{|\Gamma|}{I} \cos\left[\frac{2\pi}{\lambda_0}(D + \delta)\right] \right\} . \tag{4.80}$$

Introducing the so-called visibility of the fringes at K, defined by [4.351]

$$V = \frac{J_{max} - J_{min}}{J_{max} + J_{min}} = \frac{|\Gamma|}{I} = \frac{\left| \iint\limits_{A} \frac{|S|^2 |K|^2 C}{L_s^2 L_0^2} \exp\left(-\frac{2\pi i}{\lambda_0} \Delta D_K\right) d\bar{A} \right|}{\iint\limits_{A} \frac{|S|^2 |K|^2 C}{L_s^2 L_0^2} d\bar{A}}, \tag{4.81}$$

we finally arrive at the intensity

$$J(\varrho) = 2I \left\{ 1 + V\cos\left[ \frac{2\pi}{\lambda_0} (D + \delta) \right] \right\}. \tag{4.82}$$

If we assume the fringes so close to each other that $\cos[2\pi(D + \delta)/\lambda_0]$ is steeply varying compared to $V$, then the position of the ridge of each fringe is approximately determined by $D + \delta$, whereas the blackness of the fringe is given by $V$; thus, this latter function describes their contrast.

Our further discussion will be focussed on (4.78 and 81). It has already been said that the diameter of the aperture is generally small compared to the distances involved. We may therefore replace these quantities in the amplitude factors relative to point $\bar{P}$ by those on the reference ray through point P, so that the formulas approximately simplify to

$$\delta = -\frac{\lambda_0}{2\pi} \arg\left[ \frac{1}{\mathring{A}} \iint\limits_{A} \exp\left(-\frac{2\pi i}{\lambda_0} \Delta D_K\right) d\mathring{A} \right] \tag{4.83}$$

and to

$$V = \left| \frac{1}{\mathring{A}} \iint\limits_{A} \exp\left(-\frac{2\pi i}{\lambda_0} \Delta D_K\right) d\mathring{A} \right| \tag{4.84}$$

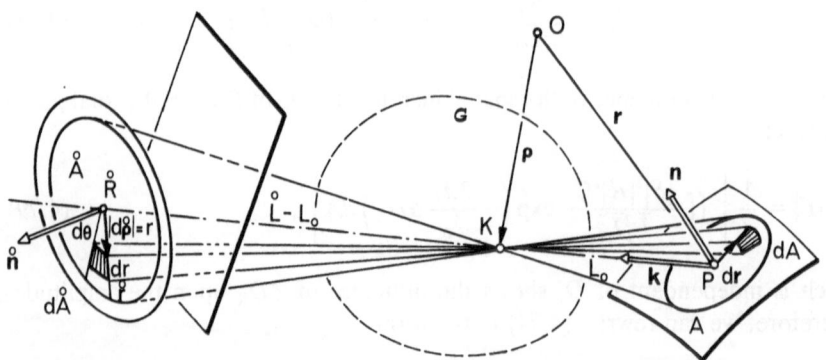

**Fig. 4.16.** Collineation between the aperture surface $\mathring{A}$ and its "image" on the diffusing surface $A$ with respect to the fringe locus K

where

$$dÅ = \frac{(\mathring{L} - L_0)^2}{L_0^2} \frac{\mathbf{n} \cdot \mathbf{k}}{\mathring{\mathbf{n}} \cdot \mathbf{k}} dA \tag{4.85}$$

is the surface element in the aperture plane with unit normal $\mathring{\mathbf{n}}$ (Fig. 4.16).

In the special case of a circular aperture, commonly used for observing instruments, we may further write the integrals with polar coordinates and more conveniently in a dimensionless form. Let $\mathring{r}$ be the radius of the aperture, $r$, $\theta$ the polar coordinates and $\sigma = r/\mathring{r}$ the dimensionless radial parameter, so that $dÅ = r\,dr\,d\theta = \mathring{r}^2 \sigma\,d\sigma\,d\theta$ and $Å = \pi \mathring{r}^2$. Then, (4.83, 84) become

$$\delta = -\frac{\lambda_0}{2\pi} \arg\left[\frac{1}{\pi}\int_0^1 \sigma\,d\sigma \int_0^{2\pi} \exp\left(-\frac{2\pi i}{\lambda_0}\Delta D_K\right) d\theta\right] , \tag{4.86}$$

$$V = \left|\frac{1}{\pi}\int_0^1 \sigma\,d\sigma \int_0^{2\pi} \exp\left(-\frac{2\pi i}{\lambda_0}\Delta D_K\right) d\theta\right| . \tag{4.87}$$

As

$$\Delta D_K = dD_K + \tfrac{1}{2}d^2 D_K ,$$

we have, see (4.63), with $\nabla_k^*$ denoting differentiation on the unit sphere,

$$dD_K = d\mathbf{k} \cdot f_K ,$$
$$d^2 D_K = d\mathbf{k} \cdot (\nabla_k^* \otimes f_K)d\mathbf{k} , \tag{4.88}$$

where it is assumed that $d^2\mathbf{k}$ is taken parallel to $\mathbf{k}$, so that $d^2\mathbf{k} \cdot f_K = 0$. Although the two increments $dD_K$ and $d^2 D_K$ are of different power in $d\mathbf{k}$, we shall see that one should consider both simultaneously for a particular reason. A corresponding theory has been outlined by *de Jong* [4.352] to which we also refer in the following.

The increment vector $d\mathbf{k}$ and the associated increment $d\mathring{\varrho}$ in the plane of the aperture are connected by the affine connection

$$d\mathbf{k} = \frac{1}{\mathring{L} - L_0} K\,d\mathring{\varrho} , \tag{4.89}$$

where $|d\mathring{\varrho}| = r$. However, by introducing a unit vector $\mathring{\mathbf{e}}$ in the direction of $d\mathring{\varrho}$ and two dimensionless quantities, namely a vector $\mathring{\mathbf{v}}$ and a tensor $\mathring{U}$ defined on the aperture plane, where $\mathring{N} = I - \mathring{\mathbf{n}} \otimes \mathring{\mathbf{n}}$, by

$$\frac{2\pi \mathring{r}}{\lambda_0(\mathring{L} - L_0)}\mathring{N}f_K = \mathring{\mathbf{v}} , \qquad \frac{2\pi \mathring{r}^2}{\lambda_0(\mathring{L} - L_0)^2}\mathring{N}(\nabla_k^* \otimes f_K)\mathring{N} = \mathring{U} , \tag{4.90}$$

Eq. (4.88) can be rewritten in the form

$$\frac{2\pi}{\lambda_0} dD_K = \sigma \mathring{e} \cdot \mathring{v} \ , \qquad \frac{2\pi}{\lambda_0} d^2 D_K = \sigma^2 \mathring{e} \cdot \mathring{U} \mathring{e} \ . \tag{4.91}$$

If the polar coordinate $\theta$ is measured from the direction of $\mathring{v}$, we have $\mathring{e}\,(\cos\theta,\ \sin\theta)$, $\mathring{e} \cdot \mathring{v} = \mathring{v}\cos\theta$. Whereas the modulus $\mathring{v}$ may be calculated relatively easily from the visibility vector (see (4.90) and the table specifying the analogy in Sect. 4.1.2), a similar detailed calculus of the tensor $\mathring{U}$ becomes very cumbersome, as can be seen from the explicit form of the analogue second derivative of $D$ given in [Ref. 4.24, p. 168, Eq. (5.49)]. However, in the quadratic form of (4.91b), only the symmetric part $(\mathring{U} + \mathring{U}^{\mathrm{T}})/2$ is relevant. Separating it into an isotropic part $-\mathring{u}\mathring{N}$ and a deviatoric part $\mathring{i}\mathring{D}$, see also (3.36), $\theta_1$ denoting the azimuth of the principal direction 1 of $\mathring{D}$, we then have

$$\frac{2\pi}{\lambda_0} dD_K = \sigma \mathring{v} \cos\theta \ , \qquad \frac{\pi}{\lambda_0} d^2 D_K = -\frac{\sigma^2}{2}\mathring{u} + \frac{\sigma^2}{2}\mathring{i}\cos 2\,(\theta - \theta_1) \ , \tag{4.92}$$

where we recall that the orders of magnitude are $\mathring{v} = O(\mathring{r}/\mathring{L})$, $\mathring{u} = O(\mathring{r}/\mathring{L})^2$ and $\mathring{i} = O(\mathring{r}/\mathring{L})^2$, respectively. Also, in order to assess the meaning of $\mathring{v}$ and $\mathring{u}$, let us look for a moment at the particular case investigated in [4.352], where the optical system is focused on the surface of an opaque deformable body. Here, we have $L_0 = 0$ (K coincides with P). For the vector $f_K$ and its derivative, $u$ is to be considered a constant vector while the projector $K$ is to be derived on the unit sphere around point K. Applying the projection rules with respect to $k$, (2.26), we have

$$f_K = Ku \ , \qquad \nabla^*_k \otimes f_K = -[K \otimes k + K \otimes k)^{\mathrm{T}}]u \ .$$

In the special case where $\mathring{n} = k$ and consequently $\mathring{N} = K$, we get

$$\mathring{v} = \frac{2\pi\mathring{r}}{\lambda_0 \mathring{L}} |Ku| \ , \qquad \mathring{u} = \frac{2\pi}{\lambda_0} \left(\frac{\mathring{r}}{\mathring{L}}\right)^2 (k \cdot u) \ , \qquad \mathring{i} = 0 \ . \tag{4.93}$$

Thus, $\mathring{v}$ and $\mathring{u}$ are the dimensionless lateral and longitudinal parts of the displacement vector $u$ of P.

Returning to the general case, we may now write the integral of (4.86, 87) into the form

$$\frac{1}{\pi}\int_0^1 \sigma \, d\sigma \int_0^{2\pi} \exp\left(-\frac{2\pi i}{\lambda_0} \Delta D_K\right) d\theta$$

$$= \frac{1}{\pi}\int_0^1 \sigma \, d\sigma \int_0^{2\pi} \exp\left\{-i\left[\sigma\mathring{v}\cos\theta - \frac{\sigma^2}{2}\mathring{u} + \frac{\sigma^2}{2}\mathring{i}\cos 2\,(\theta - \theta_1)\right]\right\} d\theta = H(\mathring{v}, \mathring{u}, \mathring{i})$$

$$\tag{4.94}$$

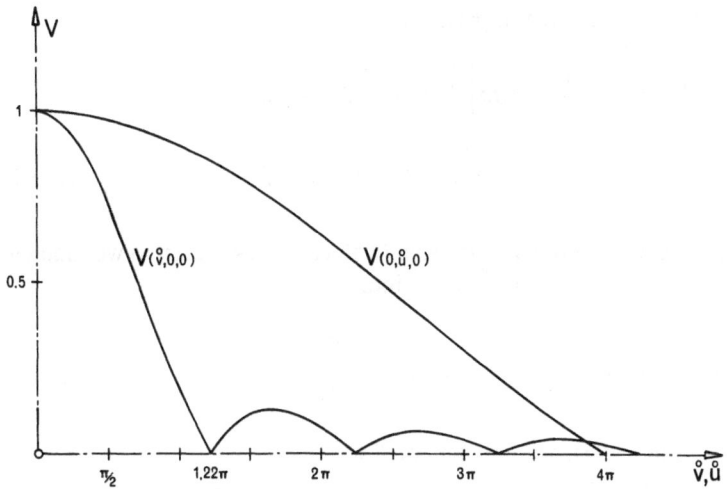

**Fig. 4.17.** Graph of the functions $V(\mathring{v}, 0, 0)$ and $V(0, \mathring{u}, 0)$ see (4.102), relevant in fringe visibility. For the definitions of $\mathring{v}$ and $\mathring{u}$, see (4.91) or, in particular, (4.93)

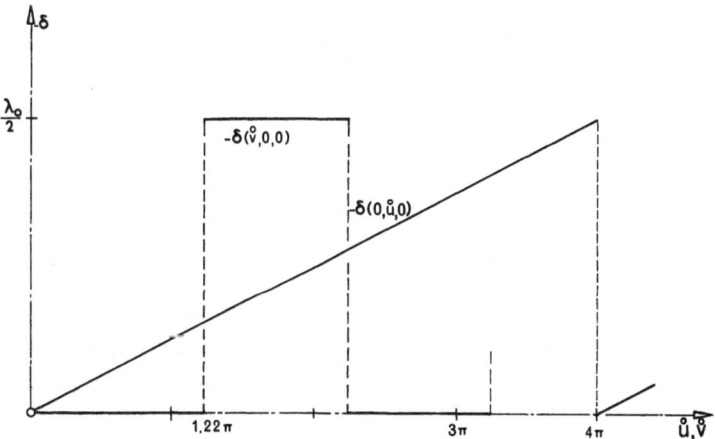

**Fig. 4.18.** Graph of the functions $\delta(\mathring{v}, 0, 0)$ and $\delta(0, \mathring{u}, 0)$, see (4.95, 97, 99) relevant to the shifting of fringes

leading to the wanted functions (Figs. 4.17, 18):

$$\delta = -\frac{\lambda_0}{2\pi} \arg H(\mathring{v}, \mathring{u}, \mathring{t}) \ , \tag{4.95}$$

$$V = |H(\mathring{v}, \mathring{u}, \mathring{t})| \ . \tag{4.96}$$

Let us first discuss $\delta$. We have in particular

$$\arg H(\hat{v}, 0, 0) = \arg \frac{1}{\pi} \int_0^1 \sigma \, d\sigma \left[ \int_0^\pi \exp(-i\sigma\hat{v}\cos\theta)d\theta \right.$$

$$\left. + \int_\pi^{2\pi} \exp(-i\sigma\hat{v}\cos\theta)d\theta \right] = 0 \quad \text{or} \quad \pi \ , \tag{4.97}$$

where this function has a zero value for small values of $\hat{v}$ ($< 1.22\pi$ as we shall see later). By a similar separation of $\int_0^{2\pi}(\dots)$ into

$$\int_{\theta_1}^{\theta_1+\pi/2} (\dots) + \int_{\theta_1+\pi/2}^{\theta_1+\pi} (\dots) + \int_{\theta_1+\pi}^{\theta_1+3\pi/2} (\dots) + \int_{\theta_1+3\pi/2}^{\theta_1+2\pi} (\dots) \ ,$$

where $\mathring{u} = \hat{v} = 0$ in (4.94), we find for small $\mathring{t}$

$$\arg H(0, 0, \mathring{t}) = 0 \ . \tag{4.98}$$

In contrast to these two relations, we get for $0 < \mathring{u} < 4\pi$ that

$$\arg H(0, \mathring{u}, 0) = \arg \frac{1}{\pi} \int_0^1 \sigma \, d\sigma \int_0^{2\pi} \exp\left(\frac{i\sigma^2}{2}\mathring{u}\right) d\theta$$

$$= \arg \left[ \frac{2i}{-\mathring{u}} \exp\left(\frac{i\sigma^2}{2}\mathring{u}\right) \right]_0^1$$

$$= \arg \left\{ \frac{2i}{-\mathring{u}} \left[ \exp\left(\frac{i\mathring{u}}{2}\right) - 1 \right] \right\} = \frac{\mathring{u}}{4} \ . \tag{4.99}$$

This shows that, in the vicinity of the origin in a $(\hat{v}, \mathring{u}, \mathring{t})$-space, the essential variable for $\delta$ is $\mathring{u}$ [although it is of the order $(\mathring{r}/\mathring{L})^2$]. Next to be considered will be $\hat{v}$, whereas $\mathring{t}$ may be neglected, since it has a behavior similar to $\hat{v}$, but is of smaller order in $\mathring{r}/\mathring{L}$ than $\hat{v}$. The map of $\delta$ in the $(\hat{v}, \mathring{u})$-plane, which *de Jong* [4.352] has given for the special case of focusing on the object surface, may therefore also be used in a more general case. It is thus worth calculating the Debye-type integral:

$$H(\hat{v}, \mathring{u}, 0) = \frac{1}{\pi} \int_0^1 \sigma \, d\sigma \int_0^{2\pi} \exp(-i\sigma\hat{v}\cos\theta)\exp\left(\frac{i\sigma^2}{2}\mathring{u}\right) d\theta$$

$$= 2 \int_0^1 J_0(\sigma\hat{v})\exp\left(\frac{i\sigma^2}{2}\mathring{u}\right) \sigma \, d\sigma \ , \tag{4.100}$$

where $J_0$ is the zero-order Bessel function. For the special cases where $\mathring{u} = 0$ and $\hat{v} = 0$, the function $H$ may also be written

$$H(\mathring{v}, 0, 0) = 2\frac{J_1(\mathring{v})}{\mathring{v}} , \qquad H(0, \mathring{u}, 0) = \frac{2i}{-\mathring{u}} \left[ \exp\left(\frac{i\mathring{u}}{2}\right) - 1 \right] , \qquad (4.101)$$

with the first-order Bessel function $J_1$.

The corresponding visibilities of these last expressions are

$$V(\mathring{v}, 0, 0) = 2\left|\frac{J_1(\mathring{v})}{\mathring{v}}\right| , \qquad V(0, \mathring{u}, 0) = \left|\frac{\sin\left(\frac{\mathring{u}}{4}\right)}{\frac{\mathring{u}}{4}}\right| \qquad (4.102)$$

with their first zeros $\mathring{v}_0 = 1.22\pi$ and $\mathring{u}_0 = 4\pi$, respectively. Beyond these values, $\delta$ changes suddenly by $\lambda_0/2$ (contrast inversion).

As $\mathring{u}$ is of higher order than $\mathring{v}$, (4.102a) has a greater importance. Varying the position of K along the ray PR on which the observing system is focused, we may seek the locus of relatively best contrast (partial localization of the fringes). This locus is determined by the condition

$$\frac{dV}{dL} = 0 \quad \text{or} \quad \frac{dV}{dL_0} = 0 , \qquad (4.103)$$

the parameter being either $L$ or $-L_0$, depending on the two cases of Sect. 4.1.1 and 4.1.2. Equations (4.103) are equivalent to

$$\frac{d}{dL}\left(\frac{|\mathring{N}f_K|}{\mathring{L} + L}\right) = 0 \quad \text{or} \quad \frac{d}{dL_0}\left(\frac{|\mathring{N}f_K|}{\mathring{L} - L_0}\right) = 0 . \qquad (4.104)$$

Two examples are of interest [4.301, 302]: (a) If $k = \mathring{n}$, $\mathring{L} + L$ or $\mathring{L} - L_0 = $ const, we find

$$L = L_p = -\frac{Mw \cdot Ku}{|Mw|^2} \quad \text{or} \quad L_p = \frac{\int\limits_{0}^{\zeta_{12}} \nabla_k n' d\zeta \cdot \int\limits_{0}^{\zeta_{12}} (L_1 + \zeta)\nabla_k n' d\zeta}{\left|\int\limits_{0}^{\zeta_{12}} \nabla_k n' d\zeta\right|^2} \qquad (4.105)$$

which, when compared to (4.71, 67), gives as locus $K_p$, the position of the shortest distance between the skewed homologous rays, or (b) if $k = \mathring{n}$, $\mathring{L} \cong L_R = $ const, we find, for instance, for the first case:

$$L = L_q = -\frac{(L_R Mw - Ku) \cdot Ku}{(L_R Mw - Ku) \cdot Mw} , \qquad (4.106)$$

which gives a point $K_q$ where the visibility vector $f_K = L_q Mw + Ku$ is perpendicular to the fringe vector $f_R = L_R Mw - Ku$. The projections of the two homologous points defined by $L = L_q$, (4.70), on the plane perpendicular to $k$ lie on the same fringe.

### 4.2.3 Time Dependent Effects

In all the cases treated in the preceding subsection, we assumed quasi-stationary configurations during short exposure times. Of course, fast time-varying config-urations are of greater interest for practical applications; as has already been mentioned in the introduction of this chapter, they involve techniques of time average [4.353–375], stroboscopy [4.376–385] and the use of derotating devices [4.386–395]. The literature referring to this field is quite extensive; nevertheless, within the scope of this book, we intend to limit ourselves to some complemen-tary aspects. For the whole theory and its applications, the reader may refer to classical works [4.6, 7, 17, 18, 396–397]. In particular, we draw the reader's attention to the book of *Ostrovsky* et al. [Ref. 4.19, p. 289ff.], where the fringe intensity function is studied in a systematic manner for the case of finite exposure times and of oscillatory processes. Other particular aspects have been actively researched as well. For instance, forms of motion differing from a simple sinusoi-dal mode [4.398–401] have been investigated, specifically motions of two or multiple sinusoidal mode [4.402–404] of periodic non-sinusoidal forms [4.405–409], the approach by density functions [4.410] and the study of three-dimensional vibrations [4.411–413]. Multiple exposures, combinations of time-average and stroboscopic techniques may disclose new information [4.414–417]. Further, for the time-average, the laser beam will be modulated in order to enhance the visibility of the higher-order fringes [4.418–421] which may also be partly achieved by an adequate choice of the holographic plate development [4.422]. The upper limit of the vibration amplitude has been analysed in [4.423–424]. The Doppler effect, produced by the moving object, has been com-pensated in stroboscopic holography by the frequency sweep of a pulsed laser beam [4.425–431]. Finally, let us mention the heterodyne method, which has led to a certain automation of the measurements [4.432]. Here, we shall discuss only the effect of time variation combined with that of spatial variation within the theory of visibility, in general situations occuring with only relatively short expo-sure times.

Let us first write the expressions for the optical path difference. In the case of phase objects (varying index of refraction), this difference is taken preferably with respect to the homogeneous configuration

$$D = n_0[L_0 - (L_1 + L_2)] - \int_0^{\zeta_{12}} n'[x(\zeta), t]d\zeta . \tag{4.107}$$

With the spatial and temporal changes of the index of refraction function $n'(x, t)$ assumed to be small, the integration is performed along the straight ray PK (Fig. 4.12) so that the influence on $D$ due to the motion of the particules in $G$ is disregarded.

In the case of deformed opaque bodies, the situation is somewhat different since the materialized point P of the rough body surface (Fig. 4.13), from which is emitted a specific light ray (not to be confused with its neighboring ray), can perform any movement defined by a displacement function $u(r, t)$. We may write, for instance, this displacement in the form

$$u(r, t) = \int_0^t v(\bar{r})d\bar{t} \ . \tag{4.108}$$

Here, $v(\bar{r})$ is the velocity of the above materialized point at an intermediate position $\bar{P}$ between P and P' with the generally unknown deformation $r, \bar{t} \rightarrow \bar{r}(r, \bar{t})$. In the special case of a constant velocity $v = v_0$, we have of course

$$u(r, t) = v_0(r)t \ . \tag{4.109}$$

In the often encountered situation of oscillation with constant direction of velocity, $v = u_0 \omega \cos \omega \bar{t}$, one gets another separation of space and time variable

$$u(r, t) = u_0(r) \sin \omega t \ . \tag{4.110}$$

Now, the optical path difference at the point K between the configuration at time $t$ and that at time $t = 0$ is similar to (4.4) ($n_0 \simeq 1$)

$$D = L_s - L - L_s'(t) + L'(t) \simeq u(r, t) \cdot g(r, k) \ , \tag{4.111}$$

where the sensitivity vector $g(r, k)$ is assumed to be independent of time $t$, because $u \ll L_s, |L|$.

In both expressions (4.107) and (4.111), the scalar function $D$ now contains three variables: $D = D(r, k, t)$; but if we fix a collineation center, for instance, at point K, so that $r = r(k)$, then $D = D_K(k, t)$ depends only on the two-dimensional (vectorial) space-variable $k$ and the one-dimensional (scalar) time-variable $t$. Since all elementary waves $U dt$ interfere with the reference wave $V$ during the total exposure time $\tau$, the relevant complex amplitude at K is not $U + U'$ as in (4.1), but is now, as in (4.60), the mean value

$$U_\tau(\varrho) = \frac{1}{\tau} \int_0^\tau U \, dt = \frac{1}{\tau} \int_0^\tau dt \iint_A \frac{SKG}{\overline{L_s}|\overline{L_0}|}$$

$$\times \ \exp \left\{ \frac{2\pi i}{\lambda_0} \left[ n_0(\overline{L}_s + \overline{L}_0) - \overline{D} \right] \right\} d\overline{A} \ . \tag{4.112}$$

For the corresponding intensity we have, instead of (4.74), the expected value

$$J(\varrho) = \tfrac{1}{2} E[U_\tau U_\tau^*] \ , \tag{4.113}$$

the asterisk again denoting the complex conjugate.

As in the cross correlation with equal argument, (4.76), we then have explicitly with (4.57):

$$J(\varrho) = \frac{1}{2\tau^2} E \left[ \int_0^\tau dt \int_0^\tau dt' \iint_A \iint_A \frac{SKG}{\bar{L}_s|\bar{L}_0|} \exp\left\{ \frac{2\pi i}{\lambda_0} [n_0(\bar{L}_s + \bar{L}_0) - \bar{D}] \right\} \right.$$

$$\times \frac{S^*K^*G^*}{\bar{L}_s|\bar{L}_0|} \exp\left\{ \frac{-2\pi i}{\lambda_0} [n_0(\bar{L}_s + \bar{L}_0) - \bar{D}'] \right\} d\bar{A} \, d\bar{A} \left. \right]$$

$$= \frac{1}{2\tau^2} \int_0^\tau dt \int_0^\tau dt' \iint_A \frac{|S|^2|K|^2 C}{\bar{L}_s^2 \bar{L}_0^2} \exp\left[ \frac{2\pi i}{\lambda_0} (\bar{D}' - \bar{D}) \right] d\bar{A} \; . \qquad (4.114)$$

In order to combine some well-known results in a preliminary step, let us assume that the aperture is very small and that the increment $\Delta D_K$ is negligible. In this case, we consider $\bar{D}$ as a constant on $A$ and a factor which may be taken out of the surface integral. Thus, with (4.75), Eq. (4.114) simplifies to

$$J(\varrho) = \frac{I}{\tau^2} \int_0^\tau dt \int_0^\tau dt' \exp\left[ \frac{2\pi i}{\lambda_0} (D' - D) \right] . \qquad (4.115)$$

In the special case (4.109) characterized by a constant velocity of the considered opaque body, we find [4.398, 433–435] the intensity function

$$J(\varrho) = \frac{I}{\tau^2} \int_0^\tau \exp\left[ \frac{2\pi i}{\lambda} (\boldsymbol{v}_0 \cdot \boldsymbol{g}) t' \right] dt' \int_0^\tau \exp\left[ -\frac{2\pi i}{\lambda} (\boldsymbol{v}_0 \cdot \boldsymbol{g}) t \right] dt$$

$$= \frac{I}{\tau^2} \frac{\exp[2\pi i(\boldsymbol{v}_0 \cdot \boldsymbol{g}) \tau/\lambda] - 1}{2\pi i \boldsymbol{v}_0 \cdot \boldsymbol{g}/\lambda} \cdot \frac{\exp[-2\pi i(\boldsymbol{v}_0 \cdot \boldsymbol{g}) \tau/\lambda] - 1}{-2\pi i \boldsymbol{v}_0 \cdot \boldsymbol{g}/\lambda}$$

$$= I \frac{\sin^2[\pi(\boldsymbol{v}_0 \cdot \boldsymbol{g}) \tau/\lambda]}{\pi^2(\boldsymbol{v}_0 \cdot \boldsymbol{g})^2 \tau^2/\lambda^2} \; . \qquad (4.116)$$

In the other special case of harmonic oscillation of an opaque body, (4.110), we get from (4.115) with an exposure duration $\tau = 2\pi/\omega$ that

$$J(\varrho) = \frac{I}{(2\pi/\omega)^2} \int_0^{2\pi/\omega} \exp\left[ \frac{2\pi i}{\lambda} (\boldsymbol{u}_0 \cdot \boldsymbol{g}) \sin \omega t' \right] dt'$$

$$\times \int_0^{2\pi/\omega} \exp\left[ -\frac{2\pi i}{\lambda} (\boldsymbol{u}_0 \cdot \boldsymbol{g}) \sin \omega t \right] dt = I J_0^2 \left( \frac{2\pi}{\lambda} \boldsymbol{u}_0 \cdot \boldsymbol{g} \right) ; \quad (4.117)$$

the intensity follows the square of the Bessel function of order zero [4.6].

Returning to the relatively general case of (4.114) for a small but finite aperture and a spatial gradient of $D$, let us consider a *double* exposure during two small but finite intervals of time $\tau/2 \ll |t_1 - t_2|$ around $t_1$ and $t_2$ [4.342]. Here the analogue of (4.114) reads

$$J(\varrho) = \frac{1}{2(\tau/2)^2}\left\{ \int\limits_{t_1-\tau/4}^{t_1+\tau/4} dt \int\limits_{t_1-\tau/4}^{t_1+\tau/4} dt' \iint\limits_{A} \frac{|S|^2|K|^2 C}{\overline{L}_s^2\overline{L}_0^2} \exp\left[\frac{2\pi i}{\lambda_0}(\overline{D}' - \overline{D})\right] d\overline{A} \right.$$

$$+ \int\limits_{t_2-\tau/4}^{t_2+\tau/4} dt \int\limits_{t_2-\tau/4}^{t_2+\tau/4} dt'(\ldots)$$

$$+ \int\limits_{t_2-\tau/4}^{t_2+\tau/4} dt \int\limits_{t_1-\tau/4}^{t_1+\tau/4} dt'(\ldots)$$

$$\left. + \int\limits_{t_1-\tau/4}^{t_1+\tau/4} dt \int\limits_{t_2-\tau/4}^{t_2+\tau/4} dt'(\ldots)\right\} . \tag{4.118}$$

As far as the intensity $|S|^2$ of the source is concerned, for simplicity, we assume it to be constant during the exposure. We may develop the function $\overline{D}$ in space around the central ray of direction $k$ and near $t_1$ and $t_2$ in time. Hence, we write the linear approximation [4.436, 437]

$$\overline{D} \cong D_\alpha + dD_\alpha + (t-t_\alpha)\partial_t D_\alpha , \quad \alpha = 1, 2 \tag{4.119}$$

where $D_1 = D(k, t_1)$ and $D_2 = D(k, t_2)$. The terms $dD_\alpha = dk \cdot \nabla_k^* D_\alpha$ denote the two first-order spatial increments, whereas $(t - t_\alpha)\partial_t D_\alpha\,(\partial_t = \partial/\partial t)$ are the two first-order temporal increments. Again with (4.75) and, since the spatial increments cancel, we get for the first term in (4.118) in a manner similar to the example of (4.115, 116), that

$$\frac{2}{\tau^2} \int\limits_{t_1-\tau/4}^{t_1+\tau/4} dt \int\limits_{t_1-\tau/4}^{t_1+\tau/4} dt' \iint\limits_{A} \frac{|S|^2|K|^2 C}{\overline{L}_s^2\overline{L}_0^2} \exp\left[\frac{2\pi i}{\lambda_0}(\overline{D}' - \overline{D})\right] d\overline{A}$$

$$= \frac{\sin^2\left(\dfrac{\pi}{2}\dfrac{\tau}{\lambda_0}\partial_t D_1\right)}{\left(\dfrac{\pi}{2}\dfrac{\tau}{\lambda_0}\partial_t D_1\right)^2} I . \tag{4.120}$$

An analogue expression can be derived for the second term of (4.118). Contrary to this, the two remaining mixed terms will contain both the temporal *and* the spatial increments in the difference of the argument, e.g.:

$$\overline{D}'(\overline{k}, t_1) - \overline{D}(\overline{k}, t_2) \cong D_1 - D_2 + dD_1 - dD_2$$
$$+ (t' - t_1)\partial_t D_1 - (t-t_2)\partial_t D_2 . \tag{4.121}$$

Therefore, we get for the third term of (4.118):

$$\frac{2}{\tau^2} \int\limits_{t_2-\tau/4}^{t_2+\tau/4} dt \int\limits_{t_1-\tau/4}^{t_1+\tau/4} dt' \iint\limits_A \frac{|S|^2 |K|^2 C}{L_s^2 L_0^2} \exp\left[\frac{2\pi i}{\lambda_0}(\bar{D}' - \bar{D})\right] d\bar{A}$$

$$= \frac{1}{2} \exp\left[\frac{2\pi i}{\lambda_0}(D_1 - D_2)\right] \iint\limits_A \frac{|S|^2 |K|^2 C}{L_s^2 L_0^2} \exp\left[\frac{2\pi i}{\lambda_0} d(D_1 - D_2)\right] d\bar{A}$$

$$\times \frac{\sin\left(\frac{\pi}{2}\frac{\tau}{\lambda_0}\partial_t D_1\right) \sin\left(\frac{\pi}{2}\frac{\tau}{\lambda_0}\partial_t D_2\right)}{\left(\frac{\pi}{2}\frac{\tau}{\lambda_0}\partial_t D_1\right)\left(\frac{\pi}{2}\frac{\tau}{\lambda_0}\partial_t D_2\right)} , \qquad (4.122)$$

which separates time variables from space variables multiplicatively [4.418]. Finally, the fourth term in (4.118) is simply the conjugate of (4.122). Defining then similarly to (4.76)

$$2\Gamma = \exp\left[\frac{2\pi i}{\lambda_0}(D_1 - D_2)\right] \iint\limits_A \frac{|S|^2 |K|^2 C}{L_s^2 L_0^2} \exp\left[\frac{2\pi i}{\lambda_0} d(D_1 - D_2)\right] d\bar{A} \quad (4.123)$$

and using the abbreviations

$$\frac{\pi\tau}{2\lambda_0}\partial_t D_1 = \mu_1 ,$$

$$\frac{\pi\tau}{2\lambda_0}\partial_t D_2 = \mu_2$$

$$(4.124)$$

we obtain for the fringe intensity

$$J(\varrho) = I\left(\frac{\sin^2\mu_1}{\mu_1^2} + \frac{\sin^2\mu_2}{\mu_2^2}\right) + (\Gamma + \Gamma^*)\frac{\sin\mu_1 \sin\mu_2}{\mu_1\mu_2}$$

$$= I\left\{\frac{\sin^2\mu_1}{\mu_1^2} + \frac{\sin^2\mu_2}{\mu_2^2} + \frac{|\Gamma|}{I}\frac{2\sin\mu_1 \sin\mu_2}{\mu_1\mu_2}\right.$$

$$\left. \times \cos\left[\frac{2\pi}{\lambda_0}(D_1 - D_2 + \delta)\right]\right\} , \qquad (4.125)$$

where $\delta$ again summarizes, as in (4.80), the influence of $d(D_1 - D_2)$ in the argument of $\Gamma$. As in (4.81), the visibility becomes

$$V = \frac{J_{max} - J_{min}}{J_{max} + J_{min}} = \frac{|\Gamma|}{I} \frac{2\left|\dfrac{\sin\mu_1 \sin\mu_2}{\mu_1\mu_2}\right|}{\dfrac{\sin^2\mu_1}{\mu_1^2} + \dfrac{\sin^2\mu_2}{\mu_2^2}} . \qquad (4.126)$$

In the particular case of a circular aperture, the relevant generalization of (4.102) is thus

$$V(\hat{v}, 0, 0, \mu_1, \mu_2) = \left| \frac{2J_1(\hat{v})}{\hat{v}} \cdot \frac{2\dfrac{\sin\mu_1 \sin\mu_2}{\mu_1\mu_2}}{\dfrac{\sin^2\mu_1}{\mu_1^2} + \dfrac{\sin^2\mu_2}{\mu_2^2}} \right| \tag{4.127}$$

with the first order Bessel function $J_1$.

We therefore conclude that time variation has no influence on the fringe visibility $V$ if $\partial_t D_1 = \partial_t D_2$ or $\partial_t D_1 = -\partial_t D_2$, since, for $|\mu_1| = |\mu_2| = \mu$, we have

$$2 \left| \frac{\sin\mu_1 \sin\mu_2}{\mu_1\mu_2} \right| = \frac{\sin^2\mu_1}{\mu_1^2} + \frac{\sin^2\mu_2}{\mu_2^2} = \frac{2\sin^2\mu}{\mu^2} \, .$$

In contrast to $V$, the intensity $J$ will be reduced by this factor. This happens, for example, in a strobo-holographic exposition of a continuously deformed opaque object [Ref. 4.438, Picture 4, p. 145] or of a vibration process in symmetric positions. However, for the variation of the index of refraction in a gas, both time derivatives are generally not known and therefore $|\mu_1| \neq |\mu_2|$. As can be expected and as has been observed, the fringe visibility decreased with increasing finite exposure time. When the first exposure is carried out with $n = n_0$, (4.125, 126) may be simplified, since $\mu_1 = 0$, i.e. $\sin\mu_1/\mu_1 = 1$, and (4.127) then becomes [4.437]

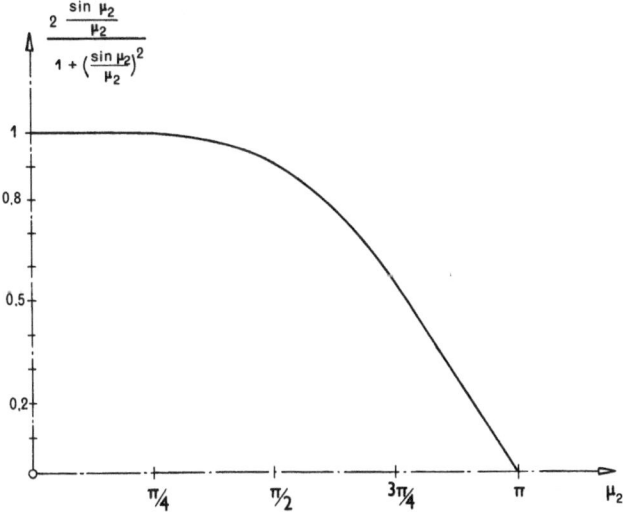

**Fig. 4.19.** Graph of the function $2(\mu_2^{-1}\sin\mu_2)(1 + \mu_2^{-2}\sin^2\mu_2)^{-1}$, relevant to fringe visibility in the case of time variation. For the definition of $\mu_2$, see (4.124)

$$V = \left| \frac{2J_1(\dot{v})}{\dot{v}} \left[ \frac{2\frac{\sin \mu_2}{\mu_2}}{1 + \left( \frac{\sin \mu_2}{\mu_2} \right)^2} \right] \right| . \tag{4.128}$$

The function in brackets is represented by Fig. 4.19. We see that, for $\partial_t D_2$ smaller than $\lambda_0/\tau$ ($\mu_2 < \pi/2$), the time effect has an influence of less than 10% on the visibility. On the other hand, we do not observe fringes ($V = 0$) for $\mu_2 = \pi$, i.e. $\partial_t D_2 = 2\lambda_0/\tau$ and this happens independently of the spatial change of $n'$. Therefore, the concept of visibility could also be applied for the determination of wave velocities in a gas [4.439], since $\partial_t D_2$ implicitely contains the velocity.

# 5. Modification at the Reconstruction in Holographic Interferometry

We have already discussed in detail the modification of the optical arrangement at reconstruction which caused deformations and aberrations of the images for single exposures in holography. Such modifications may also influence the formation of a fringe pattern in holographic interferometry. First, we shall look at some cases of interest where only one reference source and one hologram are used as in the standard set-up. In this limited case, we cannot act on both wave fields simultaneously and independently. Secondly, we shall pass to the more relevant modifications where either two reference sources or two holograms are involved. There, each wave field will be modified at the same time in a different way, but in ways which are convenient for our purposes, so that alternative measurement techniques are provided which differ from those described in Chap. 4.

## 5.1 Modifications with a Single Reference Source and a Single Hologram

This section begins with the study of two cases with double exposure; (i) the large (geometric) modifications of the image of an object together with its fringe pattern which is due to a small mechanical deformation, and (ii) the reconstruction of a fringe pattern resulting from the illumination of an object by two different wavelengths at recording. This latter case will lead us to the contour generation of a fixed object surface. Finally, we investigate the fringes formed when the actual wavefront emitted by a deformed object interferes with a diffracted wavefront coming from a single exposure hologram of the corresponding undeformed state. This is called the real-time technique.

### 5.1.1 Large Modification in the Double-Exposure Method

It was recognized long ago that, in the double-exposure method with one reference source and one wavelength at recording, the fringe order is not affected by a small geometric modification or a small wavelength change at reconstruction. However, a large modification or a large wavelength change is not trivial, necessitating a clarification of the concept of "not affected fringes", since the surface of

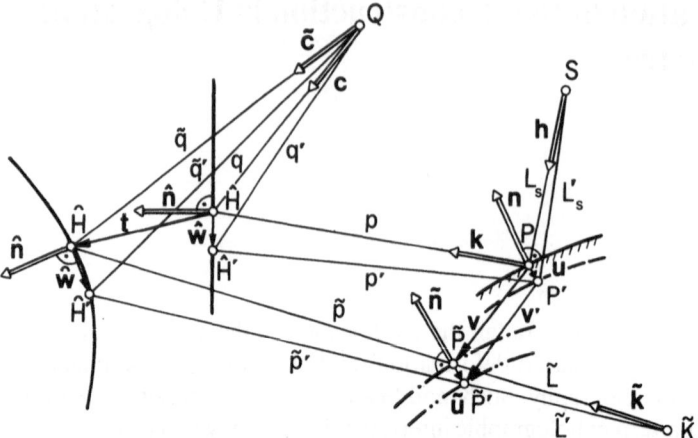

**Fig. 5.1.** Large modification at the reconstruction when using a single reference source Q and a single hologram (points $\hat{H}$). ($u$: mechanical displacement. $v$, $v'$: optical displacements. $\tilde{u}$: apparent displacement. $\mathring{n}$, $\hat{n}$, $n$, $\tilde{n}$: unit normals of the hologram and the object surface; $\mathring{w}$: collinear image of $\tilde{u}$ with respect to the fringe locus $\tilde{K}$. $\hat{w}$: corresponding vector at the recording)

the examined object becomes visibly deformed. This investigation is of particular interest in the case of dynamically deformed opaque bodies, where the recording is performed with a pulsed laser while the reconstruction is made, for convenience, with a continuous laser, each laser emitting light with different wavelengths [5.1].

Figure 5.1 shows the somewhat general situation where, besides the finite difference $\tilde{\lambda} - \lambda$ of wavelength, we assume a finite deformation of the hologram characterized by the displacement $t$ of point $\hat{H}$. Q and S are the reference and the object sources, respectively, where the latter illuminates the object surface in its two configurations {P} and {P'} at recording. The small vector $u$ describes the mechanical displacement between P and P', whereas $v$ and $v'$ denote the finite optical displacements of these points due to the large modification, so that the two images $\tilde{P}$ and $\tilde{P}'$ are separated by an apparent displacement $\tilde{u}$. However, we will not use $\tilde{u}$ at the moment because $\tilde{P}$ and $\tilde{P}'$ are, in fact, astigmatic points. Accordingly, we shall investigate the small vector $\mathring{w}$ on the hologram at reconstruction which is the "collinear image" of $\tilde{u}$ when the fringes are observed at some fixed point $\tilde{K}$. The small vector $\mathring{w}$ separates the two points $\hat{H}$, $\hat{H}'$, whose corresponding points $\hat{H}$, $\hat{H}'$ define another vector $\hat{w}$, which, however, is not a collinear image of the small displacement $u$. As before, we use the same notation for the distances: $q$ denotes the distance from $\hat{H}$ to the reference source Q, $p$ that from $\hat{H}$ to P, $\tilde{L}$ that from $\tilde{P}$ to $\tilde{K}$ and $L_s$ that from P to the source point S. Finally, the unit normals of the hologram in its two positions are $\mathring{n}$, $\hat{n}$ those of the undeformed surface of the body and its image are $n$, $\tilde{n}$.

As for the fringe order, we must consider the optical path difference referring to the modified configuration

$$\tilde{D} = \left(\frac{\tilde{\lambda}}{2\pi}\,\tilde{\phi} - \tilde{L}\right) - \left(\frac{\tilde{\lambda}}{2\pi}\,\tilde{\phi}' - \tilde{L}'\right) \tag{5.1}$$

where $\tilde{\phi}$, $\tilde{\phi}'$ are the phases of the light at reconstruction at the points $\tilde{P}$, $\tilde{P}'$ (virtual extrapolation). Referring to Sect. 3.1.2 regarding image formation, these two phases must satisfy conditions of interference identity of type (3.9) at corresponding points $\hat{H}$ and $\hat{H}$

$$\tilde{\phi} = \phi + \frac{2\pi}{\lambda}\,(p - q) - \frac{2\pi}{\tilde{\lambda}}\,(\tilde{p} - \tilde{q}) + (\tilde{\psi} - \psi)\ ,$$

$$\tilde{\phi}' = \phi' + \frac{2\pi}{\lambda}\,(p' - q') - \frac{2\pi}{\tilde{\lambda}}\,(\tilde{p}' - \tilde{q}') + (\tilde{\psi} - \psi)\ . \tag{5.2}$$

The phases $\phi$, $\phi'$ of P, P' may be expressed by the paths $L_s$, $L'_s$, since the common object source is S: $\phi = 2\pi L_s/\lambda$, $\phi' = 2\pi L'_s/\lambda$. The phases $\psi$, $\tilde{\psi}$ of Q, on the other hand, will cancel in (5.1). Thus, with (5.2), the optical path difference becomes

$$\tilde{D} = (\tilde{q} - \tilde{q}') - [(\tilde{p} + \tilde{L}) - (\tilde{p}' + \tilde{L}')]$$

$$+ \frac{\tilde{\lambda}}{\lambda}\,[(p - p') - (q - q') + (L_s - L'_s)]\ . \tag{5.3}$$

The paths involved are grouped here in such a way that the terms in parentheses represent small quantities, which allow a development with respect to unit direction vectors $\tilde{c}$, $\tilde{k}$, $k$, $c$, and $h$, namely

$$\tilde{D} = -\hat{w} \cdot \tilde{c} + \hat{w} \cdot \tilde{k} + \frac{\tilde{\lambda}}{\lambda}\,(-\hat{w} \cdot k + u \cdot k + \hat{w} \cdot c - u \cdot h)\ ; \tag{5.4}$$

$\hat{w}$ and $\hat{w}$ are corresponding small vectors lying on the hologram in its two respective positions, like two increments $d\hat{r}$ and $d\hat{r}$ defined at points $\hat{H}$ and $\hat{H}$, respectively. Therefore, we may write the linear transformation

$$\hat{w} = \hat{N}\hat{F}\hat{N}\hat{w} = \hat{F}\hat{N}\hat{w} \tag{5.5}$$

conforming to (3.39, 40). Let us recall that $\hat{F} = I - (\hat{\nabla} \otimes t)^{\mathrm{T}}$ is the deformation gradient of the hologram and $\hat{N}\hat{F}\hat{N}$ the mixed projection ($\hat{N} = I - \hat{n} \otimes \hat{n}$, $\hat{N} = I - \hat{n} \otimes \hat{n}$). Hence, (5.4) may be rewritten in the form ($g = k - h$)

$$\tilde{D} = \frac{\tilde{\lambda}}{\lambda}\,u \cdot g + \hat{w} \cdot \left[\hat{N}\hat{F}^{\mathrm{T}}(\tilde{k} - \tilde{c}) - \frac{\tilde{\lambda}}{\lambda}\,\hat{N}(k - c)\right]\ .$$

However, considering the fundamental equation (3.124) of lateral aberration,

$$\hat{N}\left[\frac{1}{\tilde{\lambda}}\,\hat{F}^{\mathrm{T}}(\tilde{k} - \tilde{c}) - \frac{1}{\lambda}\,(k - c)\right] = 0\ , \tag{5.6}$$

we see that the bracket of the second term cancels, so that we simply obtain

$$\tilde{D} = \frac{\tilde{\lambda}}{\lambda} \boldsymbol{u} \cdot \boldsymbol{g} \quad \text{or}$$

$$D = \boldsymbol{u} \cdot \boldsymbol{g}$$

$$(5.7)$$

with $D = \tilde{D}\lambda/\tilde{\lambda}$ referring to the recording (the fringe order corresponds to a multiple of $\lambda$ and $\tilde{\lambda}$). In summary, we conclude, as had been observed, that the fringe order does not change from P to $\tilde{\text{P}}$. However, for determining the sensitivity vector $\boldsymbol{g} = \boldsymbol{k} - \boldsymbol{h}$, we have to calculate the unit direction vector $\boldsymbol{k}$ from (5.6), since the observation direction at reconstruction is $\tilde{\boldsymbol{k}}$ and not $\boldsymbol{k}$. A practical application in the case of a wavelength change has been given in [5.1].

Let us now move one step forward and pass to the less trivial consideration of the derivatives of $D$ which are necessary for the determination of fringe interspace, fringe direction and fringe visibility. First, we calculate the differential of $D$ when the center $\tilde{\text{R}}$ of the observing instrument at reconstruction is the fixed point; this gives

$$dD_{\tilde{R}} = d\boldsymbol{u} \cdot \boldsymbol{g} - d\boldsymbol{h} \cdot \boldsymbol{u} + d\boldsymbol{k}' \cdot \boldsymbol{Ku} \ . \tag{5.8}$$

The prime of $d\boldsymbol{k}'$ indicates here that the neighboring rays of directions $\boldsymbol{k}$ and $\tilde{\boldsymbol{k}}$ are skew, contrary to those of direction $\boldsymbol{h}$ and $\tilde{\boldsymbol{h}}$ which are collinear at S. The first two terms in the bracket can be evaluated directly on the object surface (at recording) as in standard holographic interferometry, since

$$\nabla \otimes \boldsymbol{h} = \frac{1}{L_s} (\boldsymbol{I} - \boldsymbol{h} \otimes \boldsymbol{h}) \ ,$$

$$d\boldsymbol{u} \cdot \boldsymbol{g} - d\boldsymbol{h} \cdot \boldsymbol{u} = d\boldsymbol{r} \cdot \left[ (\nabla_n \otimes \boldsymbol{u})\boldsymbol{g} - \frac{1}{L_s} NH\boldsymbol{u} \right] \ ,$$

$$(5.9)$$

with the semi-projection of the displacement "gradient" $\nabla \otimes \boldsymbol{u}$ and the two projectors $N = \boldsymbol{I} - \boldsymbol{n} \otimes \boldsymbol{n}, H = \boldsymbol{I} - \boldsymbol{h} \otimes \boldsymbol{h}$. By using the abbreviation ($w$ should not be confused with $\hat{w}$ or $\check{w}$)

$$w = (\nabla \otimes \boldsymbol{u})\boldsymbol{g} - \frac{1}{L_s} H\boldsymbol{u} \tag{5.10}$$

and the auxiliary relations for the oblique projector $M = \boldsymbol{I} - \boldsymbol{n} \otimes \boldsymbol{k}/\boldsymbol{n} \cdot \boldsymbol{k}$, (5.8) becomes, with $d\boldsymbol{r} \cdot N\boldsymbol{w} = (M^T\boldsymbol{K} \, d\boldsymbol{r}) \cdot N\boldsymbol{w} = (\boldsymbol{K} \, d\boldsymbol{r}) \cdot MN\boldsymbol{w} = (\boldsymbol{K} \, d\boldsymbol{r}) \cdot M\boldsymbol{w}$:

$$dD_{\tilde{R}} = (\boldsymbol{K} \, d\boldsymbol{r}) \cdot M\boldsymbol{w} + d\boldsymbol{k}' \cdot \boldsymbol{Ku} \ . \tag{5.11}$$

The geometry is such as to permit the use of the basic equation of the dual ray-tracing method (3.110) for the vector $d\boldsymbol{k}'$:

$$d\boldsymbol{k}' = lT \, d\boldsymbol{k} \ , \quad \text{where} \tag{5.12}$$

$$T = - \hat{M} \left( \frac{1}{q} C - \hat{B} [\hat{n} \cdot (k - c)] \right.$$

$$- \frac{\lambda}{\tilde{\lambda}} \hat{F}^{T} \left\{ \frac{1}{\tilde{l}} \tilde{K} + \frac{1}{\tilde{q}} \tilde{C} - \hat{B} [\hat{n} \cdot (\tilde{k} - \tilde{c})] \right\} \hat{F} \right) \hat{M}^{T}$$

is the tensor describing the dual astigmatism of the origin R of $\tilde{R}$. Transforming the increment $dk$ by the sequence of relations which represent the passage from recording to reconstruction by means of an increment $d\hat{r}$ on the hologram, see (3.126),

$$l \hat{M}^{T} dk = - d\hat{r} = - \hat{F} \ d\tilde{r} = \tilde{l} \hat{F} \hat{M}^{T} d\tilde{k} \ ; \tag{5.13}$$

where $\hat{F} = \hat{F}^{-1}$ and where $\hat{M} = I - \hat{n} \otimes \tilde{k} / \hat{n} \cdot \tilde{k}$ is the dual of $\hat{M}$, (5.11) now becomes

$$dD_{\tilde{R}} = (K \ dr) \cdot Mw + \tilde{l} \ d\tilde{k} \cdot \hat{M} \hat{F}^{T} TKu \ . \tag{5.14}$$

The last step consists in using the relation of apparent image deformation (3.128) which may be written with (3.130)

$$K \ dr = \frac{p\tilde{l}}{\tilde{l} + \tilde{p}} \ \left( T + \frac{1}{p} K \right) \hat{F} \hat{M}^{T} \tilde{K} \ d\tilde{r}$$

$$= \frac{\lambda p\tilde{l}}{\tilde{\lambda} (\tilde{l} + \tilde{p})} \hat{M} \hat{F}^{T} \left( \tilde{T} + \frac{1}{\tilde{l}} \tilde{K} \right) \tilde{K} \ d\tilde{r} \ . \tag{5.15}$$

We use the auxiliary relation of collineation similar to (3.116) for the special case, or conform to Fig. 3.15 for the general case, so that

$$\frac{1}{\tilde{l} + \tilde{p}} \tilde{K} \ d\tilde{r} = - d\tilde{k} = \tilde{m} \ d\tilde{\beta} \ , \tag{5.16}$$

for an angular increment $d\tilde{\beta} = d\tilde{\phi}$ at $\tilde{R}$ and an apparent unit direction vector $\tilde{m}$ normal to $\tilde{k}$. We then insert (5.15) into (5.14) and finally obtain the derivative of $D$ for the fringe interspace and direction

$$\frac{dD_{\tilde{R}}}{d\tilde{\phi}} = \tilde{l} \tilde{m} \cdot \hat{M} \hat{F}^{T} \left[ p \left( T + \frac{1}{p} K \right) Mw - TKu \right] \ . \tag{5.17}$$

Thus, in this case there appears to be a generalization

$$f_{R} = \tilde{l} \hat{M} \hat{F}^{T} \left[ p \left( T + \frac{1}{p} K \right) Mw - TKu \right]$$

of the fringe vector given by *Stetson* in [5.2].

### 5.1.2 Two Close Wavelengths at the Recording. Holographic Contouring

Instead of changing the wavelength of the reference source at the reconstruction ($\tilde{\lambda} \neq \lambda$), one may alternatively use two different wavelengths $\lambda, \lambda' \neq \lambda$ for the two exposures at the recording. When the object is not deformed and when a single wavelength is used for the reconstruction, the interference fringes ($\tilde{\lambda} = \lambda$) give some information about the contour lines of the object for certain holographic arrangements [5.3–8]. Other means of producing contour lines also exist. For example, such a phenomenon may be produced by moving the position of the illuminating source S between the two recording exposures [5.4, 6, 9–15], or by successively immersing the considered object in glass tanks, each containing a medium with different index of refraction [5.6, 16]. Other possibilities are created by phase conjugation [5.17], by limited depth of focus [5.18], or by picosecond pulse holography [5.19, 20]. Finally, contour lines may be used as a Moiré pattern [5.11, 21]. Here, however, we shall consider the holographic contour generation by change of wavelength only.

Looking first at Fig. 5.2, we notice that the fringe order in K is given here by the optical path difference, as defined in (5.1),

$$D = \frac{\lambda}{2\pi} (\tilde{\phi} - \tilde{\phi}') - (L - \tilde{L}') , \tag{5.18}$$

where, with the phases $\psi, \psi'$ at point Q, we have

$$\tilde{\phi} = \phi + \tilde{\psi} - \psi , $$
$$\tilde{\phi}' = \phi' + \frac{2\pi}{\lambda'} (p' - q') - \frac{2\pi}{\lambda} (\tilde{p}' - q') + \tilde{\psi} - \psi' , \tag{5.19}$$

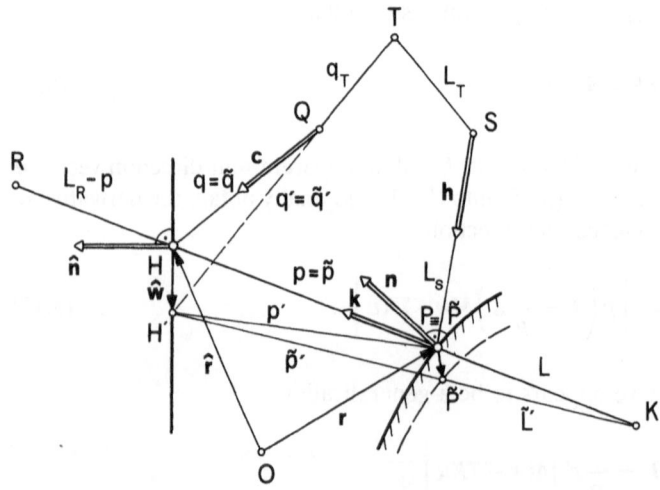

**Fig. 5.2.** Recording with two close wavelengths $\lambda, \lambda'$. Reconstruction with one wavelength $\tilde{\lambda} = \lambda$. (T: bifurcation point at the beam splitter; $\tilde{P} = P$; $\tilde{P}'$: image points; R: observation center)

and, with the phases $\chi$, $\chi'$ at point T,

$$\psi = \frac{2\pi}{\lambda} q_T + \chi , \qquad \psi' = \frac{2\pi}{\lambda'} q_T + \chi' ,$$

$$\phi = \frac{2\pi}{\lambda} (L_s + L_T) + \chi , \qquad \phi' = \frac{2\pi}{\lambda'} (L_s + L_T) + \chi' .$$

With the distances $q_T$ and $L_T$, we take into account the paths from the bifurcation point T of the laser beam to the pinhole sources Q and S. With the relations (5.19), Eq. (5.18) becomes

$$D = \left(1 - \frac{\lambda}{\lambda'}\right)(L_s + L_T - q_T)$$

$$- (p' - q') \frac{\lambda}{\lambda'} - [(p + L) - (\tilde{p}' + \tilde{L}')] + p - q' , \tag{5.20}$$

where again we may notice, as in Sect. 5.1.1, that the modified image $\tilde{P}'$ of P does not influence the optical path difference $D$, since the distances $(p + L)$ and $(\tilde{p}' + \tilde{L}')$ intervene. Furthermore, with $\lambda/\lambda' \simeq 1 - \Delta\lambda/\lambda$ and with

$$- [(p + L) - (\tilde{p}' + \tilde{L}')] \simeq \hat{w} \cdot k , \quad -(p' - p) \simeq - \hat{w} \cdot k ,$$

we obtain the basic formula [Ref. 5.22, Eq. (7.2)],

$$D \simeq \frac{\Delta\lambda}{\lambda} (L_s + p - q) + \frac{\Delta\lambda}{\lambda} (L_T - q_T) . \tag{5.21}$$

In this expression of $D$, only the first term $\Delta\lambda (L_s + p - q)/\lambda$ is important, since it varies with the two position vectors $\hat{r}$, $r$ of points H, P. Therefore, when keeping the observer's point R fixed, with (2.23 and 50) applied to the reference ray PHR, the derivative of $D$ becomes

$$\nabla_n D_R = \frac{\Delta\lambda}{\lambda} \{\nabla_n L_s + \nabla_n[L_R - (L_R - p)] - \nabla_n q\}$$

$$= \frac{\Delta\lambda}{\lambda} \left(Nh - Nk + \frac{L_R - p}{L_R} N\hat{M}k - \frac{L_R - p}{L_R} N\hat{M}c\right)$$

$$= \frac{\Delta\lambda}{\lambda} N\left[M(h - k) + \frac{L_R - p}{L_R} \hat{M}(k - c)\right] . \tag{5.22}$$

In the special case in which the observer is far away from the optical set-up ($L_R \to \infty$), with $M\hat{M} = \hat{M}$, this equation reduces to

$$M\nabla_n D_R = \frac{\Delta\lambda}{\lambda} [M(h - k) + \hat{M}(k - c)] . \tag{5.23}$$

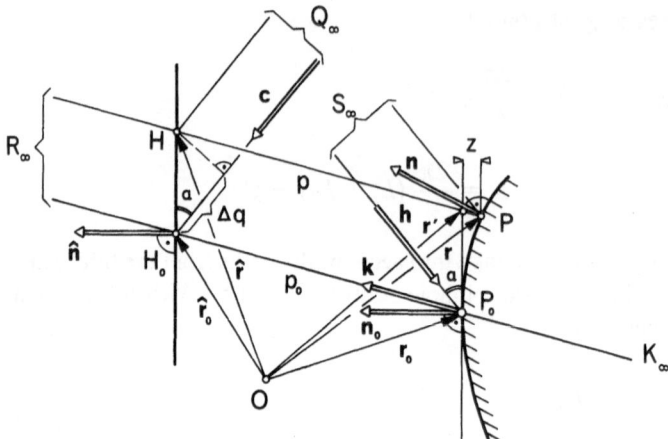

**Fig. 5.3.** Particular arrangement of Fig. 5.2, such that the fringes represent contour lines where $z = $ const on the object surface. ($P_0$: singular point; Observer $R_\infty$, reference source $Q_\infty$, and object source $S_\infty$ are all placed far away from the points considered)

Moreover, let us assume that the sources Q, S are also far away (collimated light) and that the unit direction vectors $h$ and $-c$ have the same constant inclination $\alpha$ with respect to the hologram, so that $h - c$ is parallel to $\hat{n}$ (Fig. 5.3). Accordingly, at point $P_0$ on the object surface, where the tangential plane is parallel to the hologram ($n_0 = \hat{n}$), because $M = \hat{M}$, $M(h - c) = 0$, expression (5.23) gives

$$M\nabla_n D_R(r_0) = 0 \ . \tag{5.24}$$

The vanishing gradient of $D$ shows that $P_0$ corresponds to a singular point in the fringe field. For any other point P situated at the distance $z$ from the tangential reference plane, we have

$$\Delta q = \lim_{q \to \infty} (q - q_0) = (\hat{r} - \hat{r}_0) \cdot c = (r' - r_0) \cdot c \ ,$$

$$\Delta L_s = \lim_{L_s \to \infty} (L_s - L_{s0}) = (r - r_0) \cdot h = (r' - r_0) \cdot h - z \frac{k \cdot h}{n_0 \cdot k} \ ,$$

$$\Delta p = p - p_0 = \frac{z}{n_0 \cdot k} \ ,$$

where $r'$ is the position vector of the oblique projection P' of P onto the reference plane. Since $(r' - r_0) \cdot (h - c) = 0$ and with these auxiliary relations, from (5.21) we obtain

$$\Delta D = D - D_0 = \frac{\Delta \lambda (1 - k \cdot h)}{\lambda n_0 \cdot k} z \ , \tag{5.25}$$

from which we conclude that the fringes represent *contour lines* on the object surface.

If a finite aperture is used, we often encounter contrast problems with the fringes [5.22]. Replacing $L_R$ by $-L$ in (5.22), we may write the visibility vector as

$$f_K = LM\nabla_n D_K = \frac{\Delta\lambda}{\lambda}\{L[M(h-k) + \hat{M}(k-c)] + p\hat{M}(k-c)\}, \qquad (5.26)$$

which shows that, near the ray passing through the singularity ($M \cong \hat{M}$), the fringes are almost completely localized at infinity (at the singularity point, $\nabla_n D_K = 0$ involves $L \to \infty$, but we cannot speak of a fringe at this particular point). However, far away from $P_0$, the unit normal $n$ at P generally differs much from $n_0$ so that $f_K \neq 0$. Here, conforming to (4.105), for which there also exists a modified version (compare, e.g., with [Ref. 5.23, Eq. (4.156 or 126)]), we can find a finite distance of partial localization:

$$L_p = \frac{p[M(h-k) + \hat{M}(k-c)] \cdot \hat{M}(k-c)}{|M(h-k) + \hat{M}(k-c)|^2}. \qquad (5.27)$$

At any rate, if the observed part of the object surface is strongly curved (i.e., with large variation of $z$), the fringes are not localized on a plane such that a photograph of them would show a variable contrast. In such a case, a very small aperture would be necessary. At the same time, $\Delta\lambda$ must be sufficiently small to avoid the appearance of too many close fringes. By contrast, a slightly curved object surface needs a relatively large $\Delta\lambda$ in order to create a sufficient number of fringes. On the other hand, this wavelength variation has to be considered a limited quantity by reason of visibility, as treated in the preceeding Chapter. According to Sect. 4.2.2, we have visible fringes for $|\mathring{v}| < 1.22\pi$ (first zero of the first-order Bessel function). The vector $\mathring{v}$ is defined by (4.90), which here becomes

$$\mathring{v} = \frac{2\pi\mathring{r}}{\lambda(\mathring{L} + L)}\mathring{N}f_K. \qquad (5.28)$$

For simplicity, we take the aperture plane perpendicular to the viewing direction. Let us recall that $\mathring{L}$ is the distance between the aperture plane and the object point considered. Since we want, in fact, a picture with the contour lines on the object, the camera will be focused on it and L will be zero. When using (5.26), Eq. (5.28) now reduces to

$$\mathring{v} = \frac{2\pi\mathring{r}p\Delta\lambda}{\lambda^2\mathring{L}}\hat{M}(k-c). \qquad (5.29)$$

Thus, it is possible to fix a limit for the fringe visibility

$$\left|\frac{\Delta\lambda}{\lambda}\right|_{max} = \frac{1.22\lambda\mathring{L}}{2\mathring{r}p|\hat{M}(k-c)|}. \qquad (5.30)$$

The last equation may be used to estimate the largest possible wavelength change for still visible fringes. Finally from (5.25) we can estimate the minimal "height" $z$ described by two neighboring fringes corresponding to the value $|\Delta\lambda/\lambda|_{max}$.

Since a non-negligible speckle effect may occur when greatly reducing the viewing angle $\mathring{r}/\mathring{L}$, the only way of increasing the limit $|\Delta\lambda/\lambda|_{max}$ is to place the object closer to the hologram [5.7, 22]. There is yet another way to produce an increase in fringe contrast. The reference beam can be tilted before the second exposure at the reconstruction, so as to have two reference sources Q, Q'. At the reconstruction, the source will be placed at Q only. The optical path difference then contains an additional term, similar to that which will appear in the next paragraph dealing with the real-time technique. Equation (5.25) will then not only depend on $z$, but also on the relative position vector $r' - r_0$. It must be noted that, in this case, fringes are not exact contour lines [5.4, 5].

### 5.1.3 Small Modifications in the Real-Time Technique

While referring to Fig. 5.4, let us consider some basic principles of the real-time technique, as applied to various deformation problems [5.24–36]. First, the unde-formed object (i.e., the set of points {P} on its surface), is recorded by the object source S and the reference source placed at point Q. Then, after developing and repositioning the photographic plate, the deformed object, i.e., the points {P'}, are again illuminated from S (Fig. 5.4); however, the hologram now receives light from the reference source in its new position $\tilde{Q}$. Instead of the source shift $d$ from Q to $\tilde{Q}$, one could also have produced a hologram displacement, either intentionally or not. One wave field originates from the deformed object surface (the set {P'}) the other by diffraction from the hologram. The interference of ray $\tilde{P}H$ with, for instance, the ray through P' takes place at some point $\tilde{K}$ where a fringe appears. If $\tilde{P}\tilde{K}$ is the axis of the observation system, its center $\tilde{R}$ (e.g., the observer's eye) is placed somewhere on the other side of the hologram.

As far as the formation of the image $\tilde{P}$ is concerned, the aberration $v$ may be split into a lateral part, $Kv$, and a longitudinal part, $k \cdot v$, the latter introducing an astigmatic interval $\langle \tilde{P}_1, \tilde{P}_2 \rangle$ with P being a fixed point. Of course, we assume here that the "optical displacement" $v$ is of the same order of magnitude as the "mechanical displacement" $u$. In fact, it is so small that, practically speaking, the aberration effects are invisible. Nevertheless, it is convenient to utilize all of the aberration concepts encountered in the problem of image formation, for reasons which will be evident later. When we look at the whole area around $\tilde{P}$ from the fixed point $\tilde{R}$, we may introduce the dual interval $\langle R_1 R_2 \rangle$ and the dual aberration $v_R$ decomposed into two parts $Kv_R$ and $k \cdot v_R$.

Let us now calculate the difference $D$ between the optical paths covered by the two rays coming from the corresponding points $\tilde{P}$ and P', which are at the distances $L$ and $L'$ from their point of intersection $\tilde{K}$. While the phase $\phi'_s = 2\pi L'_s/\lambda$, of the beam emitted by P', may be expressed simply by the distance $L'_s$ of this point to the object source S, the phase corresponding to point $\tilde{P}$ must be

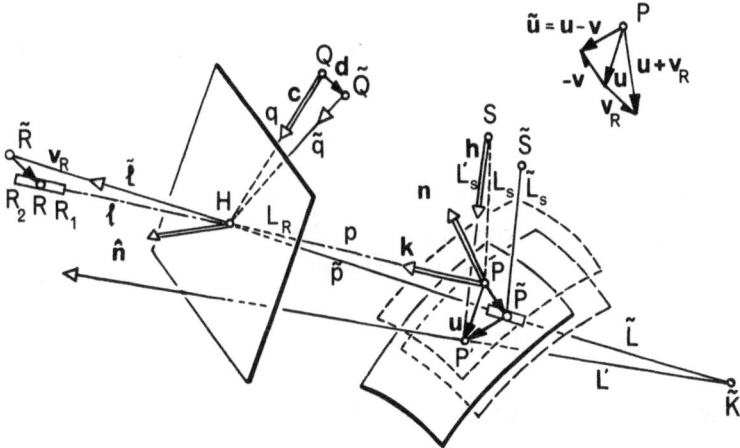

**Fig. 5.4.** Real-time holography. P: point on the undeformed object surface at the recording (single exposure). $\tilde{P}$: image of P at the reconstruction. (P': point of the illuminated deformed object surface during the reconstruction of $\tilde{P}$; $u$: mechanical displacement of P; $v$: optical displacement of P; $\tilde{u} = u - v$: apparent displacement of P; $v_R$: dual aberration of $\tilde{R}$; $u + v_R$: relative displacement of P)

calculated from the condition of interference identity at H, which is the point on the hologram where the ray $\tilde{P}\tilde{K}$ is formed at the reconstruction. This phase reads

$$\tilde{\phi}_s = \frac{2\pi}{\lambda} [L_s + (p - q) - (\tilde{p} - \tilde{q})] + (\psi - \tilde{\psi}) = \frac{2\pi}{\lambda} \tilde{L}_s , \qquad (5.31)$$

where the phases $\psi$ and $\tilde{\psi}$ would be those of Q and $\tilde{Q}$, respectively. For simplicity, we may assume that they are equal. Equation (5.31) also discloses a fictitious path $\tilde{L}_s$ to some point $\tilde{S}$ (Fig. 5.4). The path difference is then

$$D = (\tilde{L}_s - \tilde{L}) - (L'_s - L')$$
$$= (L_s - L'_s) - (\tilde{L} - L') + (p - q) - (\tilde{p} - \tilde{q}) . \qquad (5.32)$$

The linearization of $D$, in terms of displacement vectors $u$, $\tilde{u}$, $d$ and unit vectors $h$, $k$, $c$, reads ($\tilde{u} = u - v$)

$$D = -u \cdot h + \tilde{u} \cdot k + v \cdot k - d \cdot c = u \cdot g - d \cdot c . \qquad (5.33)$$

This separation of the optical path difference $D$ into the standard term $u \cdot g$, encountered in holographic interferometry, and the fringe control term $d \cdot c$ was, for instance, the starting point of investigations of the modification problem in [5.23], when the derivative of $D$ had to be calculated. This illustrates that the fringe problem may be treated independently from the image problem. We shall return to this topic later on.

However, in order to increase physical insight as well as to facilitate the determination of higher derivatives, we rewrite (5.32) by replacing $p$ and $\tilde{p}$ by $(L_R - l)$ and $(\tilde{L}_R - \tilde{l})$, respectively:

$$D = (L_s - L_s') + (L' + L_R) - (\tilde{L} + \tilde{L}_R) - (l + q) + (\tilde{l} + \tilde{q}). \qquad (5.34)$$

Instead of the phase difference $(2\pi/\tilde{\lambda})[(p - q) - (\tilde{p} - \tilde{q})] = -\theta_{\tilde{P}}$ relative to $\tilde{P}$, the phase difference function of R, without the factor $2\pi/\lambda$, has been made apparent in (5.34). The function has previously been defined as

$$\theta_R = \frac{2\pi}{\lambda}[(l + q) - (\tilde{l} + \tilde{q})] \ .$$

Accordingly, instead of (5.33), the linearization now gives separation in the form

$$D = u \cdot g + v_R \cdot k - \frac{\lambda}{2\pi} \theta_R \ . \qquad (5.35)$$

Let us then determine the fringe interspace and the fringe direction, i.e., the derivation of $D$ relative to the observation point $\tilde{R}$, which may here be approximately replaced by R

$$\frac{dD_R}{d\phi} = \frac{dr}{d\phi} \cdot \nabla_n D_R = -L_R \frac{dk}{d\phi} \cdot M\nabla_n D_R = L_R m \cdot M\nabla_n D_R \ . \qquad (5.36)$$

Referring to (5.35), we now may, on the one hand, use the condition of stationary behavior, $d\theta_R = 0$, for the origin R, see (3.14 and 108). On the other hand, we note that $v_R$ is a constant vector when a derivative $m \cdot M\nabla_n$ on the object surface is calculated. Therefore, for the derivative of $D$, we obtain with (5.10), e.g. $w = (\nabla \otimes u)g - Hu/L_s$,

$$\frac{dD_R}{d\phi} = m \cdot [L_R M\nabla_n(u \cdot g) - Kv_R] = m \cdot [L_R Mw - K(u + v_R)] \ ; \qquad (5.37)$$

that is to say, when a modification of the optical arrangement is performed, one must simply add the lateral "aberration" of $\tilde{R}$ to the term of the mechanical deformation encountered when discussing the representative case for standard holographic interferometry. Without referring to the image problem, if we replace $Kv_R$ by using (3.175), we alternatively get:

$$\frac{dD_R}{d\phi} = m \cdot \left( L_R Mw - Ku - \frac{l}{q}\hat{M}Cd \right) \ . \qquad (5.38)$$

This equation may also be deduced directly from (5.33), but we then need an affine connection between the object and hologram surfaces in order to calculate the derivative of $c$ [Ref. 5.23, p. 135]. Finally, in terms of the "optical displacement" $v$ of P, since $l = L_R - p$, we can write:

$$\frac{dD_R}{d\phi} = m \cdot \left[ L_R \left( Mw - \frac{1}{p} Kv \right) - K(u - v) \right]$$

$$= m \cdot [L_R M\tilde{w} - K\tilde{u}] . \tag{5.39}$$

This last expression is especially useful when localization problems are investigated and when point $\tilde{K}$ represents another collineation center besides $\tilde{R}$. The distance $+L_R$ has only to be exchanged by the localization distance $-L$ of the fringes, whereas the two vectors $\tilde{w}$, $\tilde{u}$ remain unchanged and replace $w$, $u$ [Ref. 5.23, p. 149]. With these modifications the visibility vector reads

$$\tilde{f}_K = L \left( Mw - \frac{1}{p} Kv \right) + K(u - v) = LM\tilde{w} + K\tilde{u} . \tag{5.40}$$

Disregarding the localization problem for the moment, we want to consider the higher derivatives of $D$. To that end, it is preferable to refer to (5.37). Let us develop $D$ along a curve on the unit sphere where the vector $k$ varies

$$\overline{D} = D + dk \cdot \nabla_k^* D_R + \tfrac{1}{2} [dk \cdot (\nabla_k^* \otimes \nabla_k^* D_R) dk + (d\phi)^2 b_k \cdot \nabla_k^* D_R] . \tag{5.41}$$

The vector $\nabla_k^* = -L_R M \nabla_n$ is the two-dimensional operator of differentiation on the unit sphere around $\tilde{R}$ and $b_k$ denotes the vector of geodesic curvature of the curve and/or the curvature of its projection on a screen perpendicular to $k$. If this curve represents a fringe, we now have from (5.41) that

$$m_t \cdot \nabla_k^* D_R = 0 , \tag{5.42}$$

$$m_t \cdot (\nabla_k^* \otimes \nabla_k^* D_R) m_t = -b_k \cdot \nabla_k^* D_R , \tag{5.43}$$

where $dk$ has been replaced by $-m_t d\phi$. The unit vector $m_t$ is situated in the viewing plane and is parallel to the fringe. The fringe vector, including the modification, here reads

$$\tilde{f}_R = -\nabla_k^* D_R = -[\nabla_k^* (u \cdot g) + Kv_R] . \tag{5.44}$$

Equation (5.43) indicates that, for this particular direction $m_t$ given by the fringe, the second derivative of $D_R$ is determined by the curvature and the interspace of the fringe. The second derivative of $D_R$, in the same direction $m_t$, generally reads (with $d^2\theta_R = 0$):

$$m_t \cdot (\nabla_k^* \otimes \nabla_k^* D_R) m_t = m_t \cdot [\nabla_k^* \otimes \nabla_k^* (u \cdot g)] m_t - (k \cdot v_R) . \tag{5.45}$$

It must be noted that the vector $v_R$ remains constant when, for $m = m_t$, the particular derivative is calculated with respect to the fixed point $\tilde{R}$. We have formally used $\nabla^* \otimes (Kv_R) = -Kv_R$, although $v_R$ would vary with $m$.

In the particular case in which the optical modification upon reconstruction (here the shift $d$) is performed *without any mechanical deformation* of the object

($u \equiv 0$), Eqs. (5.44, 45) become simply, see (5.38), (3.186 and 185)

$$\nabla_k^* D_R = K\boldsymbol{v}_R = \frac{l}{p} K\boldsymbol{v} = \frac{l}{q} \hat{M}C\boldsymbol{d} , \tag{5.46}$$

$$\boldsymbol{m}_t \cdot (\nabla_k^* \otimes \nabla_k^* D_R)\boldsymbol{m}_t = -\boldsymbol{k} \cdot \boldsymbol{v}_R = \boldsymbol{m}_t \cdot U\boldsymbol{m}_t = \frac{l^2}{q^2} (\boldsymbol{m}_t \cdot \hat{M}C\hat{M}^T\boldsymbol{m}_t)(\boldsymbol{c} \cdot \boldsymbol{d}) . \tag{5.47}$$

Considering (5.40), we note that $\tilde{f}_K$ becomes zero for $L = -p$, implying a complete localization of the fringes on the hologram plane. Equation (5.46) indicates that the fringes are perpendicular to the lateral aberration $K\boldsymbol{v}_R$ of $\tilde{R}$ or $K\boldsymbol{v}$ of P. Moreover, (5.47) reveals that the second derivative of $D_R$ is *directly related* to the longitudinal displacement $\boldsymbol{k} \cdot \boldsymbol{v}_R$ of R, to the *astigmatism tensor $U$* at R, and (since $l\hat{M}C\boldsymbol{d}/q$ is perpendicular to the fringe) to the longitudinal part $\boldsymbol{c} \cdot \boldsymbol{d}$ of the vector $\boldsymbol{d}$ describing the displacement of point source Q. Thus, as the fringe vector describes the fringe direction and fringe interspace, this interspace is inversely proportional to the *lateral displacement* of the reference source Q

$$|\Delta\phi| = \frac{\lambda}{\left| \dfrac{l}{q} \hat{M}C\boldsymbol{d} \right|} . \tag{5.48}$$

From (5.43, 47), we conclude that the curvature of the fringe is proportional to the *longitudinal displacement* of the reference source

$$\frac{1}{|\varrho|} = \frac{1}{|l\varrho_k|} = \left| \frac{\boldsymbol{c} \cdot \boldsymbol{d}}{q|\hat{M}C\boldsymbol{d}|} (\boldsymbol{m}_t \cdot \hat{M}C\hat{M}^T\boldsymbol{m}_t) \right| , \tag{5.49}$$

where $\varrho_k$ is the curvature radius of the projected fringe on the unit sphere around $\tilde{R}$.

Equations (5.48, 49) can be used in the reverse sense. Since only the relative position of the reference source with respect to the hologram is of importance, fringes produced by a repositioning error of the hologram may be compensated for in the following way [5.37]. In a first step, fringes are usually spaced by a lateral displacement of the source relative to direction $\boldsymbol{c}$. Secondly, they are straightened out by a longitudinal displacement of the source along direction $\boldsymbol{c}$.


## 5.2 Modifications with Two Reference Sources

A fringe pattern produced by double exposure in standard holographic interferometry generally does not offer sufficient flexibility for convenient analysis as needed in solving practical problems. In particular, as far as their number and contrast are concerned, inconvenient fringes may appear. Since these fringes are "frozen" for a given observation direction, no compensation is possible that would allow evaluation of fractions of their order. If we want to influence the two fields independently, as mentioned earlier, we need two separate sources or two separate holograms. The practicality of such procedures comes last, but not least, from the possibility that either the sources or the holograms may be placed very close to one another in order to get an amplification effect due to the modification. The analysis consists primarily of asymptotic considerations connecting quantities of different orders of magnitude.

### 5.2.1 General Equations. Wavelength Changes with Two Reference Sources

One of the most powerful techniques in holographic interferometry is the heterodyne method already mentioned. It was introduced by *Dändliker* et al. in 1973 [5.38] and turns out to be also of importance in other fields of optics [5.39–44]. This method has been described in various publications [5.38, 45–50]. Since the basic idea consists of using two reference sources Q, Q' with very small wavelength changes at the reconstruction, $\lambda \to \tilde{\lambda} = \lambda - \Delta\lambda$, $\lambda' \to \tilde{\lambda}' = \lambda + \Delta\lambda$, this method should be classified in the field of optical modifications, though care has been taken in many applications, so that the modification terms are negligible. In practice, these wavelength changes are produced by acousto-optical modulators, resulting in a modification factor, $\Delta\lambda/\lambda$, which is characterized by an order of magnitude varying between $10^{-7}$ and $10^{-12}$.

However, we should like to discuss here a slightly more generalized situation, including a repositioning error of the hologram upon reconstruction. This error is usually thought to be created by a rigid body motion of the hologram. Such a modification could also occur, to a greater extent, from shrinkage or from bending of the hologram (Fig. 5.5). We could also consider position changes of the reference sources since, in addition to the change in wavelength, the acousto-optical modulators may also change the direction of the reference beam, thereby creating a very small, though usually negligible, modification [5.45]. The influence of nonlinear recording effects [5.51, 52] are omitted here.

It is perhaps judicious to first explain the recording and reconstruction processes in detail, as was done in the beginning of Sect. 3.1 for the standard case. We now assume two different reference waves with complex amplitudes $V$ and $V'$, corresponding to the two successive exposures. At any hologram point H ($\hat{H}$ or $\hat{H}'$), we have the intensities

$$I = \tfrac{1}{2}(U + V)(U + V)^* = \tfrac{1}{2}(|U|^2 + |V|^2 + UV^* + U^*V)$$
$$I' = \tfrac{1}{2}(U' + V')(U' + V')^* = \tfrac{1}{2}(|U'|^2 + |V'|^2 + U'V'^* + U'^*V') \ ,$$

(5.50)

where $U$, $U'$ are the amplitudes of the waves coming from points P, P' respectively. Therefore, after development, the transmittance of the hologram at an arbitrary point with twice the same exposure times $\tau/2$ reads

$$T = T_0 - \frac{\beta\tau}{4}(UV^* + U'V'^* + U^*V + U'^*V') \ .$$

(5.51)

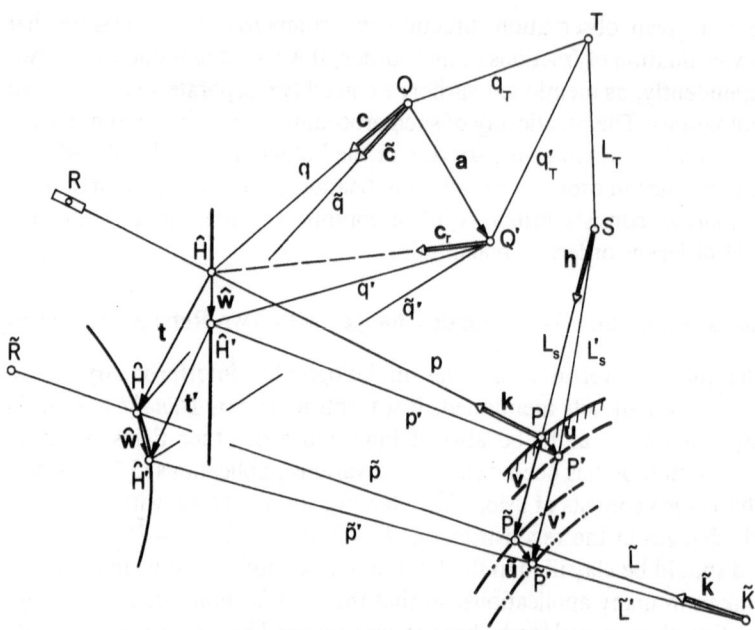

**Fig. 5.5.** Arrangement of the set-up for the heterodyne method (T: bifurcation point of the laser beam where an acousto-optical modulator is placed. Q, Q': reference sources, where spherical waves with wavelength $\lambda - \Delta\lambda$, $\lambda + \Delta\lambda$ are emitted upon reconstruction)

At the reconstruction, the hologram is simultaneously illuminated by the two waves

$$\tilde{v} = \tilde{V} \exp[-2\pi i(\nu + \Delta\nu)t] \ ,$$

$$\tilde{v}' = \tilde{V}' \exp[-2\pi i(\nu - \Delta\nu)t] \ , \tag{5.52}$$

with frequencies $\nu + \Delta\nu$, $\nu - \Delta\nu$, corresponding to $\tilde{\lambda} = \lambda - \Delta\lambda$, $\tilde{\lambda}' = \lambda + \Delta\lambda$, respectively. Just behind the hologram, we then have the transmitted wave

$$(\tilde{v} + \tilde{v}')T = (\tilde{v} + \tilde{v}')T_0$$

$$-\frac{\beta\tau}{4} \left[ U'V'^*\tilde{v} + (U'V'^*\tilde{v}' + UV^*\tilde{v}) + UV^*\tilde{v}' \right]$$

$$-\frac{\beta\tau}{4} \left[ U'^*V'\tilde{v} + (U'^*V'\tilde{v}' + U^*V\tilde{v}) + U^*V\tilde{v}' \right] \ . \tag{5.53}$$

The last line of (5.53) represents conjugate-image wave fields, whereas the primary-image wave fields are contained in the second line. If we assume, for the

**Fig. 5.6.** Images reconstructed in two-reference-beam holographic interferometry. The images on the left and right are cross-reconstructions. The center image shows interference fringes produced by simultaneous reconstruction of the undeformed and deformed configurations of the object

moment, no wavelength change ($\Delta v = 0$), and no deformation of the hologram, $\tilde{V} = V$, and $\tilde{V}' = V'$, the second line reduces to

$$-\frac{\beta\tau}{4} [U'V'^*V + (U'|V'|^2 + U|V|^2) + UV^*V'] \exp(-2\pi i vt) \ .$$

The middle term of the bracket $U'|V'|^2 + U|V|^2$ is of special interest. If $|V'| = |V|$, it represents the superposition of $U$ and $U'$, i.e., it reconstructs the undisturbed image of the object with interference fringes on it (middle part of Fig. 5.6). On the other hand, the left term $U'V'^*V$ is the same type as the term $UV^*\tilde{V}$ in (3.6), i.e., it shows a supplementary cross-image, modified as if Q' had been displaced to Q (left part of Fig. 5.6). Similarly, the right term $UV^*V'$ reveals yet another cross-image, modified as if Q had been displaced to Q' (right part of Fig. 5.6) [5.49, 53].

Let us now return to (5.53), in the case where $\Delta v \neq 0$, and, as in previous sections, use the concept of images $\tilde{P}$, $\tilde{P}'$ when reconstructing points P, P' recorded from the two configurations on the object surface, non-deformed and deformed, respectively. If the rays through these points were present *alone*, at some point $\tilde{K}$ (Fig. 5.5), the terms $UV^*\tilde{v}$ and $U'V'^*\tilde{v}'$ in (5.53) would give the wave disturbances

$$\tilde{u}(\varrho, t) = \frac{A_{\tilde{p}}}{\tilde{L}} \exp[i(\tilde{\phi} - \frac{2\pi}{\tilde{\lambda}} \tilde{L} - 2\pi\Delta vt)] \exp(-2\pi i vt) \ ,$$

$$\tilde{u}'(\varrho, t) = \frac{A_{\tilde{p}'}}{\tilde{L}'} \exp[i(\tilde{\phi}' - \frac{2\pi}{\tilde{\lambda}'} \tilde{L}' + 2\pi\Delta vt)] \exp(-2\pi i vt) \ .$$

We may isolate the common time factor $\exp(-2\pi i \nu t)$ and obtain for the resulting intensity at $\tilde{K}$ that

$$J(\varrho, t) = \frac{1}{2}\left\{\frac{|A_{\tilde{P}}|^2}{\tilde{L}^2} + \frac{|A_{\tilde{P}'}|^2}{\tilde{L}'^2} + 2\frac{|A_{\tilde{P}}||A_{\tilde{P}'}|}{\tilde{L}\,\tilde{L}'}\right.$$

$$\left.\times \cos\left[\left(\tilde{\phi} - \frac{2\pi}{\tilde{\lambda}}\tilde{L}\right) - \left(\tilde{\phi}' - \frac{2\pi}{\tilde{\lambda}'}\tilde{L}'\right) - 4\pi\Delta\nu t\right]\right\}. \qquad (5.54)$$

The argument of the cos-function in (5.54) indicates that the fringe order oscillates with the beat frequency $2\Delta\nu$. The phase contains an optical path difference which, this time, is related to the recording wavelength $\lambda$, namely,

$$D = \frac{\lambda}{2\pi}\left[\left(\tilde{\phi} - \frac{2\pi}{\tilde{\lambda}}\tilde{L}\right) - \left(\tilde{\phi}' - \frac{2\pi}{\tilde{\lambda}'}\tilde{L}'\right)\right], \qquad (5.55)$$

where $\tilde{\phi}$ and $\tilde{\phi}'$ are the phases at points $\tilde{P}$ and $\tilde{P}'$, respectively. The condition of interference identity at corresponding points $\hat{H}$ and $\hat{H}$ on the hologram here are

$$\tilde{\phi} = \frac{2\pi}{\lambda}(L_s + L_T) + 2\pi\left[\left(\frac{p}{\lambda} - \frac{\tilde{p}}{\tilde{\lambda}}\right) - \left(\frac{q}{\lambda} - \frac{\tilde{q}}{\tilde{\lambda}}\right) - \left(\frac{1}{\lambda} - \frac{1}{\tilde{\lambda}}\right)q_T\right],$$

$$\qquad (5.56)$$

$$\tilde{\phi}' = \frac{2\pi}{\lambda}(L_s' + L_T) + 2\pi\left[\left(\frac{p'}{\lambda} - \frac{\tilde{p}'}{\tilde{\lambda}'}\right) - \left(\frac{q'}{\lambda} - \frac{\tilde{q}'}{\tilde{\lambda}'}\right) - \left(\frac{1}{\lambda} - \frac{1}{\tilde{\lambda}'}\right)q_T'\right].$$

The distances $L_T$, $q_T$, $q_T'$ are again the paths between the "sources" S, Q, Q' and the common point T, where the laser light is split into three beams. Inserting (5.56) into (5.55), we obtain the general expression for the optical path difference relative to $\lambda$

$$D = (L_s - L_s') + (p - p') - (q - q')$$

$$+ \lambda\left[\frac{\tilde{L}' + \tilde{p}'}{\tilde{\lambda}'} - \frac{\tilde{L} + \tilde{p}}{\tilde{\lambda}} + \frac{\tilde{q}}{\tilde{\lambda}} - \frac{\tilde{q}'}{\tilde{\lambda}'} - \left(\frac{1}{\lambda} - \frac{1}{\tilde{\lambda}}\right)q_T\right.$$

$$\left.+ \left(\frac{1}{\lambda} - \frac{1}{\tilde{\lambda}'}\right)q_T'\right]. \qquad (5.57)$$

Using the first-order approximation for the small wavelength changes

$$\frac{1}{\tilde{\lambda}} \cong \frac{1}{\lambda}\left(1 + \frac{\Delta\lambda}{\lambda}\right), \qquad \frac{1}{\tilde{\lambda}'} \cong \frac{1}{\lambda}\left(1 - \frac{\Delta\lambda}{\lambda}\right), \qquad (5.58)$$

(5.57) becomes

$$D = (L_s - L_s') - [(\tilde{L} + \tilde{p}) - (\tilde{L}' + \tilde{p}')] + (p - p') - (q - q') + (\tilde{q} - \tilde{q}')$$

$$+ \frac{\Delta\lambda}{\lambda} [-(\tilde{L} + \tilde{p} + \tilde{L}' + \tilde{p}') + (\tilde{q} + q_T + \tilde{q}' + q_T')] . \tag{5.59}$$

Moreover, as in Sect. 5.1, we assume the mechanical displacement $u$, as well as the apparent shift $\tilde{K}\tilde{u}$, to be small. Therefore, with correspondingly small vectors $\hat{w}$, $\hat{w}$ on the hologram and unit direction vectors $h$, $k$, $\tilde{k}$, we get

$$D = -u \cdot h + \hat{w} \cdot \tilde{k} + u \cdot k - \hat{w} \cdot k - (q - q') + (\tilde{q} - \tilde{q}')$$

$$+ \frac{\Delta\lambda}{\lambda} [-2(\tilde{L} + \tilde{p}) + (q + q_T + q' + q_T')] . \tag{5.60}$$

Expression (5.60) exposes a group of small shift terms, $u \cdot k - u \cdot h + \hat{w} \cdot \tilde{k} - \hat{w} \cdot k$, and a group of small wavelength change terms, $(\Delta\lambda/\lambda) [-2(\tilde{L} + \tilde{p}) + (q + q_T) + (q' + q_T')]$. If the interfringe spaces are not too narrow, then the remaining term $(q - q') + (\tilde{q} - \tilde{q}')$ should also be small. This condition may be fulfilled in the two following limiting cases, $l_0$ denoting a finite characteristic length of the set-up:

a) When $|t| \ll l_0$, the "deformation" of the hologram is very small, i.e., of the same order of magnitude as $|u|$, whereas the distance $|a|$ between the two reference sources Q, Q' is of finite order [5.49]. With the sensitivity vector $g = k - h$, with unit direction vectors $c$, $c_r$ directed to point $\hat{H}$, and with the approximation $\hat{w} \cdot \tilde{k} \cong \hat{w} \cdot k$, we then obtain for the optical-path difference $D$, that

$$D = u \cdot g + t \cdot (c - c_r)$$

$$+ \frac{\Delta\lambda}{\lambda} [-2(\tilde{L} + \tilde{p}) + (q + q_T) + (q_r + q_T')] . \tag{5.61}$$

b) When $|a| \ll l_0$, the distance between sources Q, Q' is very small [5.54, 55], whereas the deformation of the hologram is now of finite order. With unit vector $c$, $\tilde{c}$ having its origin at Q, (5.60) takes the following linearized form

$$D = u \cdot g + \hat{w} \cdot (\tilde{k} - \tilde{c}) - \hat{w} \cdot (k - c) + a \cdot (\tilde{c} - c)$$

$$+ \frac{2\Delta\lambda}{\lambda} [-(\tilde{L} + \tilde{p}) + (q + q_T)] .$$

We may then apply (5.6) describing the image aberration for $\tilde{\lambda} \sim \lambda$, which cancels the terms in $\hat{w}$ and $\hat{w}$. This equation may now be rewritten as

$$D = u \cdot g + a \cdot (\tilde{c} - c) + \frac{2\Delta\lambda}{\lambda} [-(\tilde{L} + \tilde{p}) + (q + q_T)] . \tag{5.62}$$

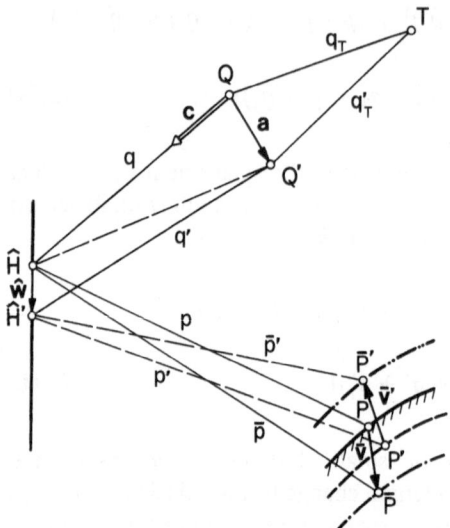

**Fig. 5.7.** Illustration of the cross-images

As far as practical applications are concerned, the wavelength change terms in (5.61, 62) may be kept negligible, since $\Delta\lambda$ is usually very small. On the other hand, it must be pointed out that, in case b, the distance $|a|$ between the two reference sources $Q$, $Q'$ should not be too small [5.54], since the image of the object (complex amplitude $UV^*V$) with the quasi-superposed cross-image (complex amplitude $UV^*V'$, Fig. 5.7) exposes supplementary fringes of interference. Since we have the same $U$ in both triple products above, this must be interpreted as a case of fringe modification without object deformation ($U' \equiv U$), where, as shown previously, the fringes are completely localized on the hologram [Ref. 5.56, p. 1097]. The optical path difference reduces to:

$$D^c = q - q' + q_T - q'_T$$
$$= a \cdot c + q_T - q'_T \ . \tag{5.63}$$

The corresponding fringe vector, see (4.19), with the relation of affine connection (2.50) becomes

$$L_R M \nabla_n D^c_R = L_R M \frac{L_R - p}{L_R} N \hat{M} \nabla_n D^c_R = \frac{L_R - p}{q} \hat{M} C a \ ,$$

so that the apparent angular fringe spacing, see (4.21), may be expressed by

$$|\Delta\phi| = \frac{\lambda q}{(L_R - p)|\hat{M}Ca|} \ .$$

Such cross-fringes should not be seen. This angle must be smaller than the angular resolution limit of the objective, given by the well-known formula $|\Delta\phi|_{\min} = 1,22\,\lambda/2\mathring{r}$; thus, we get

$$|\hat{M}Ca|_{\min} = \frac{1}{0,61} \frac{q\mathring{r}}{(L_R - p)} \tag{5.64}$$

Even more interesting than Case b are situations in which $|a|$ is only moderately small. Since such a choice necessitates a careful comparison of quantities having different orders of magnitude, we shall discuss this problem in the next paragraph where, for reasons that will become evident later, geometrical modifications are performed in lieu of wavelength changes.

## 5.2.2 Moderate Shifts of Moderately Close Reference Sources

Continuing along the same line of thought as in Sect. 5.2.1, we consider the general situation of two reference sources Q, Q'. However, instead of a wavelength change, we now discuss shifts $d, d'$ of these sources at the reconstruction (Fig. 5.8). For simplicity, we keep the hologram fixed. A rigid body motion of the hologram would be equivalent to the particular displacement of the point sources Q, Q' separated by a constant distance $|a|$. A "rigid body motion" of two sources [5.57] is very similar to the so-called sandwich holography of *Abramson* [5.58–72], where, instead of the sources, we move the two holograms, which are rigidly linked together. Consequently, the following equations have the same structure as those encountered in sandwich holography [5.73]. In the literature, the commonly discussed cases are those in which either one of the two reference sources or one of the hologram plates are displaced [5.74–89]. As in the preceeding sections, we must combine the optical path difference,

$$D = \left(\frac{\lambda}{2\pi}\tilde{\phi} - \tilde{L}\right) - \left(\frac{\lambda}{2\pi}\tilde{\phi}' - \tilde{L}'\right), \tag{5.65}$$

with the two conditions of interference identity

$$\tilde{\phi} = \phi + \frac{2\pi}{\lambda}\left[(p-q)-(\tilde{p}-\tilde{q})\right] + \tilde{\psi}-\psi , \qquad \tilde{\phi}' = \ldots . \tag{5.66}$$

If the "motion" of the sources is realized by the motion of a mirror (Fig. 5.8), the phase differences $\tilde{\psi} - \psi$, $\tilde{\psi}' - \psi'$ are zero, so that, with (5.66), Eq. (5.65) becomes

$$D = L_s - L_s' - \left[(\tilde{p}+\tilde{L})-(\tilde{p}'+\tilde{L}')\right] + (p-p')-(q-q') + (\tilde{q}-\tilde{q}') . \tag{5.67}$$

We now let $\varepsilon \ll 1$ be a very small parameter so that $|\tilde{u}|/l_0 = O(\varepsilon)$, with $l_0$ being a characteristic length of the set-up. For instance, if $\mathring{r}\,(\mathring{r} \ll l_0)$ is the radius of the

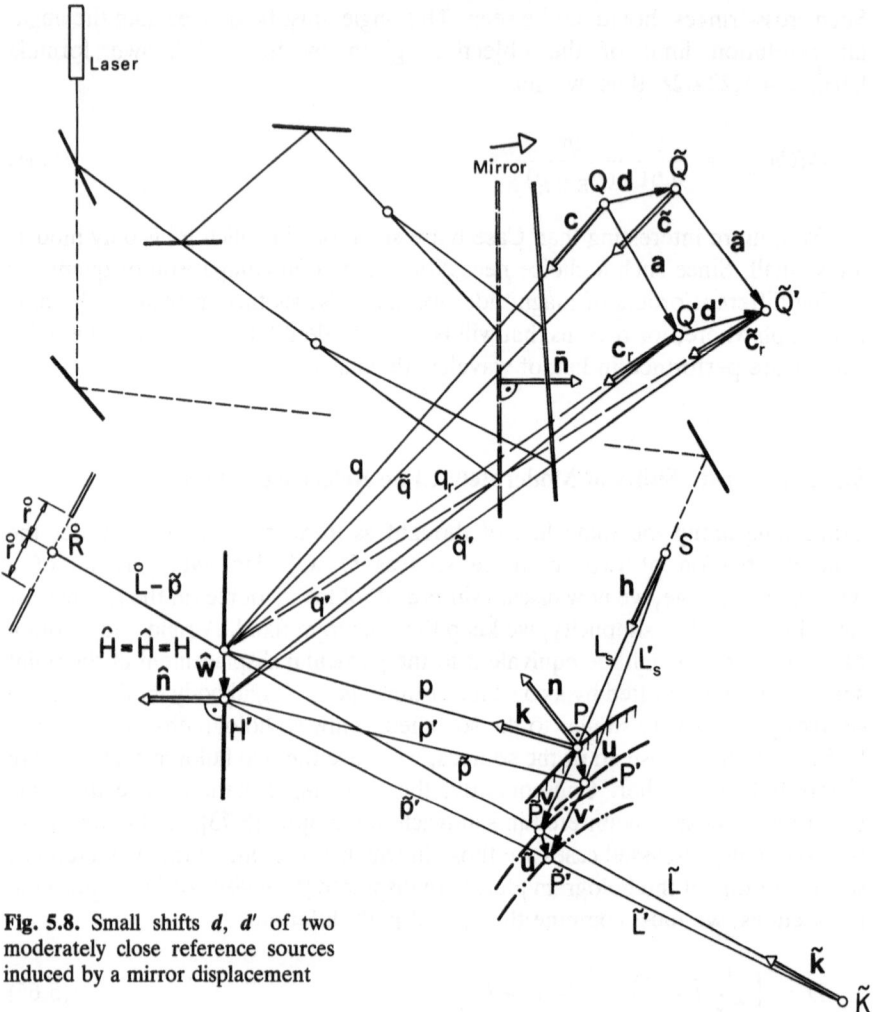

**Fig. 5.8.** Small shifts **d**, **d'** of two moderately close reference sources induced by a mirror displacement

aperture at $\mathring{R}$ and $\mathring{L}$ the distance separating point $\mathring{R}$ from point $\tilde{P}$, we have the needed condition [Ref. 5.23, Eq. (4.49)]

$$|\tilde{K}\tilde{u}| \ll \frac{\tilde{L}\mathring{r}}{\tilde{L} + \mathring{L}} \tag{5.68}$$

for the overlapping of corresponding areas on the images; otherwise, no interference fringes are visible. So, if $\kappa \ll 1$ is some other small constant, we could define $\varepsilon$ by the relation

$$\varepsilon = \kappa \frac{\mathring{r}}{\tilde{L} + \mathring{L}} \, .$$

When compared with the image displacement $\tilde{u}$, the vector $\hat{w}$ on the hologram is, of course, of the same small order. In contrast to this, the mechanical displacement $u$ on the object could be much larger, at least in principle. In such a situation, the purpose of the modification would primarily be to bring points $\tilde{P}$, $\tilde{P}'$, closer in order to let fringes appear. However, we will not continue discussing this interesting application which, instead of speckle metrology, could be useful in investigations of large deformations by holography. From now on, we assume the ratio $|u|/l_0$ to be of order $\varepsilon$. Thus the development of (5.67), with respect to unit direction vectors and small displacement vectors, and with the aberration equation (3.15) gives

$$
\begin{aligned}
D &= u \cdot g + \hat{w} \cdot (\tilde{k} - k) - \hat{w} \cdot (\tilde{c}_r - c_r) - (q - q_r) + (\tilde{q} - \tilde{q}_r) \\
&= u \cdot g - (q - q_r) + (\tilde{q} - \tilde{q}_r) \; ;
\end{aligned}
\tag{5.69}
$$

or also,

$$
D = u \cdot g - (q - \tilde{q}) + (q_r - \tilde{q}_r) \; ,
\tag{5.70}
$$

where the four lengths $q$, $\tilde{q}$, $q_r$, $\tilde{q}_r$ all meet at point H.

A development of the parentheses with respect to moderately small quantities $|a|$ or $|d|$ remains. Analogously to Sect. 5.2.1, we consider two cases. The first is characterized by a relatively small distance separating the two reference sources Q, Q':

a) where $|u| \ll |d| \ll |a| \ll l_0$; the displacement of the sources is small, while their relative distances are only moderately small (Fig. 5.8). Here we start with a refined development of (5.70), which leads to

$$
D = u \cdot g + (d' \cdot c_r - d \cdot c) - \frac{1}{2} \left( \frac{1}{q_r} d' \cdot C_r d' - \frac{1}{q} d \cdot C d \right) .
\tag{5.71}
$$

As far the vector $d'$ is concerned, we assume a rigid body motion of both sources so that $\tilde{a} = Qa$, where $Q$ is an orthogonal tensor. Thus, since $d' = \tilde{a} - a + d$, we may write

$$
d' = (Q - I)a + d = \Psi a + d \; .
\tag{5.72}
$$

If $|d|/l_0 = \delta$ is a second parameter characterizing the smallness of the shift and the rotation $\Psi$, and if $|a|/l_0 = \eta$ is a third parameter for the relative distance, then $|\Psi a|/l_0 = O(\eta \delta)$, $|d'|/l_0 = O(\delta)$; therefore,

$$
\frac{1}{2l_0} \left( \frac{1}{q_r} d' \cdot C_r d' - \frac{1}{q} d \cdot C d \right) = O(\eta \delta^2) \; .
\tag{5.73}
$$

There remains the derivation of the difference $c_r - c$. Since $a$ is only moderately small, we include the second-order terms and, accordingly, the development for $c_r$ reads

$$c_r = c - \frac{1}{q} Ca - \frac{1}{2q^2} aCa \ , \tag{5.74}$$

with the superprojection $C = c \otimes C + C \otimes c + C \otimes c)^T$. Considering (5.72, 74) and with $\Psi^T = - \Psi$, the optical path difference then becomes:

$$D = u \cdot g - a \cdot \left( \Psi c + \frac{1}{q} Cd - \frac{1}{q} \Psi Ca \right)$$

$$- \frac{1}{2q^2} (aCa) \cdot d + O(\eta \delta^2) \ . \tag{5.75}$$

If the linear term in $a$ is comparable to the standard term $u \cdot g$, then $\delta \eta = O(\varepsilon)$. Consequently, the quadratic terms are of order $\varepsilon \eta$ and larger than the neglected terms of order $(\eta \delta^2)$. These terms should not be neglected when compared to the linear terms. The following table summarizes the orders of magnitude of the different linear quantities:

| $O(1)$ | $O(\eta)$ | $O(\delta) = O(\varepsilon/\eta)$ | $O(\varepsilon)$ |
|---|---|---|---|
| $\dfrac{p}{l_0}, \dfrac{q}{l_0}, \dfrac{L}{l_0}, \dfrac{L_s}{l_0}$ | $\dfrac{\|a\|}{l_0}$ | $\dfrac{\|d\|}{l_0}, \dfrac{\|\Psi c\|}{1}, \dfrac{\|Kv\|}{l_0}$ | $\dfrac{\|u\|}{l_0}, \dfrac{\|d' \cdot c_r - d \cdot c\|}{l_0}$ |
| lengths in optical arrangement | distance between reference sources | motion of reference sources and apparent image | mechanical object deformation and relative motions of sources |

Let us now consider the second case:

b) where $\|u\| \ll \|a\| \ll \|d\| \ll l_0$, the distance of sources Q, Q' is kept small. Their displacements, however, are moderately small (Fig. 5.8). In this case, we perform a development of (5.69) as follows

$$D = u \cdot g + (\tilde{a} \cdot \tilde{c} - a \cdot c) - \frac{1}{2} \left( \frac{1}{\tilde{q}} \tilde{a} \cdot \tilde{C}\tilde{a} - \frac{1}{q} a \cdot Ca \right) \ . \tag{5.76}$$

The roles of $a$ and $d$ are now interchanged. Instead of (5.72), we use the relation

$$\tilde{a} = a + \delta \Psi_1 a + \delta^2 \Psi_2 a$$

where, like in (2.135), $\delta \Psi_1$ is only moderately small and of order $\delta \gg \eta$, so that the term $\delta^2 \Psi_2$ cannot be neglected. Accordingly,

$$\frac{1}{2l_0} \left( \frac{1}{\tilde{q}} \tilde{a} \cdot \tilde{C}\tilde{a} - \frac{1}{q} a \cdot Ca \right) = O(\eta^2 \delta) \ , \tag{5.77}$$

and, contrary to (5.74), the development of $\tilde{c}$ reads

$$\tilde{c} = c - \frac{1}{q}Cd - \frac{1}{2q^2}dCd \ . \tag{5.78}$$

The optical path difference then becomes with $\delta\Psi_1 = \Phi_1$ and $\delta^2\Psi_2 = \Phi_2$

$$D = u \cdot g + a \cdot \left[ (\Phi_2 - \Phi_1)c - \frac{1}{q}(Cd - \Phi_1 Cd) \right]$$

$$- \frac{1}{2q^2}(dCd) \cdot a + O(\eta^2\delta) \ , \tag{5.79}$$

where the quadratic term in $d$ is of order $\eta\delta^2 = \varepsilon\delta$, but still larger than the neglected terms of order $\eta^2\delta$. This case must be limited to very small deformations, otherwise a large modification has to be performed in order to compensate term $(u \cdot g)$. The observation direction $\tilde{k}$ will be much different from direction $k$ needed in (5.79). A correction according to (3.15) then becomes necessary.

### 5.2.3 Derivatives of the Path Difference for Two Moderately Close Reference Sources: Applications

Our further considerations are restricted to Case a, presented in the previous paragraph, where $|u| \ll |d| \ll |a| \ll l_0$, with the distance separating the reference sources being only moderately small. However, the conclusions also apply to the sandwich technique which reveals the same structure in the equations. Let us point out again that, as far as applications relative to deformation analysis are concerned, the calculus of the displacement vector $u$ only builds an intermediate step, whereas the main purpose is to develop a refined procedure with which to determine the strain and rotation components with sufficient precision. These quantities are related to the derivative of $D$, i.e., to the fringe vector when point $\tilde{R}$ is fixed and to the visibility parameters when point $\tilde{K}$ is fixed. However, in practice it is not easy to operate with both visibility and fringe spacing quantities, so that our attention will be mainly concentrated on an expression relative to the fringe direction. With the help of the modification parameters and different viewing directions, we intend to obtain the necessary number of independent equations and thereby determine all desired components. This method, based on fringe direction modification, has been experimentally proven in [5.57] and developed in detail in [5.90].

Starting with the basic equation (5.75) of Case a, we may form the derivative of $D$, with respect to the reference direction $k$, on the object surface instead of on the image surface. We thus have

$$\frac{dD_R}{d\phi} = \frac{dr}{d\phi} \cdot \nabla_n D_R = m \cdot L_R M \nabla_n D_R$$

$$= m \cdot \left\{ L_R M \nabla_n (u \cdot g) + L_R M \nabla_n \left[ -a \cdot \Psi c - \frac{1}{q} a \cdot Cd \right. \right.$$

$$\left. \left. + \frac{1}{q} a \cdot \Psi Ca - \frac{1}{2q^2} (aCa) \cdot d \right] \right\} . \tag{5.80}$$

The entire bracket may be called the modified fringe vector whose first term represents the fringe vector as defined in standard holography, see (4.19), $f_R = L_R M w - K u$. The second term contains derivatives of the unit vector $c$ and corresponding projectors which may be performed by means of the affine connection relating the object surface to the hologram plane

$$L_R M \nabla_n = (L_R - p) \hat{M} \nabla_{\hat{n}} . \tag{5.81}$$

In our case, when deriving with respect to direction $k$, the tensor $\Psi$ and the vectors $a$ and $d$ remain constant, so that only the derivatives of $c$, $-C/q$ and $-C/q^2$, have to be taken into account. According to (2.24, 26 and 28), for these three terms we successively get the projector $C/q$, the superprojector $C/q^2$, and the hyperprojector $C/q^3$. The derivative of the optical path difference then becomes

$$\frac{dD_R}{d\phi} = m \cdot f_R + m \cdot \left\{ \frac{L_R - p}{q} \hat{M} \left[ -a\Psi C + \frac{1}{q} aCd \right. \right.$$

$$\left. \left. - \frac{1}{q} a\Psi Ca + \frac{1}{2q^2} (aCa) d \right] \right\} , \tag{5.82}$$

where $f_R$ is the fringe vector without modification.

We have already mentioned that a rigid body motion of the sources Q, Q' may be realized by the motion of a mirror (Fig. 5.8). Let us then consider the particular and more easily performed case in which a translation $d = 2s\bar{n}$ is performed along the direction $\bar{n}$ of the unit normal of the mirror ($\Psi = 0$). Further, let $m_\xi$, $m_\eta$ be two given vectors in the plane normal to $k$. These may, for instance, represent the base vectors of the cartesian system $(\xi, \eta, \zeta)$ with the $\zeta$-axis parallel to vector $k$ (Fig. 5.9). If $s_\xi$ and $s_\eta$ are the values of two successive shifts of the mirror for which the fringes become once parallel to $m_\xi$, and once parallel to $m_\eta$ (Fig. 5.10), then the derivatives $dD_R/d\phi_\xi$ and $dD_R/d\phi_\eta$ vanish, so that we may write the system as

$$m_\xi \cdot (f_R + s_\xi t_0) = 0 ,$$

$$m_\eta \cdot (f_R + s_\eta t_0) = 0 , \qquad \text{where} \tag{5.83}$$

$$t_0 = \frac{2(L_R - p)}{q^2} \hat{M} \left[ aC\bar{n} + \frac{1}{2q} (aCa)\bar{n} \right] \tag{5.84}$$

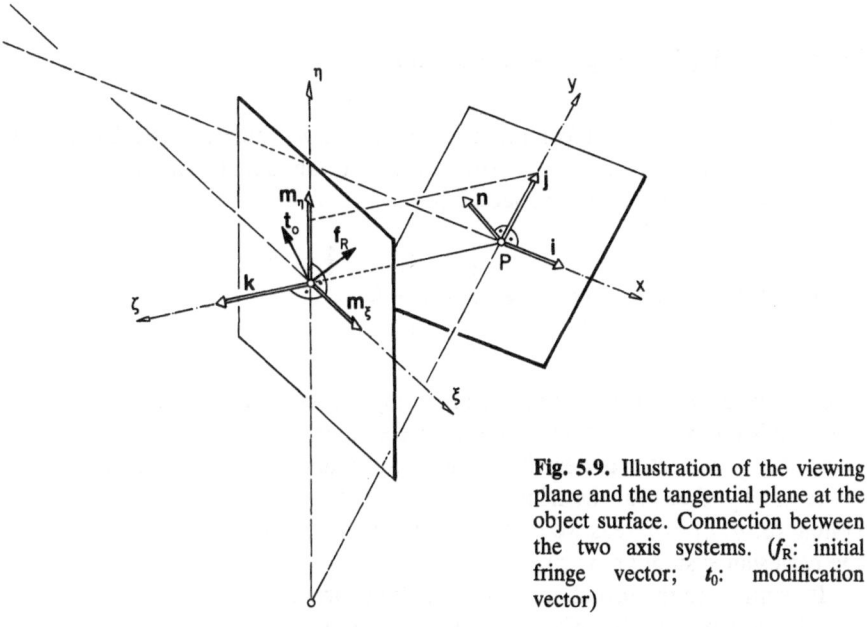

**Fig. 5.9.** Illustration of the viewing plane and the tangential plane at the object surface. Connection between the two axis systems. ($f_R$: initial fringe vector; $t_0$: modification vector)

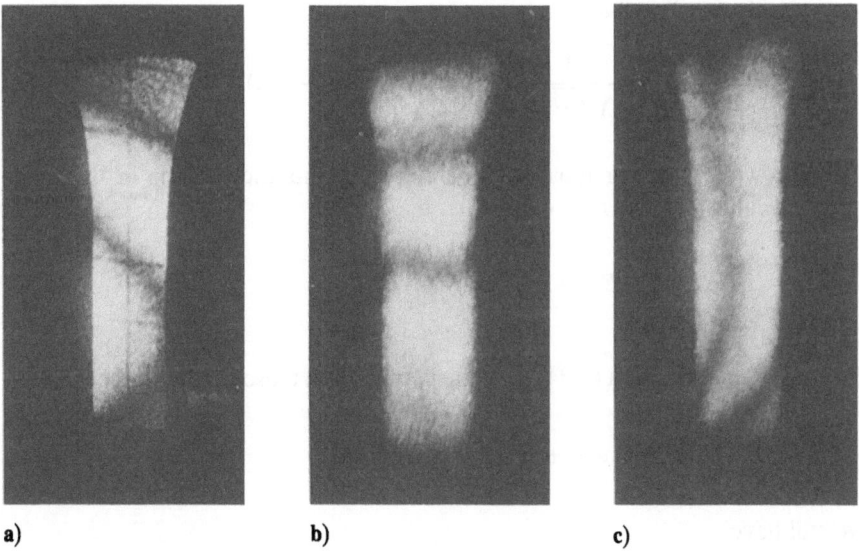

a)                              b)                              c)

**Fig. 5.10a–c.** Fringe modifications: (a) no modification; (b) the fringe is modified so that it becomes horizontal; (c) the fringe is modified so that it becomes vertical

represents the modification vector. The fringe vector

$$f_R = L_R M (\nabla_n \otimes u) g - \frac{L_R}{L_s} MHu - Ku$$

contains the unknown deformation gradient $\nabla_n \otimes u$ which in cartesian components $x$, $y$ relative to a system, situated in the tangential plane of the object surface, may be expressed by, see (2.143),

$$\nabla_n \otimes u = \gamma + \Omega E + \omega \otimes n \triangleq \begin{bmatrix} \varepsilon_x & \frac{1}{2}\gamma_{xy} + \Omega & \omega_x \\ \frac{1}{2}\gamma_{yx} - \Omega & \varepsilon_y & \omega_y \\ 0 & 0 & 0 \end{bmatrix} . \qquad (5.85)$$

The strains $\varepsilon_x$, $\varepsilon_y$ and the shear $\gamma_{xy}$ are components of the strain tensor $\gamma$. The scalar $\Omega$ is the pivot rotation of the surface element around $n$ and $\omega_x$, $\omega_y$ are the components of vector $\omega$ describing the out-of-plane inclination. The conditions expressed by (5.83) for the fringe vector concern its components in the cartesian system $(\xi, \eta, \zeta)$, whereas we look for components values of the tensor $\nabla_n \otimes u$ in the cartesian system $(x, y, z)$.

In contrast to the use of an orthogonal transformation matrix, as in [5.57], we may directly express the two equations (5.83) in the system $(x, y, z)$. In order to completely define the relative position of the two systems (we already have $\zeta \| k$, $z \| n$), we choose the $\eta$-axis to be parallel to the projection of the $y$-axis onto the viewing plane (see Fig. 5.9). With unit vectors $i$, $j$ in the directions of $x$, $y$, respectively, we then have

$$m_\eta = \frac{1}{|Kj|} Kj = \frac{1}{\sqrt{1 - k_y^2}} (j - k_y k) = \frac{1}{\alpha} (j - k_y k) . \qquad (5.86)$$

Further, with $E'$ being the two-dimensional permutation tensor in the plane perpendicular to $k$, we may express $m_\xi$ by

$$m_\xi = E' m_\eta = \frac{1}{\alpha} (E'j - k_y E'k) = \frac{1}{\alpha} E'j . \qquad (5.87)$$

With the decomposition (2.34) ($E$ is the permutation tensor in the surface plane),

$$E' = Ek , \qquad E = E \otimes n + n \otimes E - E \otimes n)^T ,$$

we still have

$$\alpha m_\xi = (Ek)j = (n \cdot k)Ej - (k \cdot Ej)n - (n \cdot j)Ek$$
$$= k_z i - k_x n . \qquad (5.88)$$

Introducing (5.88) and (5.86) in (5.83), and since $f_R$ and $t_0$ are two vectors perpendicular to $k$, we get

$$(k_z i - k_x n) \cdot (f_R + s_\xi t_0) = 0 , \tag{5.89}$$

$$j \cdot (f_R + s_\eta t_0) = 0 . \tag{5.90}$$

Using the same normality property $f_R \cdot k = t_0 \cdot k = 0$, we may also add the factor $kk_x/k_z$ to the left part of (5.89) without changing its value. With $k = k_x i + k_y j + k_z n$, (5.89) is rewritten as:

$$\left[ \left( k_z + \frac{k_x^2}{k_z} \right) i + \frac{k_x k_y}{k_z} j \right] \cdot (f_R + s_\xi t_0) = 0 .$$

Introducing (5.90) in this auxiliary relation in order to eliminate $j \cdot f_R$, the component $i \cdot f_R$ may be isolated:

$$\frac{1}{k_z} (k_z^2 + k_x^2) i \cdot (f_R + s_\xi t_0) = \frac{k_x k_y}{k_z} (s_\eta - s_\xi) t_0 \cdot j .$$

Finally, rewriting this equation and joining it to (5.90), we obtain a system of two equations for the $x,y$-components of $f_R$:

$$i \cdot f_R = (Nf_R)_x = \frac{k_x k_y}{1 - k_y^2} (s_\eta - s_\xi) t_{0y} - s_\xi t_{0x} ,$$

$$\tag{5.91}$$

$$j \cdot f_R = (Nf_R)_y = - s_\eta t_{0y} .$$

Fully expressed in components of strain, rotation, and displacement, $u \triangleq (u, v, w)$, this system reads

$$\varepsilon_x g_x + (\tfrac{1}{2}\gamma_{xy} + \Omega)g_y + \omega_x g_z = b_x , \tag{5.92}$$

$$(\tfrac{1}{2}\gamma_{xy} - \Omega)g_x + \varepsilon_y g_y + \omega_y g_z = b_y ,$$

where

$$b_x = \frac{1}{L_R} \left[ \frac{k_x k_y}{1 - k_y^2} (s_\eta - s_\xi) t_{0y} - s_\xi t_{0x} + u - (u \cdot k)k_x \right] + \frac{1}{L_s} [u - (u \cdot h)h_x] ,$$

$$\tag{5.93}$$

$$b_y = \frac{1}{L_R} [-s_\eta t_{0y} + v - (u \cdot k)k_y] + \frac{1}{L_s} [v - (u \cdot h)h_y] .$$

Similarly to the displacement determination, we need three different observation directions $k_1, k_2, k_3$ in order to get the $3 \times 2$ linear independent equations of type (5.92) for the six unknown components. Using, for instance, Cramer's "rule", we find

$$\varepsilon_x = \frac{1}{\Delta} \begin{vmatrix} b_{x1} & g_{y1} & g_{z1} \\ b_{x2} & g_{y2} & g_{z2} \\ b_{x3} & g_{y3} & g_{z3} \end{vmatrix} \quad , \quad \frac{1}{2}\gamma_{xy} + \Omega = \frac{1}{\Delta} \begin{vmatrix} g_{x1} & b_{x1} & g_{z1} \\ g_{x2} & b_{x2} & g_{z2} \\ g_{x3} & b_{x3} & g_{z3} \end{vmatrix} \quad ,$$

$$\omega_x = \frac{1}{\Delta} \begin{vmatrix} g_{x1} & g_{y1} & b_{x1} \\ g_{x2} & g_{y2} & b_{x2} \\ g_{x3} & g_{y3} & b_{x3} \end{vmatrix} \quad \ldots \tag{5.94}$$

where

$$\Delta = \begin{vmatrix} g_{x1} & g_{y1} & g_{z1} \\ g_{x2} & g_{y2} & g_{z2} \\ g_{x3} & g_{y3} & g_{z3} \end{vmatrix} \quad ,$$

and similar expressions for $\gamma_{xy}/2 - \Omega$, $\varepsilon_y$, and $\omega_y$, respectively, but with $b_x$ replaced by $b_y$. Finally the shear $\gamma_{xy}$ and the pivot rotation $\Omega$ may be separated.

In order to increase precision, measurements in more than three directions $k$ are made and the corresponding linear equations (5.92) are reduced by the least-squares method, see [5.57, 91] or (4.9) concerning the displacement $u$. In the above equations, we assume the components of the modification vector $t_0$ to be known. In fact, we distinguish two ways to determine the components of $t_0$:

a) this vector may be *calculated* from its definition (5.84), for which the convenient form for computing reads

$$t_0 = \frac{2(L_R - p)}{q^2} \hat{M} \bar{t}_0 = \frac{2(L_R - p)}{q^2} \left( \bar{t}_0 - \frac{k \cdot \bar{t}_0}{\hat{n} \cdot k} \hat{n} \right) ,$$

where

$$\bar{t}_0 = a C \bar{n} + \frac{1}{2q} (a C a) \bar{n}$$

$$= a \left[ c \cdot \bar{n} + \frac{3}{q} (c \cdot \bar{n})(a \cdot c) - \frac{1}{q} a \cdot \bar{n} \right]$$

$$+ \bar{n} \left[ a \cdot c + \frac{3}{2q} (a \cdot c)^2 - \frac{1}{2q} |a|^2 \right]$$

$$+ c \left[ a \cdot \bar{n} - 3(a \cdot c)(c \cdot \bar{n}) + \frac{3}{q} (a \cdot \bar{n})(a \cdot c) - \frac{15}{2q} (a \cdot c)^2 (c \cdot \bar{n}) \right.$$

$$\left. + \frac{3}{2q} |a|^2 (c \cdot \bar{n}) \right] ; \tag{5.95}$$

b) the vector $t_0$ may be *measured* from a double-exposure hologram of the undeformed object ($u = 0$ and $f_R = 0$). The fringes produced by the modifica-

a)                                                                    b)

**Fig. 5.11a, b.** Fringes produced by modification only ($u = 0, f_R = 0$). The camera is focused on the hologram (locus of the fringes). The modification vector $t_0$ is the same in both pictures; thus, the direction of the fringes is the same. The displacement $|s|$ of the mirror in (b) is twice the value of (a)

tion alone are completely localized on the hologram (Fig. 5.11) and perpendicular to the vector $t_0$. Their interspaces depend on the length of $t_0$ as well as on the displacement $s$ of the mirror, since the modified fringe vector $\hat{f}_R$ is here equal to $st_0$. The vector $t_0$ varies slightly and continously over the whole hologram, maintaining the fringes nearly straight and equally spaced. These fringes present a finite but small curvature which could be studied by the second derivative of $D$, as was done in Sect. 5.1.3. The fringes are very similar to those produced by a rigid body motion of the object (see end of Sect. 4.1.1), but care should be taken when making this comparison, since the equations governing these two fringe fields are completely different. The length of $t_0$ is then given within a good approximation by

$$|t_0| = \frac{\lambda |\Delta n|}{|s| |\Delta \phi|} ,$$

where $\Delta n$ is the number of fringes delimited by the viewing angle $\Delta \phi$. This approximation further shows that the fringe number is linear in $s$, as is visible in Fig. 5.11a, b.

The following table illustrates a simple application and contains numerical results from two tensile tests with an aluminium bar (Fig. 5.12a, b). In the second test, the bar was guided in order to influence the out-of-plane rotation, though the same force was applied. Without any imposed restrictions, all six components of ($\nabla_n \otimes u$) were supposed unknown. The strain components should not have changed from the first to the second test. These components may be well separated from the rotation components which differ much in the two tests. The

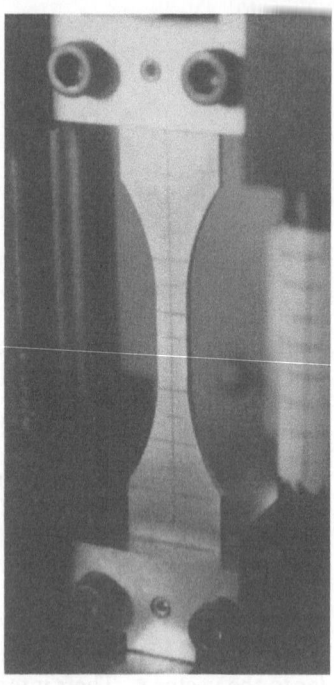

a)                                                  b)

**Fig. 5.12.** (a) Holographic set-up. *Upper left:* two microscopic objectives with two spatial filters. *Upper right:* mirror placed on a translating stage. The images of the two reference sources Q, Q′ (two bright points) are visible. *Lower left:* holographic plate holder for precise repositioning (after Abramson). *Lower right:* aluminium bar being tested. (b) Magnified view of the aluminium bar

|  |  | 1st test | 2nd test |
|---|---|---|---|
| $\varepsilon_x$ | $\times\,10^4$ | −0.20 | −0.18 |
| $\varepsilon_y$ | $\times\,10^4$ | 0.56 | 0.61 |
| $\gamma_{xy}/2$ | $\times\,10^4$ | 0.02 | −0.04 |
| $\Omega$ | $\times\,10^4$ | −0.04 | −0.03 |
| $\omega_x$ | $\times\,10^4$ | −0.59 | −0.83 |
| $\omega_y$ | $\times\,10^4$ | −1.37 | 0.03 |

results were in good agreement with the expected values, in particular with $\varepsilon_x = -\nu\varepsilon_y$, $\nu \cong 0.33$, and $\gamma_{xy} \cong 0$. The system presented here requires a precise determination of $t_0$. The upper limit of a measurable deformation depends mainly on the fringe visibility (see [5.90] or Sect. 4.2.2). Also, (5.93 and 84) together show that, for large $L_s$ and $L_R$, the measurements of the displacements $u$, $v$, $w$ need not be precise compared to those of the modification parameters.

In the case of a large rotation, the strain components must be corrected by using (2.147) instead of (2.143). Further, (5.92) may be replaced by an equation similar to that deduced at the end of Sect 4.1.1, with a modification that would be insensitive to the out-of-plane components of the rotation by inclusion of holographic Moiré. Here, the sensitivity vector $\Delta h$ may be chosen in a manner such that $\Delta h_z = 0$ (see end of Sect. 4.1.1 or [5.21]). Another method consists of fixing the holographic plate onto the object. For $p \ll L_R$, $c \cong h$, and $\hat{n} \cong n$, the rotation components $\Omega$, $\omega_x$, $\omega_y$ then represent only the relative rotation between the holographic plate and the object surface element [5.92–95].

In summary, we conclude that in holography the modification at the reconstruction represents a flexible tool with which to make suitable quantitative applications of industrial interest.

# References

## Chapter 1

1.1    D. Gabor: Microscopy by reconstructed wave-fronts, Proc. Roy. Soc. (London) A**197**, 454–487 (1949)

1.2    D. Gabor: Microscopy by reconstructed wave-fronts II, Proc. Phys. Soc. (London) B**64**, 449–469 (1951)

1.3    E. N. Leith, J. Upatnieks: Reconstructed wavefronts and communication theory, J. Opt. Soc. Am. **52**, 1123–1130 (1962)

1.4    Y. N. Denisyuk: On the reproduction of the optical properties of an object by the wave field of its scattered radiation, Opt. Spektrosk. **15**, 522–532 (1963)

1.5    P. J. van Heerden: Theory of optical information storage in solids, Appl. Opt. **2**, 393–400 (1963)

1.6    R. L. Powell, K. A. Stetson: Interferometric vibration analysis by wavefront reconstruction, J. Opt. Soc. Am. **55**, 1593–1598 (1965)

1.7    J. M. Burch: The application of lasers in production engineering, Prod. Eng. **44**, 431–443 (1965)

1.8    R. E. Brooks, L. O. Heflinger, R. F. Wuerker: Interferometry with a holographically reconstructed comparison beam, Appl. Phys. Lett. **7**, 248–249 (1965)

1.9    H. Kogelnik: Coupled wave theory for thick hologram gratings, Bell Syst. Techn. J. **48**, 2909–2947 (1969)

1.10   K. A. Stetson: A rigorous treatment of the fringes of hologram interferometry, Optik **29**, 386–400 (1969)

1.11   K. A. Stetson: The argument of the fringe function in hologram interferometry of general deformations, Optik **31**, 576–591 (1970)

1.12   S. Walles: Visibility and localization of fringes in holographic interferometry of diffusely reflecting surfaces, Ark. Fys. **40**, 299–403 (1970)

1.13   C. Froehly, J. Monneret, J. Pasteur, J. C. Viénot: Etude des faibles déplacements d'objets opaques et de la distorsion optique dans les lasers à solide par interférométrie holographique, Opt. Acta **16**, 343–362 (1969)

1.14   C. M. Vest: *Holographic Interferometry* (Wiley, New York 1979)

1.15   D. W. Sweeney, C. M. Vest: Reconstruction of three-dimensional refractive index fields by holographic interferometry, Appl. Opt. **11**, 205–207 (1972)

1.16   S. Cha, C. M. Vest: Tomographic reconstruction of strongly refracting fields and its application to interferometric measurement of boundary layers, Appl. Opt. **20**, 2787–2794 (1981)

1.17   Y. I. Ostrovsky, M. M. Butusov, G. V. Ostrovskaya: *Interferometry by Holography*, Springer Ser. Opt. Sci., Vol. 20 (Springer Berlin, Heidelberg 1980)

1.18   R. Dändliker: Heterodyne holographic interferometry, in *Progress in Optics,* **17**, Chap. 1 (North-Holland, Amsterdam 1980)

1.19   R. Dändliker, B. Ineichen, F. M. Mottier: High resolution hologram interferometry by electronic phase measurement, Opt. Commun. **9**, 412–416 (1973)

# Chapter 2

2.1    R. Abraham: *Linear and Multilinear Algebra* (Benjamin, New York 1966)
2.2    A. E. Green, W. Zerna: *Theoretical Elasticity*, 2nd ed. (Clarendon, Oxford 1968)
2.3    L. Brillouin: *Les Tenseurs en Mécanique et en Elasticité*, 2nd ed. (Masson, Paris 1949)
2.4    I. S. Sokolnikoff: *Tensor Analysis: Theory and Applications* (Wiley, New York 1951)
2.5    W. Blaschke, K. Leichtweiss: *Elementare Differentialgeometrie* (Springer, Berlin, Heidelberg 1973)
2.6    W. Schumann, M. Dubas: *Holographic Interferometry. From the Scope of Deformation Analysis of Opaque Bodies,* Springer Ser. Opt. Sci., Vol. 16 (Springer, Berlin, Heidelberg 1979)
2.7    D. C. Leigh: *Nonlinear Continuum Mechanics* (McGraw-Hill, New York 1968)
2.8    C. C. Wang, C. Truesdell: *Introduction to Rational Elasticity* (Noordhoff, Leyden 1973)
2.9    W. Prager: *Einführung in die Kontinuumsmechanik* (Birkhäuser, Basel 1961)
2.10   H. Bremmer: Propagation of Electromagnetic Waves, in Handbuch der Physik, ed. by S. Flügge, Vol. 16: *Elektrische Felder und Wellen* (Springer, Berlin 1958) pp. 423–639
2.11   M. Kline, I. W. Kay: *Electromagnetic Theory and Geometrical Optics* (Wiley, New York 1965)
2.12   M. Born, E. Wolf: *Principles of Optics,* 5th ed. (Pergamon, Oxford 1975)
2.13   C. L. Andrews: *Optics of the Electromagnetic Spectrum* (Prentice-Hall, Englewood Cliffs, NJ 1960)
2.14   J. R. Meyer-Arendt: *Introduction to Classical and Modern Optics* (Prentice-Hall, Englewood Cliffs, NJ 1972)
2.15   E. B. Brown: *Modern Optics* (Reinhold, New York 1965)
2.16   F. A. Jenkins, H. E. White: *Fundamentals of Optics,* 4th ed. (McGraw-Hill, New York 1976)
2.17   L. Levi: *Applied Optics. A Guide to Optical System Design,* Vol. 1 (Wiley, New York 1968)
2.18   S. G. Lipson, H. Lipson: *Optical Physics,* 2nd ed. (Cambridge U. Press, London 1981)
2.19   W. Lukosz: "Physikalische Optik I und II. Vorlesung an der ETH Zürich" (Lecture notes 1973, 1979)
2.20   A. E. Green, J. E. Adkins: *Large Elastic Deformations* (Clarendon, Oxford 1960)
2.21   W. A. Nash: *Bibliography on Shells and Shell-like Structures,* Dep. of the Navy, David Taylor Model Basin, Dep. of Eng. Mech. (1954–1956)
2.22   W. Z. Wlassov: *Allgemeine Schalentheorie und ihre Anwendung in der Technik,* translated from Russian (Akademie Verlag, Berlin 1958)
2.23   S. Timoshenko, S. Woinowsky-Krieger: *Theory of Plates and Shells* (McGraw-Hill, New York 1959)
2.24   W. Flügge: *Stresses in Shells* (Springer, Berlin, Göttingen 1960)
2.25   A. L. Gol'denveizer: *Theory of Elastic Thin Shells* (Pergamon, Oxford 1961)
2.26   P. M. Naghdi: The theory of shells and plates, in Encyclopedia of Physics, ed. by S. Flügge, Vol. 6a (Springer, Berlin, Heidelberg 1972) pp. 425–640
2.27   P. Seide: *Small Elastic Deformations of Thin Shells* (Noordhoff, Leyden 1975)
2.28   J. L. Sanders: "An improved first approximation theory of shells" (NASA Rpt. 24, 1959)
2.29   W. T. Koiter: A consistent first approximation in the general theory of thin elastic shells, in *The Theory of Thin Elastic Shells,* Proc. Symp. Delft 1959, ed. by W. T. Koiter (North-Holland, Amsterdam 1960) pp. 12–33
2.30   C. R. Steele: A geometric optics solution for the thin shell equation. Int. J. Eng. Sci. **9,** 681–704 (1971)
2.31   G. Teichmann: "Quelques aspects généraux de la théorie linéaire des coques orthotropes et inhomogènes, en particulier en vue d'utiliser un principe de variation mixte". Thesis No. 6301 ETH Zürich (1979)
2.32   J.-P. Zürcher, W. Schumann: Some intrinsic considerations on the nonlinear theory of thin shells, Acta Mech. **40,** 123–140 (1981)

2.33   J.-P. Zürcher: "Quelques aspects généraux de la théorie non-linéaire des coques minces. Utilisation de fonctions scalaires à partir d'un principe de variation mixte", Thesis No. 7250, ETH Zürich (1983)

2.34   W. Pietraszkiewicz: "On consistent approximations in the geometrically non-linear theory of shells". Mitteilungen aus dem Institut für Mechanik No. 26, Ruhr-Universität, Bochum (Juni 1981)

2.35   V. L. Berdichevskii: Variational-asymptotic method of constructing the nonlinear shell theory, in *Theory of Shells,* ed. by W. T. Koiter, G. K. Mikhailov (North-Holland, Amsterdam 1980) pp. 137–161

2.36   M. Sayir, C. Mitropoulos: On elementary theories of linear elastic beams, plates and shells, J. of Appl. Math. Phys. (ZAMP) **31**, 1–55 (1980)

2.37   P. Pleus: "Eine Theorie zweiter Ordnung für endliche Deformationen des schlanken Balkens", Thesis No. 6956, ETH Zürich (1982)

# Chapter 3

3.1   G. W. Stroke: *An Introduction to Coherent Optics and Holography* (Academic, New York 1966)

3.2   J. B. De Velis, G. O. Reynolds: *Theory and Applications of Holography* (Addison-Wesley, Reading, MA 1967)

3.3   J. W. Goodman: *Introduction to Fourier Optics* (McGraw-Hill, San Francisco 1968)

3.4   H. Kiemle, D. Roess: *Einführung in die Technik der Holographie* (Akad. Verlagsgesellschaft, Frankfurt/Main 1969)

3.5   H. Lenk: *Holographie* (VEB Georg Thieme, Leipzig 1969)

3.6   M. Françon: *Holographie* (Masson, Paris 1969)

3.7   H. J. Caufield, S. Lu: *The Applications of Holography* (Wiley, New York 1970)

3.8   J. N. Butters: *Holography and Its Technology* (Institution of Electrical Engineers, Peregrinus Ltd., London 1971)

3.9   J.-C. Viénot, P. Smigielski, H. Royer: *Holographie optique* (Dunod, Paris 1971)

3.10  R. Dändliker: *Laser-Kurzlehrgang* (Fachschriftenverlag Aargauer Tagblatt, Aarau 1971)

3.11  R. J. Collier, C. B. Burckhardt, L. H. Lin: *Optical Holography* (Academic, New York 1971)

3.12  G. Wernicke, W. Osten: *Holografische Interferometrie* (VEB Fachbuchverlag, Leipzig 1982)

3.13  G. Groh: *Holographie, physikalische Grundlagen und Anwendungen* (Berliner Union, Stuttgart; W. Kohlhammer, Stuttgart 1973)

3.14  E. Menzel, W. Mirandé, I. Weingärtner: *Fourier-Optik und Holographie* (Springer, Wien 1973)

3.15  F. T. S. Yu: *Introduction to Diffraction, Information Processing, and Holography* (MIT Press, Cambridge 1973)

3.16  O. Bryngdahl: Evanescent waves in optical imaging, in *Progress in Optics,* **11**, Chap. 4 (North-Holland, Amsterdam 1973)

3.17  R. K. Erf (ed.): *Holographic Nondestructive Testing* (Academic, New York 1974)

3.18  W. E. Kock: *Engineering Applications of Lasers and Holography* (Plenum, New York 1975)

3.19  K. Biedermann: Information storage materials for holography and optical data processing, Opt. Acta **22**, 103–124 (1975)

3.20  H. M. Smith: *Principles of Holography,* 2nd ed. (Wiley, New York 1975)

3.21  P. Greguss (ed.): *Holography in Medicine: Applications in the Biomedical Sciences* (IPC Science and Technology Press, Guildford, UK 1976)

3.22  T. Okoshi: *Three-Dimensional Imaging Techniques* (Academic, New York 1976)

3.23  Y. I. Ostrovsky: *Holography and Its Applications* (Mir Publishers, Leningrad 1977)

3.24  P. Hariharan: *Optical Holography. Principles, techniques and applications* (Cambridge U. Press, Cambridge 1984)

3.25  G. Schmahl, D. Rudolph: Holographic diffraction gratings, in *Progress in Optics* **14**, Chap. 5 (North-Holland, Amsterdam 1977)

3.26  G. von Bally (ed.): *Holography in Medicine and Biology,* Springer Ser. Opt. Sci., Vol. 18 (Springer, Berlin, Heidelberg 1979)

3.27  T. Okoshi: Projection holography, in *Progress in Optics* **15**, Chap. 3 (North-Holland, Amsterdam 1978)

3.28  W. H. Lee: Computer-generated holograms, in *Progress in Optics* **16**, Chap. 3 (North-Holland, Amsterdam 1978)

3.29  C. M. Vest: *Holographic Interferometry* (Wiley, New York 1979)

3.30  H. J. Caufield (ed.): *Handbook of Optical Holography* (Academic, New York 1979)

3.31  W. Schumann, M. Dubas: *Holographic Interferometry. From the Scope of Deformation Analysis of Opaque Bodies,* Springer Ser. Opt. Sci., Vol. 16 (Springer, Berlin, Heidelberg 1979)

3.32  L. M. Soroko: *Holography and Coherent Optics* (Plenum, New York 1980)

3.33  Y. I. Ostrovsky, M. M. Butusov, G. V. Ostrovskaya: *Interferometry by Holography,* Springer Ser. Opt. Sci., Vol. 20 (Springer, Berlin, Heidelberg 1980)

3.34  L. P. Yaroslavskii, N. S. Merzlyakov: *Methods of Digital Holography* (Consultants Bureau, Plenum, New York 1980)

3.35  R. Dändliker: Heterodyne holographic interferometry in *Progress in Optics* **17**, Chap. 1 (North-Holland, Amsterdam 1980)

3.36  N. Abramson: *Making and Evaluation of Holograms* (Academic, London 1981)

3.37  K. Leonhardt: *Optische Interferenzen, Theorie der Auswertbarkeit und Atlas wichtiger Interferometertypen* (Wissenschaftliche Verlagsgesellschaft, Stuttgart 1981)

3.38  L. Solymar, D. J. Cooke: *Volume Holography and Volume Gratings* (Academic, London 1981)

3.39  C. Bonjour, M. Matthey (eds.): *Lasers et Applications Industrielles* (Presses Polytechniques Romandes, Lausanne 1982)

3.40  R. J. Jones, C. Wykes: *Holographic and Speckle Interferometry* (Cambridge U. Press, Cambridge 1983)

3.41  D. Gabor: A new microscopic principle, Nature **161**, 777–778 (1948)

3.42  D. Gabor: Microscopy by reconstructed wave-fronts, Proc. Roy. Soc. (London) A **197**, 454–487 (1949)

3.43  D. Gabor: Microscopy by reconstructed wave-fronts II, Proc. Phys. Soc. (London) B **64**, 449–469 (1951)

3.44  E. N. Leith, J. Upatnieks: Reconstructed wavefronts and communication theory, J. Opt. Soc. Am. **52**, 1123–1130 (1962)

3.45  E. N. Leith, J. Upatnieks: Wavefront reconstruction with continuous-tone objects, J. Opt. Soc. Am. **53**, 1377–1381 (1963)

3.46  E. N. Leith, J. Upatnieks: Wavefront reconstruction with diffused illumination and three-dimensional objects, J. Opt. Soc. Am. **54**, 1295–1301 (1964)

3.47  Y. N. Denisyuk: On the reproduction of the optical properties of an object by the wave field of its scattered radiation, Opt. Spektrosk. **15**, 522–532 (1963)

3.48  P. J. van Heerden: Theory of optical information storage in solids, Appl. Opt. **2**, 393–400 (1963)

3.49  Y. N. Denisyuk: Holographic art with recording in three-dimensional media on the basis of Lippmann photographic plates, Opt. Appl. **8**, 49–53 (1978)

3.50  H. M. Smith (ed.): *Holographic Recording Materials,* Topics Appl. Phys., Vol. 20 (Springer, Berlin, Heidelberg 1977)

3.51  A. Kozma: Photographic recording of spatially modulated coherent light, J. Opt. Soc. Am. **56**, 428–432 (1966)

3.52  R. F. VanLigten: Influence of photographic film on wavefront reconstruction. I: plane wavefronts, J. Opt. Soc. Am. **56**, 1–9 (1966)

3.53  R. F. VanLigten: Influence of photographic film on wavefront reconstruction. II: "cylindrical" wavefronts, J. Opt. Soc. Am. **56**, 1009–1014 (1966)

3.54  M. J. Landry, G. S. Phipps: Holographic characteristics of 10E75 plates for single- and multiple-exposure holograms, Appl. Opt. **14**, 2260–2266 (1975)

3.55  H. F. Dietrich, R. J. Raine, R. N. O'Brien: A 5-minute monobath for KODAK 649-F plates used in holography and holographic interferometry, J. Phot. Sci. **24**, 120–123 (1976)

3.56  J. J. A. Couture, R. A. Lessard: Intermittent characteristic curves for KODAK 649F plates at 514,5 nm, Appl. Opt. **18**, 3644–3651 (1979)

3.57  J. J. A. Couture, R. A. Lessard: Diffraction efficiency of specular multiplexed holograms recorded on KODAK 649F plates, Appl. Opt. **18**, 3652–3660 (1979)

3.58  B. J. Chang, C. D. Leonard: Dichromated gelatin for the fabrication of holographic optical elements, Appl. Opt. **18**, 2407–2417 (1979)

3.59  T. W. Hou, M. S. Chang: Holographic evaluation of resolution in amorphous chalcogenide inorganic photoresists, Appl. Opt. **18**, 1753–1756 (1979)

3.60  W. R. Graver, J. W. Gladden, J. W. Eastes: Phase holograms formed by silver halide (sensitized) gelatin processing, Appl. Opt. **19**, 1529–1536 (1980)

3.61  G. S. Phipps, C. E. Robertson, F. M. Tamashiro: Reprocessing of nonoptimally exposed holograms, Appl. Opt. **19**, 802–811 (1980)

3.62  D. A. Woodbury, F. Davidson, T. A. Rabson, F. K. Tittel: Hologram characterization in an optical memory experiment using photorefractive $LiNbO_3$, Appl. Opt. **19**, 812–817 (1980)

3.63  J. W. Gladden: Grating formation in diazo salt (sensitized) gelatin, Appl. Opt. **19**, 1537–1540 (1980)

3.64  T. Saito, S. Oshima, T. Honda, J. Tsujiuchi: An improved technique for holographic recording on a thermoplastic photoconductor, Opt. Commun. **16**, 90–95 (1976)

3.65  T. Saito, T. Imamura, T. Honda, J. Tsujiuchi: Solvent vapour method in thermoplastic photoconductor media, J. Opt. **11**, 285–292 (1980)

3.66  B. Ineichen, C. Liegeois, P. Meyrueis: Thermoplastic film camera for holographic recording of extended objects in industrial applications, Appl. Opt. **21**, 2209–2214 (1982)

3.67  P. Hariharan, G. S. Kaushik, C. S. Ramanathan: Simplified, low-noise processing technique for photographic phase holograms, Opt. Commun. **6**, 75–76 (1972)

3.68  R. Dändliker, B. Ineichen: Nonlinear cross-talk in two-reference-beam holographic interferometry, Opt. Commun. **19**, 365–369 (1976)

3.69  D. Dameron, C. M. Vest: Fringe sharpening and diffraction in nonlinear two-exposure holographic interferometry, J. Opt. Soc. Am. **66**, 1418–1421 (1976)

3.70  J. Katz: "Non-linear effects in holographic interferometry and their influence on measurement accuracy", M.S. Thesis, Tel Aviv Univ. (1976)

3.71  E. Marom, J. Katz: Unconventional interferometric realizations based on holographic nonlinear effects, Appl. Opt. **16**, 1400–1403 (1977)

3.72  A. V. Alekseev-Popov, I. I. Komissarova, G. V. Ostrovskaya: Wave intensities of higher orders reconstructed by a nonlinearly recorded hologram, Opt. Spektrosk. **37**, 1143–1149 (1974)

3.73  S. Toyooka: Elimination of wavefront aberration of optical elements used in phase difference amplification, Appl. Opt. **13**, 2014–2018 (1974)

3.74  S. Toyooka: Holographic interferometry with increased sensitivity for diffusely reflecting objects, Appl. Opt. **16**, 1054–1057 (1977)

3.75  K. Chalasinska-Macukow, J. Slaby, T. Szoplik: Comparison of nonlinear effects in amplitude and phase holograms recorded in photographic materials, Opt. Commun. **35**, 332–336 (1980)

3.76  H. Ghandeharian, W. M. Boerner: Degradation of holographic images due to depolarization of reflected light, J. Opt. Soc. **68**, 931–934 (1978)

3.77  H. Ghandeharian: Multiple-reference-beam nonlinear holography with applications in suppression of intermodulation background noise, J. Opt. Soc. Am. **70**, 835–842 (1980)

3.78  T. Lipowiecki: Optical noises in amplitude holography, developed by nonlinear properties of recording media, Opt. Appl. **11**, 105–122 (1981)

3.79  J. C. Charmet, A. Vareille: Formalisme pour la description de la lumière polarisée. Application à l'holographie en lumière polarisée, Rev. Phys. Appl. **13**, 317–327 (1978)

3.80  G. Pirard: Influence de la polarisation en holographie, Scien. Techn. Armement **53**, 35–78 (1979)

3.81  P. C. Mehta, R. Hradaynath: Elimination of depolarization effects in holography, Appl. Opt. **21**, 4549–4552 (1982)

3.82  A. A. Bugaev: Holography based on the correlation of intensities, Opt. Spektrosk. **50**, 627–630 (1981)

3.83  V. V. Kazankova, V. I. Protasevich, Y. A. Pryakhin: Superposition of holograms taking into account the limits of the dynamic range of the photographic layer, Opt. Spektrosk. **44**, 561–565 (1978)

3.84  E. N. Leith: Fundamentals of Modern Holography (SPIE 1983 Conference Geneva, Tutorial T9)

3.85  E. N. Leith, J. Upatnieks, K. A. Haines: Microscopy by wavefront reconstruction, J. Opt. Soc. Am. **55**, 981–986 (1965)

3.86  I. A. Abramowitz, J. M. Ballantyne: Evaluation of hologram aberrations by ray tracing, J. Opt. Soc. Am. **57**, 1522–1526 (1967)

3.87  I. A. Abramowitz: Evaluation of hologram imaging by ray tracing, Appl. Opt. **8**, 403–410 (1969)

3.88  F. I. Diamond: Magnification and resolution in wavefront reconstruction, J. Opt. Soc. Am. **57**, 503–508 (1967)

3.89  V. A. Vanin, G. I. Greisukh: Aberration properties of Fresnel copies of holograms, Opt. Spektrosk. **48**, 326–329 (1980)

3.90  C. J. Budhiraja, S. C. Som: Improvement of image quality in holographic microscopy, Appl. Opt. **20**, 1848–1853 (1981)

3.91  Y. Shono, T. Inuzuka: Elimination of aberrations due to a wavelength shift in holographic microscopy, Appl. Phys. Lett. **33**, 111–112 (1978)

3.92  J. Ojeda-Castaneda, J. H. Altamirano, S. Guel-Sandoval: Generalized holographic formation of a lensless Fourier transform. 2: Wavelength variation, Appl. Opt. **19**, 485–487 (1980)

3.93  M. Nazarathy, J. Shamir: Wavelength variation in Fourier optics and holography described by operator algebra, Isr. J. Techn. **18**, 224–231 (1980)

3.94  G. Ade: Die holographischen Rekonstruktionsbedingungen bei nichtisoplanatischen Abbildungen, Optik **58**, 321–339 (1981)

3.95  A. Offner: Ray tracing through a holographic system, J. Opt. Soc. Am. **56**, 1509–1512 (1966)

3.96  K. A. Stetson: An analysis of the properties of total internal reflection holograms, Optik **29**, 520–536 (1969)

3.97  R. W. Smith: The s and t formulae for holographic lens elements, Opt. Commun. **21**, 106–109 (1977)

3.98  V. V. Aristov, G. A. Ivanova: Some consequences of the diffraction theory of holography, Opt. Spektrosk. **39**, 563–570 (1975)

3.99  E. B. Champagne, N. G. Massey: Resolution in holography, Appl. Opt. **8**, 1879–1885 (1969)

3.100  I. Přikryl: A contribution to hologram imagery, Opt. Acta **21**, 517–528 (1974)

3.101  I. Přikryl: Studying hologram imagery by a ray-tracing method, Opt. Acta **19**, 623–631 (1972)

3.102  M. Born, E. Wolf: *Principles of Optics,* 5th ed. (Pergamon, Oxford 1975)

3.103  H. H. Hopkins: *Wave Theory of Aberrations* (Clarendon, Oxford 1950)

3.104  A. Maréchal: Optique Géométrique Générale, in Encyclopedia of Physics, ed. by S. Flügge, Vol. 24 (Springer, Berlin 1956) pp. 44–170

3.105  D. B. Neumann: Geometrical relationships between the original object and the two images of a hologram reconstruction, J. Opt. Soc. Am. **56**, 858–861 (1966)

3.106  C. W. Helstrom: Image luminance and ray tracing in holography, J. Opt. Soc. Am. **56**, 433–441 (1966)

3.107  W. Schumann, M. Dubas: On the motion of holographic images caused by movements of the reconstruction light source, with the aim of application to deformation analysis, Optik **46**, 377–392 (1976)

3.108  E. N. Leith, J. Upatnieks: Imagery with coherent optics, SPIE J. **3**, 123–126 (1965)

3.109  E. N. Leith, G. J. Swanson: Holographic aberration compensation with partially coherent light, Opt. Lett. **7**, 596–598 (1982)

3.110  V. L. Afanaseva, L. T. Mustafina, V. A. Seleznev: Method of compensating aberrations in holographic interferometry of increased sensitivity, Opt. Spektrosk. **37**, 788–789 (1974)

3.111  H. Kogelnik: Holographic image projection through inhomogeneous media, Bell Syst. Techn. J. **44**, 2451–2455 (1965)

3.112  H. Madjidi-Zolbanine, C. Froehly: Holographic correction of both chromatic and spherical aberrations of single glass lenses, Appl. Opt. **18**, 2385–2393 (1979)

3.113  T. Tsuruta, Y. Itoh: Image correction using holography, Appl. Opt. **7**, 2139–2140 (1968)

3.114  J. E. Ward, D. C. Auth, F. P. Carlson: Lens aberration correction by holography, Appl. Opt. **10**, 896–900 (1971)

3.115  Y. N. Denisyuk, S. I. Soskin: Holographic correction of deformational aberrations of the main mirror of a telescope, Opt. Spektrosk. **31**, 992–999 (1971)

3.116  M. Miler: Off-axis paraxial interpretation of holography, Opt. Appl. **7**, 41–45 (1977)

3.117  T. Jannson: Effective pupil model in plane holography, Optik **60**, 225–235 (1982)

3.118  E. B. Champagne: Nonparaxial imaging, magnification, and aberration properties in holography, J. Opt. Soc. Am. **57**, 51–55 (1967)

3.119  J. N. Latta: Computer-based analysis of hologram imagery and aberrations. I: Hologram types and their nonchromatic aberrations. II: Aberrations induced by a wavelength shift, Appl. Opt. **10**, 599–608 and 609–618 (1971)

3.120  J. N. Latta: Computer-based analysis of holography using ray tracing, Appl. Opt. **10**, 2698–2710 (1971)

3.121  D. G. McCauley, C. E. Simpson, W. J. Murbach: Holographic optical element for visual display applications, Appl. Opt. **12**, 232–242 (1973)

3.122  K. A. Winick, J. R. Fienup: Optimum holographic elements recorded with nonspherical wave fronts, J. Opt. Soc. Am. **73**, 208–217 (1983)

3.123  W. T. Welford: Isoplanatism and holography, Opt. Commun. **8**, 239–243 (1973)

3.124  W. T. Welford: A vector raytracing equation for hologram lenses of arbitrary shape, Opt. Commun. **14**, 322–323 (1975)

3.125  H. W. Holloway, R. A. Ferrante: Computer analysis of holographic systems by means of vector ray tracing, Appl. Opt. **20**, 2081–2084 (1981)

3.126  W. Lukosz, A. Wüthrich: Holography with evanescent waves. I: Theory of the diffraction efficiency for s-polarized light, Optik **41**, 191–211 (1974)

3.127  A. Wüthrich, W. Lukosz: Holographie mit quergedämpften Wellen. II: Experimentelle Untersuchungen der Beugungswirkungsgrade, Optik **42**, 315–334 (1975)

3.128  J. Woznicki: Geometry of recording and color sensitivity for evanescent wave holography using a Gaussian beam, Appl. Opt. **19**, 631–637 (1980)

3.129  J. F. Miles: Imaging and magnification properties in holography, Opt. Acta **19**, 165–186 (1972)

3.130  J. F. Miles: Evaluation of the wavefront aberration in holography, Opt. Acta **20**, 19–31 (1973)

3.131  V. A. Dombrovskii, S. A. Dombrovskii, E. F. Pen: Effect of a Gaussian reference beam on the distortion of images reconstructed from holograms, Opt. Spektrosk. **45**, 974–981 (1978)

3.132  J. Ojeda-Castaneda, S. Guel-Sandoval: Generalized holographic formation of a lensless Fourier transform. 1: Different geometries, Appl. Opt. **18**, 3550–3552 (1979)

3.133  J. Ojeda-Castaneda: Generalized holographic formation of a lensless Fourier transform. 3: Spherical wave illumination, Appl. Opt. **19**, 1386–1388 (1980)

3.134  H. H. Arsenault: Geometrical optics of holograms, J. Opt. Soc. Am. **65**, 903–908 (1975)

192    References

3.135  Y. Ishii, J. Maeda, K. Murata: Holographic display of diffraction patterns suffering from third- and fifth-order aberrations, Opt. Acta **26**, 969–983 (1979)
3.136  R. Dändliker, K. Hess, T. Sidler: Astigmatic pencils of rays reconstructed from holograms, Israel J. of Techn. **18**, 240–246 (1980)
3.137  A. N. Zaborov, G. N. Pavlygin: Method of calculating the image in an astigmatic holographic system, Opt. Spektrosk. **50**, 197–199 (1981)
3.138  V. V. Smirnov: Analysis of the anamorphotic properties of hologram optical elements when astigmatism is present, Opt. Spektrosk. **45**, 1153–1157 (1978)
3.139  R. Dändliker: Private Communication
3.140  M. Matsumura: Analysis of wave-front aberrations caused by deformation of hologram media, J. Opt. Soc. Am. **64**, 677–681 (1974)
3.141  M. Matsumura: Evaluation of deformation tolerance of the hologram medium, J. Opt. Soc. Am. **64**, 928–933 (1974)
3.142  O. Ersoy: Virtual holography: A method of source and channel encoding and decoding of information, Appl. Opt. **18**, 2543–2554 (1979)
3.143  W. T. Cathey: Effect of reference and illuminating sources size in holography, J. Opt. Soc. Am. **69**, 273–277 (1979)
3.144  M. Nazarathy, J. Shamir: Holography described by operator algebra, J. Opt. Soc. Am. **71**, 529–541 (1981)
3.145  J.-P. Zürcher, W. Schumann: Some intrinsic considerations on the nonlinear theory of thin shells, Acta Mech. **40**, 123–140 (1981)
3.146  J.-P. Zürcher: "Quelques aspects généraux de la théorie non-linéaire des coques minces. Utilisation de fonctions scalaires à partir d'un principe de variation mixte", Thesis No. 7250, ETH Zürich (1983)
3.147  W. Schumann: Duality property in holographic imaging, J. Opt. Soc. Am. **71**, 525–528 (1981)
3.148  K. S. Mustafin: Astigmatism of holographic diffraction grating on toroidal substrates, Opt. Spektrosk. **47**, 588–590 (1979)
3.149  K. S. Mustafin: Aberrations of thin holograms produced on a spherical substrate, Opt. Spektrosk. **37**, 1158–1162 (1974)
3.150  W. T. Welford: Practical design of an aplanatic hologram lens of focal length 50 mm and numerical aperture 0.5, Opt. Commun. **15**, 46–49 (1975)
3.151  R. W. Smith: Astigmatism free holographic lens elements, Opt. Commun. **21**, 102–105 (1977)
3.152  S. Morozumi: Aberration theory of diffraction gratings, Optik **53**, 75–88 (1979)
3.153  M. Gaj, A. Kijek: Aberrations of third and fifth orders of holograms made on rotational surfaces of second degree, Opt. Appl. **10**, 341–349 (1980)
3.154  B. J. Brown, I. J. Wilson: Holographic grating aberration correction for a Rowland circle mount I, Opt. Acta **28**, 1587–1599 (1981)
3.155  B. J. Brown, I. J. Wilson: Holographic grating aberration correction for a Rowland circle mount II, Opt. Acta **28**, 1601–1610 (1981)
3.156  C. H. F. Velzel: A general theory of the aberrations of diffraction gratings and grating-like optical instruments, J. Opt. Soc. Am. **66**, 346–353 (1976), errata **67**, 1695 (1977)
3.157  C. H. F. Velzel: On the imaging properties of holographic gratings, J. Opt. Soc. Am. **67**, 1021–1027 (1977)
3.158  H. Noda, T. Namioka, M. Seya: Geometric theory of the grating, J. Opt. Soc. Am. **64**, 1031–1036 (1974)
3.159  T. Namioka, H. Noda, M. Seya: Possibility of using the holographic concave grating in vacuum monochromators, Sci. Light **22**, 77–99 (1973)
3.160  M. P. Chrisp: Aberrations of holographic toroidal grating systems, Appl. Opt. **22**, 1508–1518 (1983)
3.161  D. Lepère: Monochromateur à simple rotation du réseau, à réseau holographique sur support torique pour l'ultraviolet lointain, Nouv. Rev. Opt. **6**, 173–178 (1975)
3.162  M. Pouey: Second-order focusing conditions for ruled concave gratings, J. Opt. Soc. Am. **64**, 1616–1622 (1974)

3.163  W. C. Sweatt: Drescribing holographic optical elements as lenses, J. Opt. Soc. Am. **67**, 803–808 (1977)

3.164  M. R. Latta, R. V. Pole: Design techniques for forming 488-nm holographic lenses with reconstruction at 633 nm, Appl. Opt. **18**, 2418–2421 (1979)

3.165  N. G. Kiselev: Calculation of ray paths through a holographic diffraction grating arbitrarily oriented in space, Opt. Spektrosk. **48**, 352–357 (1980)

3.166  A. P. Yakimovich: Selective properties of 3-D holographic gratings using spherical wave fronts, Opt. Spektrosk. **47**, 960–967 (1979)

3.167  L. L. Kolyshkina, E. F. Zhigalko: Analysis of a colored shadow pattern obtained from a hologram, Zh. Tekh. Fiz. **51**, 613–617 (1981)

3.168  G. Hesse, R. Kowarschik, A. Richter: Volume holograms as frequency-selective elements, Opt. Quant. Elect. **11**, 87–96 (1979)

3.169  K. Winick: Designing efficient aberration-free holographic lenses in the presence of a construction-reconstruction wavelength shift, J. Opt. Soc. Am. **72**, 143–148 (1982)

3.170  S. K. Case, V. Gerbig: Efficient and flexible laser beam scanners constructed from volume holograms, Opt. Commun. **36**, 94–100 (1981)

3.171  Y. Ono, N. Nishida: Holographic laser scanners using generalized zone plates, Appl. Opt. **21**, 4542–4548 (1982)

3.172  T. Jannson: Structural information in volume holography, Opt. Appl. **9**, 169–177 (1979)

3.173  Y. Belvaux: Influence de divers paramètres d'enregistrement lors de la restitution d'un hologramme, Nouv. Rev. Opt. **6**, 137–147 (1975)

3.174  C. Durou, J.-P. Hot, R. Lefèvre: Effets simultanés du tassement et de la diminution d'indice de réfraction de l'émulsion photographique en holographie, Nouv. Rev. Opt. **7**, 87–94 (1976)

3.175  J. Ben Uri: Holography – a different approach, Optik **51**, 397–415 (1978)

3.176  K. Goto, Y. Kato, K. Togawa: Design of holographic gratings, Opt. Acta **26**, 841–861 (1979)

3.177  M. G. Moharam, T. K. Gaylord, R. Magnusson: Criteria for Bragg regime diffraction by phase gratings, Opt. commun. **32**, 14–18 (1980)

3.178  M. R. B. Forshaw: Explanation of the "venetian blind" effect in holography using the Ewald sphere concept, Opt. Commun. **8**, 201–206 (1973)

3.179  R. Alferness: Equivalence of the thin-grating decomposition and coupled-wave analysis of thick holographic gratings, Opt. Commun. **15**, 209–212 (1975)

3.180  O. V. Konstantinov, M. M. Panakhov, Y. F. Romanov, A. Y. Tropchenko: Difference in the widths of two Bragg peaks from a three-dimensional phase grating with inclined layers, Opt. Spektrosk. **47**, 591–597 (1979)

3.181  V. B. Konstantinov, Y. F. Romanov, A. F. Ryklilov, A. Y. Tropchenko: Secondary structures formed during the inscribing of diffractive three-dimensional phase gratings, Zh. Tekh. Fiz. **52**, 1849–1853 (1982)

3.182  V. Kondilenko, V. Markov, S. Odulov, M. Soskin: Diffraction of coupled waves and determination of phase mismatch between holographic grating and fringe pattern, Opt. Acta **26**, 239–251 (1979)

3.183  R. R. A. Syms, L. Solymar: The effect of angular selectivity on the monochromatic imaging performance of volume holographic lenses, Opt. Acta **30**, 1303–1318 (1983)

3.184  T. Todorov, P. Markovski, M. Mazakova, M. Miteva, V. Razsolkov, M. Pancheva: Spectral characteristics of thick-phase holographic gratings, Opt. Acta **28**, 379–388 (1981)

3.185  H. Kogelnik: Coupled wave theory for thick hologram gratings, Bell Syst. Tech. J. **48**, 2909–2947 (1969)

3.186  R. Petit (ed.): *Electromagnetic Theory of Gratings,* Topics Current phys., Vol. 22 (Springer, Berlin, Heidelberg 1980)

3.187  S. I. Ragnarsson: Holograms recorded in extremely thick photographic emulsions, Opt. Commun. **14**, 39–41 (1975)

3.188  M. R. B. Forshaw: Explanation of the diffraction finestructure in overexposed thick holograms, Opt. Commun. **15**, 218–221 (1975)

3.189  T. Kubota: The diffraction efficiency of hologram gratings recorded in an absorptive medium, Opt. Commun. **16**, 347–349 (1976)

3.190  T. Kubota: Characteristics of thick hologram grating recorded in absorptive medium, Opt. Acta **25**, 1035–1053 (1978)

3.191  T. Kubota: The bending of interference fringes inside a hologram, Opt. Acta **26**, 731–743 (1979)

3.192  S. Reich: Photodielektrische Polymere für die Holographie, Angew. Chem. **89**, 467–474 (1977)

3.193  M. I. Dzyubenko, A. P. Pyatikop, V. V. Shevchenko: Increasing the diffraction efficiency of reflecting three-dimensional holograms, by preventing emulsion shrinkage, Zh. Tekh. Fiz. **45**, 1522–1524 (1975)

3.194  B. Janowska, J. Szydlowska: Improved efficiency reflection holograms of diffusely reflecting objects, Opt. Appl. **9**, 3–6 (1979)

3.195  G. I. Lashkov, V. I. Sukhanov: Use of dispersion photorefraction due to processes in which triplet states participate to record 3-D phase holograms, Opt. Spektrosk. **44**, 1008–1015 (1978)

3.196  A. V. Alekseev-Popov, N. G. Dyachenko, V. E. Mandel, A. V. Tyurin: Dispersion of the optical parameters in thick amplitude-phase holograms, Opt. Spektrosk. **47**, 583–587 (1979)

3.197  A. V. Alekseev-Popov: Limiting diffraction efficiency of three-dimensional amplitude holograms, Zh. Tekh. Fiz. **51**, 1275–1278 (1981)

3.198  A. V. Alekseev-Popov, S. A. Gevelyuk: Contributions of amplitude and phase modulation to diffraction efficiency in three-dimensional reflective holograms, Zh. Tekh. Fiz. **52**, 2100–2102 (1982)

3.199  A. P. Yakimovich: Dynamic self-amplification of scattering noise in volume-hologram recording, Opt. Spektrosk. **49**, 354–358 (1980)

3.200  J. Růžek, P. Fiala: Reflection holographic portraits, Opt. Acta **26**, 1257–1264 (1979)

3.201  D. A. Woodbury, T. A. Rabson, F. K. Tittel: Hologram indexing in $LiNbO_3$ with a tunable pulsed laser source, Appl. Opt. **18**, 2555–2558 (1979)

3.202  J. P. Huignard, B. Ledu: Collinear Bragg diffraction in photorefractive $Bi_{12}SiO_{20}$, Opt. Lett **7**, 310–312 (1982)

3.203  M. P. Owen, A. A. Ward, L. Solymar: Internal reflections in bleached reflection holograms, Appl. Opt. **22**, 159–163 (1983)

3.204  R. R. A. Syms, L. Solymar: Planar volume phase holograms formed in bleached photographic emulsions, Appl. Opt. **22**, 1479–1496 (1983)

3.205  R. Dändliker: Lecture notes: "Holographie und Phasenkonjugation" at the ETH Zürich (1983)

3.206  Y. Ninomiya: Recording characteristics of volume holograms, J. Opt. Soc. Am. **63**, 1124–1130 (1973)

3.207  B. Benlardi, L. Solymar: The effect of the relative intensity of the reference beam on the reconstructing properties of volume phase holograms, Opt. Acta **26**, 271–278 (1979)

3.208  R. Ferrante, M. P. Owen, L. Solymar: Conjugate diffraction order in a volume holographic off-axis lens, J. Opt. Soc. Am. **71**, 1385–1389 (1981)

3.209  R. R. A. Syms, L. Solymar: Localized one-dimensional theory for volume holograms, Opt. Quant. Elect. **13**, 415–419 (1981)

3.210  R. R. A. Syms, L. Solymar: Higher diffraction orders in on-axis holographic lenses, Appl. Opt. **21**, 3263–3268 (1982)

3.211  M. P. Owen, L. Solymar: Efficiency of volume phase reflection holograms recorded in an attenuating medium, Opt. Commun. **34**, 321–326 (1980)

3.212  R. Magnusson, T. K. Gaylord: Analysis of multiwave diffraction of thick gratings, J. Opt. Soc. Am. **67**, 1165–1170 (1977)

3.213  M. G. Moharam, T. K. Gaylord: Rigorous coupled-wave analysis of planar-grating diffraction, J. Opt. Soc. **71**, 811–818 (1981)

3.214  Z. Zylberberg, E. Marom: Rigorous coupled-wave analysis of pure reflection gratings, J. Opt. Soc. Am. **73**, 392–398 (1983)

3.215  K. C. Johnson: Coupled scalar wave diffraction theory, Appl. Phys. **24**, 249–260 (1981)

3.216  E. I. Krupitskii, B. K. Chernov: Light diffraction by a sinusoidally inhomogeneous dielectric layer, Opt. Spektrosk. **39**, 571–578 (1975)

3.217  J. Čtyroký: Coupled-mode theory of Bragg diffraction in the presence of multiple internal reflections, Opt. Commun. **16**, 259–261 (1976)

3.218  A. A. Leshchyev, V. G. Sidorovich: Theory of the transformation of light waves by reflection 3-D holograms, Opt. Spektrosk. **44**, 302–308 (1978)

3.219  N. D. Vorzobova, A. A. Leshchev, P. M. Semenov, V. G. Sidorovich, D. I. Staselko: Method of optimizing 3-D hologram recording conditions, Opt. Spektrosk. **45**, 779–787 (1978)

3.220  V. L. Vinetsky, N. V. Kukhtarev: Geometrical factors in dynamic holographic conversion of light beams, Kvan Elektr. **5**, 405–411 (1978)

3.221  V. G. Sidorovich, V. V. Shkunov: Spectral selectivity of 3-D holograms, Opt. Spektrosk. **44**, 1001–1007 (1978)

3.222  A. P. Yakimovich: Multilayer three-dimensional holographic gratings, Opt. Spektrosk. **49**, 158–164 (1980)

3.223  M. Gaj: The influence of the layer thickness on the coupling efficiency of plane waveguide with periodically variable refractive index, Opt. Appl. **8**, 29–33 (1978)

3.224  P. Markovski, N. Koleva, T. Todorov: Some characteristics of phase holographic gratings in the intermediate regime of diffraction, Opt. Quant. Elect. **13**, 515–518 (1981)

3.225  N. Tsukada, R. Tsujinishi, K. Tomishima: Effects of the relative phase relationships of gratings on diffraction from thick holograms, J. Opt. Soc. Am. **69**, 705–711 (1979)

3.226  P. Sheng, W. J. Burke, H. A. Weakliem: Signal distortion noise in volume phase holograms, Opt. Quant. Elect. **9**, 427–436 (1977)

3.227  S. Kessler, R. Kowarschik: Diffraction efficiency of volume holograms. Part 1: Transmission holograms, Opt. Quant. Elect. **7**, 1–14 (1975)

3.228  R. Kowarschik, S. Kessler: Zum Beugungswirkungsgrad von Volumenhologrammen. Teil II. Reflexionshologramme, Opt. Quant. Elect. **7**, 399–411 (1975)

3.229  F. Lederer, U. Langbein: Attenuated thick hologram gratings. Part I: Diffraction efficiency, Opt. Quant. Elect. **9**, 473–485 (1977)

3.230  F. Lederer, U. Langbein: Attenuated thick hologram gratings. Part II: Anomalous absorption, Opt. Quant. Elect. **9**, 487–491 (1977)

3.231  W. H. Lee, W. Streifer: Diffraction efficiency of evanescent-wave holograms. I. TE-polarization, J. Opt. Soc. Am. **68**, 795–801 (1978)

3.232  W. H. Lee, W. Streifer: Diffraction efficiency of evanescent-wave holograms. II. TM-polarisation, J. Opt. Soc. Am. **68**, 802–806 (1978)

3.233  T. Jannson: Informations capacity of Bragg holograms in planar optics, J. Opt. Soc. Am. **71**, 342–347 (1981)

3.234  T. Jannson, J. Jannson: Cylindrical Bragg holograms, J. Opt. Soc. Am. **72**, 1062–1067 (1982)

3.235  O. V. Konstantinov, M. M. Panakhov, Y. F. Romanov: Theory of phase reflection gratings, Opt. Spektrosk. **44**, 1016–1024 (1978)

3.236  O. V. Konstantinov, M. M. Panakhov, Y. F. Romanov: Electrodynamic perturbation theory for light diffraction from 3-D phase gratings, Opt. Spektrosk. **46**, 979–985 (1979)

3.237  M. G. Moharam, T. K. Gaylord: Rigorous coupled-wave analysis of grating diffraction – E-mode polarization and losses, J. Opt. Soc. Am. **73**, 451–455 (1983)

3.238  L. Solymar, C. J. R. Sheppard: A two-dimensional theory of volume holograms with electric polarization in the plane of the grating, J. Opt. Soc. Am. **69**, 491–495 (1979)

3.239  P. St. J. Russell, L. Solymar: The properties of holographic overlap gratings, Opt. Acta **26**, 329–347 (1979)

3.240  L. Solymar, B. Benlardi, D. J. Cooke: A two-dimensional theory of higher order modes in volume phase holograms, Opt. Quant. Elect. **11**, 558–560 (1979)

3.241  W. E. Parry, L. Solymar: A general solution for two-dimensional volume holograms, Opt. Quant. Elect. **9**, 527–531 (1977)

3.242  L. Solymar, M. P. Jordan: Two-dimensional transmission type volume holograms for incident plane waves of arbitrary amplitude distribution, Opt. Quant. Elect. **9**, 437–444 (1977)

3.243  R. R. A. Syms, L. Solymar: Analysis of volume holographic cylindrical lenses, J. Opt. Soc. Am. **72**, 179–186 (1982)

3.244  P. St. J. Russell: Reconstruction fidelity from volume holograms of finite width and variable index modulation, J. Opt. Soc. Am. **69**, 496–503 (1979)

3.245  I. Y. Brusin: Dependence of diffraction efficiency on hologram thickness, Opt. Spektrosk. **39**, 750–758 (1975)

3.246  A. Korpel: Two-dimensional plane wave theory of strong acousto-optic interaction in isotropic media, J. Opt. Soc. Am. **69**, 678–683 (1979)

3.247  J. Van Roey, P. E. Lagasse: Coupled wave analysis of obliquely incident waves in thin film gratings, Appl. Opt. **20**, 423–429 (1981)

3.248  M. G. Moharam, T. K. Gaylord: Coupled-wave analysis of reflection gratings, Appl. Opt. **20**, 240–244 (1981)

3.249  M. G. Moharam, T. K. Gaylord, R. Magnusson: Diffraction characteristics of three-dimensional crossed-beam volume gratings, J. Opt. Soc. Am. **70**, 437–442 (1980)

3.250  U. Langbein, F. Lederer: Spatial filtering properties of volume holograms, Opt. Quant. Elect. **11**, 29–42 (1979)

3.251  D. J. Cooke, L. Solymar, C. J. R. Sheppard: A three-dimensional vectorial theory for volume holograms, Int. J. Elect. **46**, 337–356 (1979)

3.252  W. E. Parry, D. J. Cooke, L. Solymar: Solutions of the vector differential equations of volume holography, Int. J. Elect. **46**, 357–365 (1979)

3.253  V. V. Shkunov, B. Y. Zeldovich: Recording and reconstruction of an object's wave state of polarization by a volume hologram, Appl. Opt. **18**, 3633–3643 (1979)

3.254  M. G. Moharam, T. K. Gaylord: Three-dimensional vector coupled-wave analysis of planar-grating diffraction, J. Opt. Soc. Am. **73**, 1105–1112 (1983)

3.255  A. W. Lohmann: Reconstruction of vectorial wavefronts, Appl. Opt. **4**, 1667–1668 (1965)

3.256  F. A. Sattarov: Polarizing properties of thick-film hologram gratings, Opt. Spektrosk. **47**, 764–768 (1979)

3.257  V. M. Serdyuk, A. P. Khapalyuk: Diffraction of an arbitrarily polarized plane electromagnetic wave by a phase hologram in the coupled-wave approximation, Zk. Tekh. Fiz. **51**, 2537–2540 (1981)

3.258  V. M. Serdyuk, A. P. Khapalyuk: Diffraction of arbitrarily polarized plane electromagnetic waves by vector holograms, Zh. Tekh. Fiz. **51**, 2541–2545 (1981)

3.259  H. Eklund, A. Roos, S. T. Eng: Rotation of laser beam polarization in acousto-optic devices, Opt. Quant. Elect. **7**, 73–79 (1975)

3.260  K. C. Johnson: General coupled-wave diffraction theory, Appl. Phys. **25**, 169–178 (1981)

3.261  K. C. Johnson: Image reconstruction from a dielectric volume hologram, Appl. Phys. **25**, 357–360 (1981)

3.262  T. Wilson, D. K. Saldin, L. Solymar: Phase conjugation by degenerate four-wave mixing. A general two-dimensional theory, Opt. Acta **29**, 1041–1047 (1982)

3.263  R. W. Meier: Magnification and third-order aberrations in holography, J. Opt. Soc. Am. **55**, 987–992 (1965)

3.264  M. Marquet, H. Royer: Etude des aberrations géométriques des images reconstituées par holographie, C. R. Acad. Sc. Paris **260**, 6051–6053 (1965)

3.265  J. A. Armstrong: Fresnel holograms: their imaging properties and aberrations, IBM J. Res. and Dev. **9**, 171–178 (1965)

3.266  W. Lukosz: Equivalent-lens theory of holographic imaging, J. Opt. Soc. Am. **58**, 1084–1091 (1968)

3.267  P. Hariharan: Longitudinal distortion in images reconstructed by reflection holograms, Opt. Commun. **17**, 52–54 (1976)

3.268  M. A. Gan: Third-order aberrations and the fundamental parameters of axisymmetrical holographic elements, Opt. Spektrosk. **47**, 759–763 (1979)

3.269 S. T. Bobrov, G. I. Greisukh, N. A. Prokhorov, Y. G. Turkevich, V. G. Shitov: Third-order monochromatic aberrations of axial holographic lenses, Opt. Spektrosk. **46**, 153–157 (1979)

3.270 S. T. Bobrov, Y. G. Turkevich: Method of calculating the wave aberrations of complex holographic systems, Opt. Spektrosk. **46**, 986–991 (1979)

3.271 G. Mulak: Hologram aberrations outside the binomial expansion, Opt. Appl. **8**, 139–144 (1978)

3.272 G. Mulak: An analysis of the hologram aberration in the intermediate and far regions, Opt. Appl. **9**, 257–265 (1979)

3.273 G. Mulak: Higher order aberrations in holograms, Opt. Appl. **10**, 421–433 (1980)

3.274 A. Pulka: The aberration coefficients of Fourier holograms, Opt. Appl. **10**, 451–464 (1980)

3.275 J. Nowak: Holograms of corrected spherical and comatic aberrations, Opt. Appl. **8**, 145–148 (1978)

3.276 J. Nowak: A contribution to the hologram aberration correction, Opt. Appl. **9**, 121–124 (1979)

3.277 J. Nowak, G. Mulak: Field curvature and astigmatism in holographic imaging, Opt. Appl. **11**, 161–168 (1981)

3.278 J. Nowak: Third-order aberrations of holograms, Opt. Appl. **10**, 245–251 (1980)

3.279 J. Nowak: Estimation of aberration coefficients in holography, J. Opt. **11**, 121–122 (1980)

3.280 J. Nowak, M. Zajac: Influence of the entrance pupil position on the hologram aberration correction, Opt. Appl. **10**, 285–293 (1981)

3.281 P. C. Mehta, K. Syam Sunder Rao, R. Hradaynath: Higher order aberrations in holographic lenses, Appl. Opt. **21**, 4553–4558 (1982)

3.282 G. N. Buinov, I. E. Kit, K. S. Mustafin, M. I. Savrasova: Compensation of spherical aberration of hologram lenses by a plane-parallel plate, Opt. Spektrosk. **38**, 159–162 (1975)

3.283 G. N. Buinov, K. S. Mustafin: Method of compensating for the spherical aberration of axial hologram lenses, Opt. Spektrosk. **41**, 341–342 (1976)

3.284 K. S. Mustafin: Possibility of reducing astigmatism and coma in a hologram lens system, Opt. Spektrosk. **47**, 390–394 (1979)

3.285 G. N. Buinov, F. A. Sattarov, N. F. Eiken: Monochromatic aberrations of an axial holographic lens obtained using spherical aberration of the wave-fronts, Opt. Spektrosk. **49**, 398–400 (1980)

3.286 V. G. Shitov, G. I. Greisukh: Compensation of aberrations in the simplest refraction-diffraction optical systems, Opt. Spektrosk. **50**, 786–792 (1981)

3.287 W. H. Swantner, W. H. Lowrey: Zernike-Tatian polynomials for interferogram reduction, Appl. Opt. **19**, 161–163 (1980)

3.288 J. Y. Wang, D. E. Silva: Wave-front interpretation with Zernike polynomials, Appl. Opt. **19**, 1510–1518 (1980)

3.289 L. N. Vagin, L. G. Nazarova, T. M. Arseneva, V.A. Vanin: Holographic miniaturization of scientific and technical documents, Opt. Spektrosk. **38**, 994–998 (1975)

3.290 V. G. Mityakov, V. D. Fedorov: Aberrations of the image field and information capacity of microholograms in recording and readout by light sources with different wavelengths, Opt. Spektrosk. **39**, 951–955 (1975)

3.291 E. Müller: "Auswertung holographischer Interferogramme unter Berücksichtigung der durch Rekonstruktion mit geänderter Lichtwellenlänge erzeugten Bildmodifikationen", Thesis No. 7246, ETH Zürich (1983)

3.292 W. Schumann: On the deformation of holographic images as a result of an optical modification at the reconstruction, Opt. Acta **27**, 241–250 (1980)

3.293 H. Rüll, E. Storck: Systematic analysis of the invariance properties of plane holograms, Opt. Commun. **15**, 38–41 (1975)

3.294 H. Rüll, E. Storck: Experimental investigation of the translation properties of some holographic configurations, Opt. Commun. **15**, 42–45 (1975)

3.295  A. N. Zaborov, G. N. Pavlygin: Properties of a virtual holographic image when its scale is altered, Opt. Spektrosk. **48**, 808–814 (1980)

3.296  A. W. Lohmann, N. Streibl: On the fundamentals of 3-D display, SPIE **402**, 6–12 (1983)

3.297  W. Schumann, D. Cuche: Deformation of a holographic image in space, J. Opt. Soc. Am. **72**, 136–142 (1982)

3.298  J. D. Redman: Novel applications of holography, J. Sci. Inst. **1**, 821–822 (1968)

3.299  J. D. Redman, W. P. Wolton, E. Shuttleworth: Use of holography to make truly three-dimensional X-ray images, Nature **220**, 58–60 (1968)

3.300  T. F. Krile, M. O. Hagler, W. D. Redus, J. F. Walkup: Multiplex holography with chirp-modulated binary phase-coded reference-beam masks, Appl. Opt. **18**, 52–56 (1979)

3.301  R. Kasturi, T. F. Krile, J. F. Walkup: Multiplex holography for space-variant processing: a transfer function sampling approach, Appl. Opt. **20**, 881–886 (1981)

3.302  D. J. De Bitetto: Holographic panoramic stereograms synthesized from white light recordings, Appl. Opt. **8**, 1740–1741 (1969)

3.303  N. George, J. T. McCrickerd: Holography and stereoscopy: the holographic stereogram, Phot. Sci. Eng. **13**, 342–350 (1969)

3.304  T. Kasahara, Y. Kimura, M. Kawai: 3-D construction of imaginary objects by the method of holographic stereogram, in *Applications of Holography,* ed. by E. S. Barrekette et al. (Plenum, New York 1971) pp. 19–34

3.305  G. Groh, M. Kock: 3-D display of X-ray images by means of holography, Appl. Opt. **9**, 775–777 (1970)

3.306  M. C. King, A. M. Noll, D. H. Berry: A new approach to computer-generated holography, Appl. Opt. **9**, 471–475 (1970)

3.307  I. Glaser: Anamophic imagery in holographic stereograms, Opt. Commun. **7**, 323–326 (1973)

3.308  I. Glaser, A. A. Friesem: Imaging properties of holographic stereograms, SPIE **120**, 150–162 (1977)

3.309  S. A. Benton: Holographic displays – a review, Opt. Eng. **14**, 402–407 (1975)

3.310  S. A. Benton: Holographic displays: 1975–1980, Opt. Eng. **19**, 686–690 (1980)

3.311  S. A. Benton: White light transmission/reflection holographic imaging, in *Applications of Holography and Optical Data Processing,* ed. by E. Marom, A. A. Friesem, E. Wiener-Avnear (Pergamon, Oxford 1977) pp. 401–409

3.312  K. Okada, T. Honda, J. Tsujiuchi: Distortions of reconstructed images from cylindrical holographic stereograms, SPIE **212**, 28–33 (1979)

3.313  T. Honda, K. Okada, J. Tsujiuchi: 3-D distortion of observed images reconstructed from a cylindrical holographic stereogram. (1) Laser light reconstruction type, Opt. Commun. **36**, 11–16 (1981)

3.314  K. Okada, T. Honda, J. Tsujiuchi: 3-D distortion of observed images reconstructed from a cylindrical holographic stereogram. (2) White light reconstruction type, Opt. Commun. **36**, 17–21 (1981)

3.315  K. Okada, T. Honda, J. Tsujiuchi: Image blur of multiplex holograms, Opt. Commun. **41**, 397–402 (1982)

3.316  K. Okada, T. Honda, J. Tsujiuchi: Multiplex holograms made of computer processed images, SPIE **402**, 33–37 (1983)

3.317  J. Tsujiuchi: Holographic stereograms as a tool of non-destructive testing, SPIE **370**, 17–19 (1983)

3.318  R. L. Fusek, L. Huff: Use of a holographic lens for producing cylindrical holographic stereograms, SPIE **215**, 32–38 (1980)

3.319  L. Huff, R. L. Fusek: Optical techniques for increasing image width in cylindrical holographic stereograms, SPIE **215**, 39–45 (1980)

3.320  L. Huff, R. L. Fusek: Color holographic stereograms, Opt. Eng. **19**, 691–695 (1980)

3.321  R. L. Fusek, L. Huff: Use of a holographic lens for producing cylindrical holographic stereograms, Opt. Eng. **20**, 236–240 (1981)

3.322  L. Huff, R. L. Fusek: Optical techniques for increasing image width in cylindrical holographic stereograms, Opt. Eng. **20**, 241–245 (1981)

3.323  L. Huff, R. L. Fusek: Three-dimensional imaging with holographic stereograms, SPIE **402**, 38–50 (1983)

3.324  I. R. Utyamishev, R. I. Utyamishev: Holographic methods of 3-D representation from a number of plane images for medical X-ray diagnostics, in [Ref. 3.26, pp. 102–109]

3.325  M. Grosmann, P. Meyrueis, J. Fontaine: Three dimensional holographic synthesis (T.H.I.S.) of X-ray pictures in [Ref. 3.26, pp. 110–116]

3.326  K. Iwata, S. Wantanabe, M. Suzuki, T. Saito: Holographic viewing of neuroradiograms – An attempt of a new method for 3-dimensional and dynamic observation –, SPIE **211**, 59–61 (1979)

3.327  V. M. Antonov, R. M. Mamilayev, I. P. Nalimov, A. H. Shakirov: 3-D holographic pictures of human lungs, Opt. Commun. **38**, 81–84 (1981)

3.328  K. Sugimura, M. Matsuo, S. Ikeda: Clinical application of the multiplex holography, SPIE **370**, 20–25 (1983)

3.329  R. J. Perlmutter, J. W. Goodman, A. Macovski: Digital holographic display of medical CT images, SPIE **367**, 109–116 (1982)

3.320  K. M. Johnson, L. Hesselink, J. W. Goodman: Multiple exposure holographic display of CT medical data, SPIE **367**, 149–154 (1982)

3.331  S. M. Jaffey, K. Dutta: Digital perspective correction for cylindrical holographic stereograms, SPIE **367**, 130–140 (1982)

## Chapter 4

4.1   R. E. Brooks, L. O. Heflinger, R. F. Wuerker: Interferometry with a holographically reconstructed comparison beam, Appl. Phys. Lett. **7**, 248–249 (1965)

4.2   J. M. Burch: The application of lasers in production engineering, Product. Engineer **44**, 431–443 (1965)

4.3   R. J. Collier, E. T. Doherty, K. S. Pennington: Application of moiré techniques to holography, Appl. Phys. Lett. **7**, 223–225 (1965)

4.4   D. Gabor, G. W. Stroke, D. Brumm, A. Funkhouser, A. Labeyrie: Reconstruction of phase objects by holography, Nature **208**, 1159–1162 (1965)

4.5   M. H. Horman: An application of wavefront reconstruction to interferometry, Appl. Opt. **4**, 333–336 (1965)

4.6   R. L. Powell, K. A. Stetson: Interferometric vibration analysis by wavefront reconstruction, J. Opt. Soc. Am. **55**, 1593–1598 (1965)

4.7   K. A. Stetson, R. L. Powell: Interferometric hologram evaluation and real-time vibration analysis of diffuse objects, J. Opt. Soc. Am. **55**, 1694–1695 (1965)

4.8   J. M. Burch, A. E. Ennos, R. J. Wilton: Dual- and multiple-beam interferometry by wavefront reconstruction, Nature **209**, 1015–1016 (1966)

4.9   K. A. Haines, B. P. Hildebrand: Surface-deformation measurement using the wavefront reconstruction technique, Appl. Opt. **5**, 595–602 (1966)

4.10  L. O. Heflinger, R. F. Wuerker, R. E. Brooks: Holographic interferometry, J. Appl. Phys. **37**, 642–649 (1966)

4.11  B. P. Hildebrand, K. A. Haines: Interferometric measurements using the wavefront reconstruction technique, Appl. Opt. **5**, 172–173 (1966)

4.12  H. Nassenstein: Holographische Interferometrie diffus reflektierender Objekte, Phys. Lett. **21**, 290–291 (1966)

4.13  K. A. Stetson, R. L. Powell: Hologram interferometry, J. Opt. Soc. Am. **56**, 1161–1166 (1966)

4.14  G. W. Stroke, A. E. Labeyrie: Two-beam interferometry by successive recording of intensities in a single hologram, Appl. Phys. Lett. **8**, 42–44 (1966)

4.15  G. Wernicke, W. Osten: *Holografische Interferometrie* (VEB Fachbuchverlag, Leipzig 1982)

4.16  P. Hariharan: *Optical Holography* (Cambridge U. Press, Cambridge 1984)

4.17  C. M. Vest: *Holographic Interferometry* (Wiley, New York 1979)

200     References

4.18   H. J. Caufield (ed.): *Handbook of Optical Holography* (Academic, New York 1979)
4.19   Y. I. Ostrovsky, M. M. Butusov, G. V. Ostrovskaya: *Interferometry by Holography,* Springer Ser. Opt. Sci., Vol. 20 (Springer, Berlin, Heidelberg 1980)
4.20   R. Dändliker: Heterodyne holographic interferometry, in *Progress in Optics,* 17 Chap. 1 (North-Holland, Amsterdam 1980)
4.21   N. Abramson: *Making and Evaluation of Holograms* (Academic, London 1981)
4.22   C. Bonjour, M. Matthey (eds): *Lasers et Applications Industrielles* (Presses Polytechniques Romandes, Lausanne 1982)
4.23   R. J. Jones, C. Wykes: *Holographic and Speckle Interferometry* (Cambridge U. Press, Cambridge 1983)
4.24   W. Schumann, M. Dubas: *Holographic Interferometry. From the Scope of Deformation Analysis of Opaque Bodies,* Springer Ser. Opt. Sci. Vol. 16 (Springer, Berlin, Heidelberg 1979)
4.25   C. M. Vest: Status and future of holographic nondestructive evaluation, SPIE **349,** 186–198 (1982)
4.26   C. A. Sciammarella, S. K. Chawla: Multiplication of holographic fringes, its application to crack detection, VDI-Berichte Nr. 313, 775–779 (1978)
4.27   W. L. Haworth: Holographic study of fatigue deformation and crack growth in metals, Fatig. Eng. Mat. Struct. **1,** 351–361 (1979)
4.28   S. Amadesi, A. D'Altorio, D. Paoletti: Real-time holography for microcrack detection in ancient golden paintings, Opt. Eng. **22,** 660–662 (1983)
4.29   G. Cadoret, P. Bertho: Etude des déformations plastiques en fond de fissure, Rev. Franç. Méc. **71,** 5–10 (1979)
4.30   J. T. Ke, H. Q. Zhang, C. I. Zhao: Measurement of residual stresses by modern optical methods, ICO-13 Conference digest, Optics in Modern Science and Technology, Sapporo (1984) pp. 656–657
4.31   V. I. Arkhipov: Field of elastoplastic strains in the mouth zone of a crack, J. Appl. Mech. Tech. Phys. **21,** 576–579 (1981)
4.32   D. C. Holloway: Application of holographic interferometry to stress wave and crack propagation problems, Opt. Eng. **21,** 468–473 (1982)
4.33   D. W. Park, D. A. Summers: Model Studies of subsidence and ground movement using laser holographic interferometry, Int. J. Rock Mech. Mining Scien. Geomech. Abst. **14,** 235–245 (1977)
4.34   V. J. Parks, R. J. Sanford: Photoelastic and holographic analysis of a turbine-engine component, Exp. Mech. **18,** 328–334 (1978)
4.35   E. Archbold, A. E. Ennos, M. S. Virdee: The deformation of steel bars in a four-point bending machine, measured by holographic interferometry, VDI Berichte Nr. 313, 517–522 (1978)
4.36   A. E. Ennos, M. S. Virdee: Application of reflection holography to deformation measurement problems, Exp. Mech. **22,** 202–209 (1982)
4.37   W. H. Sheldon: Comparative evaluation of potential NDE techniques for inspection of advanced composite structures, Mat. Evaluation **36,** 41–46 (1978)
4.38   H. G. Leis: Messung des Zylinderverzugs, VDI Berichte Nr. 313, 629–635 (1978)
4.39   H. G. Leis, F. W. Leipold: Kombination der holografischen Interferometrie mit der DMS-Messtechnik zur Spannungsanalyse, Tech. Messen **49,** 271–276 (1982)
4.40   V. P. Chernov, B. B. Gorbatenko: Color holographic interferogram, Opt. Spektrosk. **39,** 963–966 (1975)
4.41   F. T. S. Yu, A. Tai, H. Chen: One-step rainbow holographic interferometry, SPIE **192,** 178–183 (1979)
4.42   E. Müller, W. Looser, P. E. Gygax: Comparison of modal analysis and holography, Annals of the CIRP **29,** 397–402 (1980)
4.43   N. I. Prigovorskii, N. S. Cherpakova: Holographic methods in mechanical tests, Indus. Lab. **44,** 830–842 (1978)
4.44   V. P. Shchepinov, V. V. Yakovlev: Measurement of elastoplastic strain by hologram interferometry, Zh. Tekh. Fiz. **49,** 1005–1007 (1979)

4.45  V. P. Shchepinov et al.: Measuring the elastic and permanent deflections of a gear tooth by holographic interferometry, Russian Eng. J. **60**, 3–5 (Dec. 1980)

4.46  V. P. Shchepinov, V. V. Yakovlev: Investigation of deformation of parts by the method of holographic interferometry, J. Appl. Mech. Tech. Phys. **20**, 776–778 (1980)

4.47  H. Wachutka, H. Kordisch, B. Fischer: Holographisch-Interferometrische Verformungs-messungen zur Bewertung lokaler Inhomogenitäten im Inneren dickwandiger Bauteile, VDI-Berichte Nr. 366, 71–77 (1980)

4.48  T. A. Il'inskaya, V. L. Kazak, A. M. Kudryashov, I. M. Nagibina: Universal apparatus for holographic interferometry, Sov. J. Opt. Techn. **48**, 82–83 (1981)

4.49  V. A. Zhilkin: Study on plane problems by the method of holographic interferometry, Strength of Mat. **12**, 916–920 (1981)

4.50  G. P. Monakhov-il'in et al.: Analyzing the behavior of composite structures by the finite element and holographic interferometry methods, Russian Eng. J. **60**, 3–5 (June 1980)

4.51  A. G. Khadakkar, N. R. Iyer, T. V. S. R. Appa Rao: Determination of the torsional stiffness of an azimuth tube with a stiffened opening by holographic interferometry, Opt. Lasers Eng. **1**, 147–159 (1980)

4.52  J. A. Gilbert, D. T. Vedder: Development of holographic techniques to study thermally induced deformation, Exp. Mech. **21**, 138–144 (1981)

4.53  P. C. Metha, D. Mohan, C. Bhan, P. Lal, R. Hradaynath: Holographic visualization of flaws using magnetic stressing, Opt. Laser Techn. **14**, 269–271 (1982)

4.54  M. J. Marchant, M. B. Snell: Determination of the flexural stiffness of thin plates from small deflection measurements using optical holography, J. Strain Analy. Eng. Design **17**, 53–61 (1982)

4.55  F. Erdmann-Jesnitzer, T. Winkler: Application of the holographic nondestructive testing method to the evaluation of disbondings in sandwich plates, Int. J. Adhesion and Adhesives **1**, 189–194 (1981)

4.56  M. Yonemura, T. Nishisaka, H. Machida: Endoscopic hologram interferometry using fiber optics, Appl. Opt. **20**, 1664–1667 (1981)

4.57  D. M. Rowley: Interferometry with miniature format volume reflection holograms, Opt. Acta **28**, 907–915 (1981)

4.58  D. M. Rowley: A novel form of holocamera, Opt. Lasers Eng. **2**, 67–70 (1981)

4.59  P. Paulet: Holographic measuring of the deformations of various internal combustion engine parts, SPIE **398**, 30–34 (1983)

4.60  G. Schönebeck: New holographic means to exactly determine coefficients of elasticity, SPIE **398**, 130–136 (1983)

4.61  P. Jacquot, P. K. Rastogi, L. Pflug: Holographic interferometry applied to external osteosynthesis: comparative analysis of the performances of external fixation prototypes, SPIE **398**, 149–158 (1983)

4.62  J. A. Gilbert, T. D. Dudderar, M. E. Schultz, A. J. Boehnlein: The monomode filter – a new tool for holographic interferometry, Exp. Mech. **23**, 190–195 (1983)

4.63  H. W. Chang: Out-of-plane deformation of balanced and symmetric composites as measured by holographic interferometry, Exp. Tech. **1**, 30–34 (1983)

4.64  S. Toyooka: Contour mapping of the first and the second derivatives of plate deflection by using modulated diffraction gratings made by double-exposure holography, Opt. Acta **25**, 991–1000 (1978)

4.65  D. C. Holloway, A. M. Patacca, W. L. Fourney: Direction-sensitive displacement analysis by multiple frequency holographic interferometry, Appl. Opt. **17**, 1213–1219 (1978)

4.66  J. A. Gilbert, J. W. Herrick: Holographic displacement analysis with multimode-fiber optics, Exp. Mech. **21**, 315–320 (1981)

4.67  D. C. Holloway, A. J. Durelli: Interpretation of fringes in holographic interferometry, Mech. Res. Commun. **10**, 97–103 (1983)

4.68  B. P. Hildebrand, J. D. Trolinger: Statistical analysis of a holographic system intended for the Space Shuttle, Appl. Opt. **22**, 2124–2127 (1983)

4.69  D. Groves, M. J. Lalor, N. Cohen, J. T. Atkinson: A holographic technique with computer aided analysis for the measurement of wear, J. Phys. E, Sci. Instr. **13**, 741–746 (1980)

4.70    J. P. Hot, C. Durou: System for the automatic analysis of interferograms obtained by holographic interferometry, SPIE **210**, 144–151 (1980)

4.71    D. W. Robinson: Automatic fringe analysis with a computer image-processing system, Appl. Opt. **22**, 2169–2176 (1983)

4.72    Y. Xu, C. M. Vest: Holographic technique for simultaneous measurement of displacement and tilt, Appl. Opt. **22**, 2137–2140 (1983)

4.73    D. A. Tichenor, V. P. Madsen: Computer analysis of holographic interferograms for nondestructive testing, Opt. Eng. **18**, 469–472 (1979)

4.74    B. C. R. Ewan: Particle velocity distribution measurement by holography, Appl. Opt. **18**, 3156–3160 (1979)

4.75    E. R. Freniere, O. E. Toler, R. Race: Interferogram evaluation program for the HP-9825A calculator, Opt. Eng. **20**, 253–255 (1981)

4.76    W. R. J. Funnell: Image processing applied to the interactive analysis of interferometric fringes, Appl. Opt. **20**, 3245–3250 (1981)

4.77    R. Peralta-Fabi: Measurements of microdisplacements by holographic digital image processing, SPIE **398**, 169–173 (1983)

4.78    Y. Katzir, I. Glaser, A. A. Friesem, B. Sharon: On-line acquisition and analysis for holographic nondestructive evaluation, Opt. Eng. **21**, 1016–1021 (1982)

4.79    Y. Katzir, I. Glaser: Separation of in-plane and out-of-plane motions in holographic interferometry, Appl. Opt. **21**, 678–683 (1982)

4.80    Y. Katzir, A. A. Friesem, I. Glaser: Orthogonal in-plane and out-of-plane fringe maps in holographic interferometry, Opt. Lett. **8**, 163–165 (1983)

4.81    D. R. Axelrad, K. Rezai: Determination of surface displacements by holographic electrooptical processing, Appl. Opt. **21**, 2001–2005 (1982)

4.82    T. M. Kreis, H. Kreitlow: Quantitative evaluation of holographic interference patterns under image processing aspects, SPIE **210**, 196–202 (1980)

4.83    W. Jüptner, T. M. Kreis, H. Kreitlow: Automatic evaluation of holographic interferograms by reference beam phase shifting, SPIE **398**, 22–29 (1983)

4.84    S. Nakadate, N. Magome, T. Honda, J. Tsujiuchi: Hybrid holographic interferometer for measuring three-dimensional deformations, Opt. Eng. **20**, 246–252 (1981)

4.85    S. Nakadate, T. Yatagai, H. Saito: Computer-aided speckle pattern interferometry, SPIE **370**, 180–188 (1983)

4.86    N. A. Massie: Real-time digital heterodyne interferometry: a system, Appl. Opt. **19**, 154–160 (1980)

4.87    R. Dändliker, B. Eliasson: Accuracy of heterodyne holographic strain and stress determination, Exp. Mech. **19**, 93–101 (1979)

4.88    R. Dändliker, J. F. Willemin: Measuring microvibrations by heterodyne speckle interferometry, Opt. Lett. **6**, 165–167 (1981)

4.89    P. Hariharan, B. F. Oreb, N. Brown: A digital system for real-time holographic stress analysis, SPIE **370**, 189–194 (1983)

4.90    P. Hariharan, B. F. Oreb, N. Brown: Real-time holographic interferometry: a microcomputer system for the measurement of vector displacements, Appl. Opt. **22**, 876–880 (1983)

4.91    A. E. Ennos: Measurement of in-plane surface strain by hologram interferometry, J. Sc. Instrum. **1**, 731–734 (1968)

4.92    D. Bijl, R. Jones: On tri-hologram interferometry and the measurement of elastic anisotropies using an adapted form of Cornu's technique, in *Applications de l'Holographie*, Proc. Symp. Besançon 1970, ed. by J. C. Viénot, J. Bulabois, J. Pasteur. Paper 5.3

4.93    M. R. Wall: Zero motion fringe identification, in *Applications de l'Holographie*, Proc. Symp. Besançon 1970, ed. by J. C. Viénot, J. Bulabois, J. Pasteur. Paper 4.9

4.94    K. Shibayama, H. Uchiyama: Measurement of three-dimensional displacements by hologram interferometry, Appl. Opt. **10**, 2150–2154 (1971)

4.95    U. Kopf: Fringe order determination and zero motion fringe identification in holographic displacement measurements, Opt. Laser Techn. **5**, 111–113 (1973)

4.96   Y. Y. Hung, C. P. Hu, D. R. Henley, C. E. Taylor: Two improved methods of surface-displacement measurements by holographic interferometry, Opt. Commun. **8**, 48–51 (1973)

4.97   J. M. Burch: Holographic measurement of displacement and strain – an introduction, J. Strain Anal. **9**, 1–3 (1974)

4.98   T. Matsumoto, K. Iwata, R. Nagata: Measurement of deformation in a cylindrical shell by holographic interferometry, Appl. Opt. **13**, 1080–1084 (1974)

4.99   J. P. Sikora: A three-dimensional displacement analysis from an image-plane hologram, Exp. Mech. **18**, 101–106 (1978)

4.100  H. Ukita: Measurement of three-dimensional displacement distribution on a twisted belt by holographic interferometry, Rev. Electr. Comm. Lab. **25**, 478–486 (1977)

4.101  L. Ek, K. Biedermann: Implementation of hologram interferometry with a continuously scanning reconstruction beam, Appl. Opt. **17**, 1727–1732 (1978)

4.102  Z. Füzessy: Methods of holographic interferometry for industrial measurements, Per. Poli ME **21**, 257–263 (1977)

4.103  Z. Füzessy: Improved interference-hologram method for measurement of displacement, Zh. Tekh. Fiz. **49**, 399–403 (1979)

4.104  Z. Füzessy: Measurement of 3-D displacement by incoherent superposition of interferograms, Isr. J. Tech. **18**, 251–254 (1980)

4.105  Z. Füzessy, N. Abramson: Measurement of 3-D displacement: sandwich holography and regulated path length interferometry, Appl. Opt. **21**, 260–264 (1982)

4.106  Z. Füzessy: Measurement of 3D – displacement by regulated path length interferometry, SPIE **398**, 17–21 (1983)

4.107  G. Schönebeck: Eine allgemeine holografische Methode zur Bestimmung räumlicher Verschiebungsfelder, VDI-Berichte Nr. 313, 155–162 (1978)

4.108  G. Schönebeck: „Eine allgemeine holografische Methode zur Bestimmung räumlicher Verschiebungen", Diss. Tech. Univ. München (1979)

4.109  V. G. Seleznev: Holographic attachment for a machine for determining the field of displacements, Indus. Lab. **46**, 850–853 (1981)

4.110  H. Kohler: Holografische Interferometrie: IV. Alternative Auswertung mit absoluten Ordnungen, Optik **58**, 193–199 (1981)

4.111  H. Kohler: Holografisch-interferometrische Triangulation: I. Ein Messverfahren für Entfernungen und Oberflächenkonturen unter Verwendung von Ordnungen, Optik **60**, 61–72 (1981)

4.112  H. Kohler: Holografische Interferometrie: V. Zur Problematik der Ordnungsbestimmung, Optik **60**, 411–425 (1982)

4.113  A. Lev, J. Politch: Measuring the displacement vector by holographic interferometry, Opt. Laser Techn. **11**, 45–47 (1979)

4.114  J. Ebbeni, A. Huybrecht, S. Orloff: Holographic determination of demineralization of bones, SPIE **211**, 84–89 (1979)

4.115  J. Petkovsek, K. Rankel: Measurement of three-dimensional displacement by four small holograms, SPIE **210**, 173–177 (1980)

4.116  E. Müller, V. Hrdlicka, D. Cuche: Computer based evaluation of holographic interferograms, SPIE **398**, 46–52 (1983)

4.117  J. E. Sollid: Holographic interferometry applied to measurements of small static displacements of diffusely reflecting surfaces, Appl. Opt. **8**, 1587–1595 (1969)

4.118  J. E. Sollid: A comparison of two methods of measuring inplane surface displacements using holographic interferometry, in *Applications de l'Holographie*, Proc. Symp. Besançon 1970, ed. by J. C. Viénot, J. Bulabois, J. Pasteur. Paper 4.8

4.119  J. E. Sollid: Translational displacements versus deformation, displacements in double-exposure holographic interferometry, Opt. Commun. **2**, 282–288 (1970)

4.120  C. M. Vest: Comment on: holographic interferometry applied to measurements of small static displacements of diffusely reflecting surfaces, Appl. Opt. **12**, 612–613 (1973)

4.121  R. J. Pryputniewicz, W. W. Bowley: Techniques of holographic displacement measurement: an experimental comparison, Appl. Opt. **17**, 1748–1756 (1978)

4.122  W. Schreiber, L. Wenke, W. Osten: Bestimmung von Verschiebungsvektoren mittels holographischer Interferometrie bei reduzierter Anzahl der zu vermessenden geometrischen Größen, Opt. Acta **28**, 1163–1167 (1981)

4.123  B. Harnisch, W. Schreiber, L. Wenke: A method to avoid sign ambiguity in holographic interferometric displacement measurements, Opt. Acta **30**, 699–703 (1983)

4.124  E. B. Aleksandrov, A. M. Bonch-Bruevich: Investigation of surface strains by the hologram technique, Sov. Phys.-Tech. Phys. **12**, 258–265 (1967)

4.125  J. W. C. Gates: Holographic measurement of surface distortion in three dimensions, Opt. Technol. **1**, 247–250 (1969)

4.126  S. K. Dhir, J. P. Sikora: Holographic analysis of a general displacement field, Eng. Appl. of Holography (Proceedings), Los Angeles (1972) pp. 147–153

4.127  H. E. Gascoigne: Fringe formation and interpretation in holographic interferometry, Final Report, Grant GK-5423 (University of Utah 1972)

4.128  H. Kohler: Untersuchungen zur quantitativen Analyse der holografischen Interferometrie, Optik **39**, 229–235 (1974)

4.129  P. W. King III: Holographic interferometry technique utilizing two plates and relative fringe orders for measuring microdisplacements, Appl. Opt. **13**, 231–233 (1974)

4.130  H. Kohler: Interferenzliniendynamik bei der quantitativen Auswertung holografischer Interferogramme, Optik **47**, 9–24 (1977)

4.131  H. Kohler: Ein neues Verfahren zur quantitativen Auswertung holografischer Interferogramme, Optik **47**, 135–152 (1977)

4.132  H. Kohler: Zur Vermessung holografisch-interferometrischer Verformungsfelder, Optik **47**, 271–282 (1977)

4.133  H. Kohler: General formulation of the holographic- interferometric evaluation methods, Optik **47**, 469–475 (1977)

4.134  J. Petkovšek, K. Rankel, B. Cencič: An approach to define the zero-order fringe for reducing errors in interpreting holographic interferograms, in *Optoelektronik in der Technik,* ed. by W. Waidelich (Springer, Berlin, Heidelberg 1982) pp. 68–72

4.135  A. A. Voevodin, V. L. Kazak, I. M. Nagibina: Reading holographic interferograms used in surface deformation measurements, Zh. Tekh. Fiz. **52**, 703–708 (1982)

4.136  S. K. Dhir, J. P. Sikora: An improved method for obtaining the general-displacement field from a holographic interferogram, Exp. Mech. **7**, 323–327 (1972)

4.137  C. A. Sciammarella, J. A. Gilbert: Strain analysis of a disk subjected to diametral compression by means of holographic interferometry, Appl. Opt. **12**, 1951–1956 (1973)

4.138  T. Matsumoto, K. Iwata, R. Nagata: Measuring accuracy of three-dimensional displacements in holographic interferometry, Appl. Opt. **12**, 961–967 (1973)

4.139  W. M. Ewers, W. Fritzsch, K. Grünewald, H. Wachutka: Bestimmung dreidimensionaler Verformungsfelder mit Hilfe der holographischen Interferometrie, Optik **40**, 57–68 (1974)

4.140  V. Fossati-Bellani, A. Sona: Measurement of three-dimensional displacements by scanning a double-exposure hologram, Appl. Opt. **13**, 1337–1341 (1974)

4.141  L. Ek, K. Biedermann: Analysis of a system for hologram interferometry with a continuously scanning reconstruction beam, Appl. Opt. **16**, 2535–2542 (1977)

4.142  D. Nobis, C. M. Vest: Statistical analysis of errors in holographic interferometry, Appl. Opt. **17**, 2198–2204 (1978)

4.143  V. S. Pisarev, V. V. Yakovlev, V. P. Shchepinov, V. O. Indisov: Use of the C-optimality criterion in designing a holographic interferometer for strain determination, Zh. Tekh. Fiz. **51**, 869–871 (1981)

4.144  O. G. Lisin: Accuracy with which spatial displacements of diffuse objects can be measured using holographic interferogram data, Opt. Spektrosk. **50**, 521–531 (1981)

4.145  J. B. Schemm, C. M. Vest: Fringe pattern recognition and interpolation using nonlinear regression analysis, Appl. Opt. **22**, 2850–2853 (1983)

4.146  R. J. Pryputniewicz: Determination of the sensitivity vectors directly from holograms, J. Opt. Soc. Am. **67**, 1351–1353 (1977)

4.147 R. J. Pryputniewicz, K. A. Stetson: Determination of sensitivity vectors in hologram interferometry from two known rotations of the object, Appl. Opt. **19**, 2201–2205 (1980)

4.148 H. Kohler: Interferometric instead of geometric measurement of object points in holographic-interferometric deformation analysis, Opt. Acta **29**, 275–280 (1982)

4.149 S. K. Dhir, J. P. Sikora: Holographic determination of surface stresses in an arbitrary three-dimensional object, in digest of papers, *Int. Opt. Computing Conf.* Zürich 1974 (IEEE, New York 1974) pp. 79–83

4.150 R. E. Rowlands, J. A. Jensen, K. D. Winters: Differentiation along arbitrary orientations, Exp. Mech. **18**, 81–86 (1978)

4.151 R. Dändliker: Extrapolation of strain and stress from holographically measured surface displacement, VDI-Berichte Nr. 313, 163–169 (1978)

4.152 R. Dändliker: Holographic interferometry and speckle photography for strain measurement: a comparison, Opt. Lasers Eng. **1**, 3–19 (1980)

4.153 R. Dändliker, R. Thalmann: Determination of 3-D displacement and strain by holographic interferometry for non-plane objects, SPIE **398**, 11–16 (1983)

4.154 J. Spörri: Spannungsbestimmung in Bauteilen auf Grund von Oberflächenverschiebungsmessungen (Holographie), VDI-Berichte Nr. 313, 171–174 (1978)

4.155 J. E. Sollid, K. A. Stetson: Strains from holographic data, Exp. Mech. **18**, 208–214 (1978)

4.156 W. J. Beranek, A. J. A. Bruinsma: Determination of displacement and strain fields using dual-beam holographic-moiré interferometry, Exp. Mech. **22**, 317–323 (1982)

4.157 J. Rendl: „Holographisch interferometrische Untersuchungen zur Schallausbreitung in einem transparenten Festkörper", Diplomarbeit, Inst. Med. Optik, Univ. München (1982)

4.158 J. L. Goldberg: A method of three-dimensional strain measurement on non-ideal objects using holographic interferometry, Exp. Mech. **23**, 59–73 (1983)

4.159 K. A. Stetson: Fringe interpretation for hologram interferometry of rigid-body motions and homogeneous deformations, J. Opt. Soc. Am. **64**, 1–10 (1974)

4.160 K. A. Stetson: Fringe vectors and observed-fringe vectors in hologram interferometry, Appl. Opt. **14**, 272–273 (1975)

4.161 K. A. Stetson: Homogeneous deformations: determination by fringe vectors in hologram interferometry, Appl. Opt. **14**, 2256–2259 (1975)

4.162 R. J. Pryputniewicz, K. A. Stetson: Holographic strain analysis: extension of fringe-vector method to include perspective, Appl. Opt. **15**, 725–728 (1976)

4.163 R. J. Pryputniewicz: „Holographic analysis of body deformation", Thesis (University of Connecticut, Storrs 1976)

4.164 R. J. Pryputniewicz: Holographic strain analysis: an experimental implementation of fringe-vector theory, Appl. Opt. **17**, 3613–3618 (1978)

4.165 K. A. Stetson: Use of projection matrices in hologram interferometry, J. Opt. Soc. Am. **69**, 1705–1710 (1979)

4.166 F. Lamy, C. Liegeois: Perspective corrections for measurement of deformations by holographic interferometry, J. Opt. Soc. Am. **72**, 972–974 (1982)

4.167 K. A. Stetson: The relationship between strain and derivatives of observed displacement in coherent optical metrology, Exp. Mech. **21**, 273–275 (1981)

4.168 K. A. Stetson: Speckle and its application to strain sensing, SPIE **353**, 12–18 (1982)

4.169 D. Bijl, R. Jones: A new theory for the practical interpretation of holographic interference patterns resulting from static surface displacements, Opt. Acta **21**, 105–118 (1974)

4.170 R. Jones: An experimental verification of a new theory for the interpretation of holographic interference patterns resulting from static surface displacements, Opt. Acta **21**, 257–266 (1974)

4.171 W. Schumann: Some aspects of the optical techniques for strain determination, Exp. Mech. **13**, 225–231 (1973)

4.172 I. Přikril: Evaluation of rigid body displacement by differential holographic interferometry, Opt. Appl. **10**, 3–11 (1980)

4.173   R. J. Pryputniewicz: Holographic determination of rigid-body motions and application of the method to orthodontics, Appl. Opt. **18**, 1442–1444 (1979)

4.174   N. Bolognini, H. J. Rabal, E. E. Sicre, M. Garavaglia: Rigid body motion measurements with Fourier lensless holography, Appl. Opt. **20**, 2342–2344 (1981)

4.175   B. V. Feduleev, V. P. Ryabukho, V. B. Rabkin: Measurement of the thermal coefficient of linear expansion by holographic interferometry, Zh. Tekh. Fiz. **52**, 324–329 (1982)

4.176   A. R. Luxmoore, C. House: In-plane strain measurements by a three beam holographic-method, in *Applications de l'Holographie*, Proc. Symp. Besançon 1970, ed. by J. C. Viénot, J. Bulabois, J. Pasteur. Paper 5.2

4.177   J. Ebbeni: Combinaison d'une méthode de moiré et d'une méthode holographique pour déterminer l'état de déformation d'un objet diffusant, in Preprints *5th Inter. Conf. Exp. Stress Anal. Udine 1974* (Udine 1974) pp. 4.20–25

4.178   C. A. Sciammarella, S. K. Chawla: Holographic-moiré technique to obtain displacement components and derivatives, Mech. Res. Commun. **4**, 333–338 (1977)

4.179   C. A. Sciammarella, S. K. Chawla: A lens holographic-moiré technique to obtain components of displacements and derivatives, Exp. Mech. **18**, 373–381 (1978)

4.180   C. A. Sciammarella: Holographic moiré, an optical tool for the determination of displacements, strains, contours, and slopes of surfaces, Opt. Eng. **21**, 447–457 (1982)

4.181   J. A. Gilbert: Differentiation of holographic-moiré patterns, Exp. Mech. **18**, 436–440 (1978)

4.182   J. A. Gilbert, J. W. Herrick: Dual-beam holographic deflection measurement, Exp. Mech. **21**, 349–354 (1981)

4.183   P. K. Rastogi, M. Spajer, J. Monneret: In-plane deformation measurement using holographic moiré, Opt. Lasers Eng. **2**, 79–103 (1981)

4.184   S. Toyooka: Determination of in-plane and out-of-plane components of deformation of an object from one double-exposure hologram, Opt. Acta **26**, 429–438 (1979)

4.185   S. Toyooka: Determination of in-plane strain using a holographic modulated diffraction grating, Jap. J. Appl. Phys. **18**, 1289–1293 (1979)

4.186   L. H. Tanner: The scope and limitations of three-dimensional holography of phase objects, J. Sci. Instrum. **44**, 1011–1014 (1967)

4.187   S. M. Fraser, K. A. R. Kinloch: Large viewing angle holograms, J. Phys. E. **7**, 774–776 (1974)

4.188   E. L. Nodov: Moiré-holography method for measuring inhomogeneous distributions of refractive index, Appl. Opt. **13**, 1551–1553 (1974)

4.189   R. L. Kurtz, L. M. Perry: A holographic optical Schlieren system (HOSS), Opt. Eng. **18**, 243–248 (1979)

4.190   C. J. Reinheimer, C. E. Wiswall, R. A. Schmiege, R. J. Harris, J. E. Dueker: Holographic subsonic flow visualization, Appl. Opt. **9**, 2059–2065 (1970)

4.191   A. Hirth, P. Smigielski, A. Stimpfling: Use of holography for visualization of the wake of projectiles in hypersonic flight at Mach 6, Opt. Laser Tech. **3**, 195–199 (1971)

4.192   A. B. Witte, J. Fox, H. Rungaldier: Localized measurements of wake density fluctuations using pulsed laser holographic interferometry, AIAA J. **10**, 481–487 (1972)

4.193   V. T. Chernykh, I. N. Zelinskii: A method of obtaining a multifrequency hologram element and its use in the holographic interferometry of 3-D phase objects, Opt. Spektrosk. **46**, 795–799 (1979)

4.194   D. C. Prince: Three-dimensional shock structures for transonic/supersonic compressor rotors, J. Aircraft **17**, 28–37 (1980)

4.195   A. G. Havener, R. J. Radley: Turbulent boundary-layer flow separation measurements using holographic interferometry, AIAA J. **12**, 1071–1075 (1974)

4.196   W. A. Benser, E. E. Bailey, T. F. Gelder: Holographic studies of shock waves within transonic fan rotors, J. Eng. Power **97**, 75–84 (1975)

4.197   D. T. Hove, A. A. Smith: Holographic analysis of particle-induced hypersonic bow-shock distortions, AIAA J. **13**, 947–949 (1975)

4.198   J. D. Trolinger: Flow visualization holography, Opt. Eng. **14**, 470–481 (1975)

4.199  A. G. Havener: Detection of boundary-layer transition using holography, AIAA J. **15**, 592–593 (1977)

4.200  K. Iwata, T. Hakoshima, R. Nagata: Measurement of flow velocity distribution by means of double-exposure holographic interferometry, J. Opt. Soc. Am. **67**, 1117–1121 (1977)

4.201  A. Ozkul: Investigation of acoustic radiation from supersonic jets by double-pulse holographic interferometry, AIAA J. **17**, 1068–1075 (1979)

4.202  J. D. Trolinger, G. D. Simpson: Diagnostics of turbulence by holography, Opt. Eng. **18**, 161–166 (1979)

4.203  I. S. Zeilikovich, V. A. Komissaruk, I. I. Komissaruk, N. P. Mende: Holograms with displacement interferometers and a broad light source, Zh. Tekh. Fiz. **49**, 597–600 (1979)

4.204  I. N. Zelinskiĭ, V. T. Chernykh, A. G. Belyaev: Use of a holographic interferometer for visualizing three-dimensional gas flows along an aeroballistic trajectory, Sov. J. Opt. Techn. **47**, 629–630 (1980)

4.205  A. K. Beketova, V. I. Lakhtionov, L. E. Legu: Holographic interferometer for aerodynamic studies, Sov. J. Opt. Techn. **47**, 403–404 (1980)

4.206  P. J. Bryanston-Cross, T. Lang, M. L. Oldfield, R. J. G. Norton: Interferometric measurements in a turbine cascade using image-plane holography, J. Eng. Power **103**, 124–130 (1981)

4.207  F. W. Spaid, W. D. Bachalo: Experiments on the flow about a supercritical airfoil including holographic interferometry, J. Aircraft **18**, 287–294 (1981)

4.208  L. T. Clark, D. C. Koepp, J. J. Thykkuttathil: A three dimensional density field measurement of transonic flow from a square nozzle using holographic interferometry, J. Fluids Eng. **99**, 737–744 (1977)

4.209  R. Pawluczyk, Z. Kraska: Holographic investigation of shock wave diffraction, SPIE **370**, 39–43 (1983)

4.210  T. A. Dullforce, R. E. Faw: High-speed cine recording of realtime holographic interference fringes, Opt. Commun. **31**, 111–113 (1979)

4.211  I. N. Il'in, V. P. Grivtsov, A. D. Amelin, S. R. Yaundalders: Investigation of the boiling mechanism by means of holographic interferometry, Heat Transfer – Soviet Res. **12**, 51–56 (1980)

4.212  W. Aung, R. O'Regan: Precise measurement of heat transfer using holographic interferometry, Rev. Sc. Intr. **42**, 1755–1759 (1971)

4.213  R. F. Stevens: Three-dimensional time resolved measurements from holographic records, Opt. Laser Techn. **8**, 167–173 (1976)

4.214  G. M. Neumann, U. Müller, E. Schmidt: Holographie an gasgefüllten Glühlampen I, Lichttechnik **27**, 214–218 (1975)

4.215  G. M. Neumann, U. Müller, E. Schmidt: Holographie an gasgefüllten Glühlampen II, Lichttechnik **27**, 282–283 (1975)

4.216  G. M. Neumann, U. Müller, E. Schmidt: Holographische Interferometrie zeitlich stationärer und instationärer Zustände, Z. Naturforsch. **30a**, 1164–1165 (1975)

4.217  G. Antonini, G. Guiffant, P. Perrot: The effect of transverse oscillations on heat transfer from a horizontal hot-wire to a liquid: holographic visualization, Int. J. Heat Mass Transfer **20**, 88–92 (1977)

4.218  G. Guiffant, G. Antonini: A propos d'interférométrie holographique, Rev. Polytech. No. 1372, 431–435 (1978)

4.219  G. Larcher, D. Allano: Application de l'interférométrie holographique à trois ondes à l'étude d'un sillage thermique turbulent, J. Opt. **8**, 159–164 (1977)

4.220  R. J. Parker, J. W. C. Gates: An investigation of the instabilities in photometric standard lamps by holographic interferometry, J. Phys. E. **12**, 18–20 (1979)

4.221  F. Mayinger: Strömung und Wärmetransport in Natur und Technik, sichtbar gemacht, Naturwissensch. **66**, 300–306 (1979)

4.222  S. Kuroda, M. Kimura, K. Kubota: Dynamical study of heat conduction in liquid crystals by high-speed optical holography, Mol. Cryst. Liq. Cryst. **33**, 235–246 (1976)

4.223  A. Choudry: Digital holographic interferometry of convective heat transport, Appl. Opt. **20**, 1240–1244 (1981)

4.224  P. J. Walklate: A two wavelength holographic technique for the study of two-dimensional thermal boundary layers, Int. J. Heat Mass Transfer **24**, 1051–1057 (1981)

4.225  R. E. Faw, T. A. Dullforce: Holographic interferometry measurement of convective heat transport beneath a heated horizontal plate in air, Int. J. Heat Mass Transfer **24**, 859–869 (1981)

4.226  J. Guerry, J. P. Hot, C. Durou: Analysis by real-time holographic interferometry of heat transfer at the surface of cold solar collectors, SPIE **210**, 178–186 (1980)

4.227  J. H. Masliyah, T. T. Nguyen: Qualitative study in mass transfer by laser holography, Can. J. Ch. Eng. **52**, 664–665 (1974)

4.228  D. N. Kapur, N. Macleod: The determination of local mass-transfer coefficients by holographic interferometry – I, Int. J. Heat Mass Transfer **17**, 1151–1162 (1974)

4.229  D. N. Kapur, N. Macleod: The use of holographic interferometry for the measurement of local mass transfer coefficients, J. Phot. Sci. **23**, 81–84 (1975)

4.230  D. N. Kapur, N. Macleod: Determination of local mass transfer coefficients by holography, Nature Phys. Sci. **237**, 57–59 (1972)

4.231  M. Ueda, K. Kagawa, K. Yamada, C. Yamaguchi, Y. Harada: Flow visualization of Bénard convection using holographic interferometry, Appl. Opt. **21**, 3269–3272 (1982)

4.232  N. Bochner, J. Pipman: A simple method of determining diffusion constants by holographic interferometry, J. Phys. D. **9**, 1825–1830 (1976)

4.233  H. Grosse-Wilde, J. Uhlenbusch: Measurement of local mass-transfer coefficients by holographic interferometry, Int. J. Heat Mass Transfer **21**, 677–682 (1978)

4.234  J. H. Masliyah, T. T. Nguyen: Holographic determination of mass transfer due to impinging square jet, Can. J. Ch. Eng. **54**, 299–304 (1976)

4.235  K. Srimannarayana, V. Venkateswara Rao: Holographic formation of equal inclination and equal thickness fringes and estimation of refractive index of solution, Current Sci. **44**, 340–341 (1975)

4.236  K. Srimannarayana, V. Venkateswara Rao: Holographic method of determination of refractive index of solutions, Opt. Appl. **7**, 19–22 (1977)

4.237  R. N. O'Brien: Concentration gradients within electrodialysis membranes by holographic interferometry, Electrochimica Acta **20**, 447–449 (1975)

4.238  G. Carini, M. Cutroni, F. Wanderlingh: Holographic measurement of high viscosities, Opt. Laser Techn. **10**, 241–242 (1978)

4.239  K. D. Hinsch: Holographic interferometry of surface deformations of transparent fluids, Appl. Opt. **17**, 3101–3107 (1978)

4.240  D. H. McQueen: Holographic refractometer, J. Phys. E. **12**, 111–114 (1979)

4.241  O. N. Ertanova, I. A. Lepeshinskii: Holographic method for measurement of liquid film size at a nozzle mouth, Fluid Dynamics **13**, 119–121 (1978)

4.242  W. R. Debler, C. M. Vest: Observations of a stratified flow by means of holographic interferometry, Proc. Royal Soc. London A, Math. Phys. Sci. **358**, 1692, 1–16 (1977)

4.243  M. Clifton, V. Sanchez: Optical errors encountered in using holographic interferometry to observe liquid boundary layers in electrochemical cells, Electrochimica Acta **24**, 445–450 (1979)

4.244  L. Gabelmann-Gray, H. Fenichel: Holographic interferometric study of liquid diffusion, Appl. Opt. **18**, 343–345 (1979)

4.245  E. M. Gates, J. Bacon: A note on the determination of cavitation nuclei distributions by holography, J. Ship Research **22**, 29–31 (1978)

4.246  C. Bräuchle, D. M. Burland, G. C. Bjorklund: Hydrogen abstraction by benzophenone studied by holographic photochemistry, J. Phys. Chem. **85**, 123–127 (1981)

4.247  C. Bräuchle, D. M. Burland, G. C. Bjorklund: Study of the photolysis of dimethyl-s-tetrazine using a holographic technique, J. Am. Chem. Soc. **103**, 2515–2519 (1981)

4.248  J. D. Trolinger: Holographic interferometry, as a diagnostic tool for reactive flows, Comb. Sci. Techn. **13**, 229–244 (1976)

4.249  A. Bellmann, M. Palócz, L. von Wolfersdorf: Hologramm-interferometrische Untersuchung der Vermischungsvorgänge in Flammensystemen, Chem. Techn. **30**, 482–484 (1978)

4.250  A. E. Davydov, V. S. Abrukov, S. A. Abrukov: Holographic interferometry of a singing flame, Comb. Expl. Shock Waves **14**, 519–521 (1978)

4.251  H. Grosse-Wilde, J. Uhlenbusch: Holographic interferometry of arc discharges displaced by magnetic fields, IEEE Trans. PS-1, 55–61 (1973)

4.252  A. Bernard, A. Jolas, J. Launspach, J. P. Watteau: Etude d'une décharge linéaire non-cylindrique par interférométrie holographique, Plasma-Phys. **15**, 1019–1030 (1973)

4.253  A. H. Guenther, W. K. Pendleton, C. Smith, C. H. Skeen, S. Zivi: Pulsed interferometric holography of laser-produced air breakdown, Opt. Laser Techn. **5**, 20–23 (1973)

4.254  P. T. Rumsby, M. M. Michaelis: holographic interferometry of colliding laser produced plasmas, Phys. Lett. A **48**, 11–12 (1974)

4.255  J. L. Seftor: Two-wavelength holographic interferometry of exploding wires, J. Appl. Phys. **45**, 2903–2906 (1974)

4.256  R. A. Jeffries: Two-wavelength holographic interferometry of partially ionized plasmas, Phys. Fluids **13**, 210–212 (1970)

4.257  R. Fedosejevs, M. C. Richardson: Subnanosecond microscopic holographic interferometry of plasmas produced by 1-nsec $CO_2$ laser pulses, Appl. Phys. Lett. **27**, 115–117 (1975)

4.258  R. J. Radley: Two-wavelength holography for measuring plasma electron density, Phys. Fluids **18**, 175–179 (1975)

4.259  V. M. Ginzburg, B. M. Stepanov, Y. I. Filenko: An investigation of the discharge in flash lamps by a holographic method, Radio Eng. Electr. Phys. **17**, 1781–1782 (1972)

4.260  R. Hines, A. Gaudet, K. Chen, A. Pallone, E. W. Heinonen: Application of holography to in-situ roughness measurements in plasma arc, ISA Trans. **17**, 89–96 (1978)

4.261  D. K. Koopman, H. J. Siebeneck, G. Jellison, W. G. Niessen: Resonant holography of plasma flow phenomena, Rev. Sci. Instrum. **49**, 524–525 (1978)

4.262  P. D. Rockett, D. R. Bach: Holographic interferometry of a high-energy-density exploding lithium wire plasma, J. Appl. Phys. **50**, 2670–2674 (1979)

4.263  E. L. Pierce: Designing a probe beam and an ultraviolet holographic microinterferometer for plasma probing, Appl. Opt. **19**, 952–961 (1980)

4.264  L. V. Sutter, P. K. Baily, G. Wakalopulos, R. A. Hill, E. R. Peressini, F. Dolezal: Holographic interferometry of a large-bore cw high-energy gas-laser medium during laser-power extraction, Appl. Opt. **18**, 3835–3837 (1979)

4.265  W. Tiemann: Arc-gas flow interactions in a double-nozzle flow system, IEEE Trans. PS-**8**, 368–375 (1980)

4.266  O. Willi, A. Raven: Holographic microinterferometer to study laser-produced plasmas, Appl. Opt. **19**, 192–194 (1980)

4.267  K. P. Alum, Y. V. Koval'chuk, G. V. Ostrovskaya: Holographic interferometry of a plasma with frequency conversion of radiation passing through the plasma, Zh. Tekh. Fiz. **51**, 1618–1623 (1981)

4.268  A. P. Burmakov, G. M. Novik: Holographic interferometry of a supersonic plasma jet in a pulsed discharge, Zh. Tekh. Fiz. **51**, 68–72 (1981)

4.269  V. G. Kulkarni, P. N. Puntambekar: Holographic interferometry for testing large phase objects, Opt. Commun. **27**, 33–36 (1978)

4.270  P. L. Chu, D. Peri: Holographic measurement of refractive-index profile in the transition region of an optical fiber preform, Appl. Opt. **20**, 1418–1423 (1981)

4.271  R. J. Sanford, V. J. Parks: A holographic method for matching and measuring the refractive index of a photoelastic material, Exp. Mech. **18**, 112–114 (1978)

4.272  W. L. Howes, D. R. Buchele: Optical interferometry of inhomogeneous gases, J. Opt. Soc. Am. **56**, 1517–1528 (1966)

4.273  R. A. Kosakoski, D. J. Collins: Application of holographic interferometry to density field determination in transonic corner flow, AIAA J. **12**, 767–770 (1974)

4.274  P. V. Farrell, G. S. Springer, C. M. Vest: Heterodyne holographic interferometry: concentration and temperature measurements in gas mixtures, Appl. Opt. **21**, 1624–1627 (1982)

4.275  I. I. Komissarova, G. V. Ostrovskaya: Diagnostics of a laser spark with a hologram interferometer, Zh. Tekh. Fiz. **48**, 2062–2067 (1978)

4.276  J. Surget: Etude quantitative d'un écoulement aérodynamique par interférométrie holographique, Rech. Aerosp. No. 1973–3, 161–171 (1973)

4.277  J. Délery, J. Surget, J. P. Lacharme: Interférométrie holographique quantitative en écoulement transonique bidimensionel, Rech. Aérosp. 1977–2, 89–101 (1977)

4.278  W. Schumann: Fringe localization in holographic interferometry in the case of a transparent object with a nonuniformly varying index of refraction, Opt. Lett. **7**, 119–121 (1982)

4.279  C. M. Vest, D. W. Sweeney: Holographic interferometry of transparent objects with illumination derived from phase gratings, Appl. Opt. **9**, 2321–2325 (1970)

4.280  K. A. Stetson: The use of heterodyne speckle photogrammetry to measure high-temperature strain distributions, SPIE **370**, 46–55 (1983)

4.281  D. W. Sweeney, C. M. Vest: Reconstruction of three-dimensional refractive index fields by holographic interferometry, Appl. Opt. **11**, 205–207 (1972)

4.282  D. W. Sweeney, C. M. Vest: Reconstruction of three-dimensional refractive index fields from multidirectional interferometric data, Appl. Opt. **12**, 2649–2664 (1973)

4.283  C. M. Vest: Formation of images from projections: Radon and Abel transforms, J. Opt. Soc. Am. **64**, 1215–1218 (1974)

4.284  C. M. Vest: Interferometry of strongly refracting axisymmetric phase objects, Appl. Opt. **14**, 1601–1606 (1975)

4.285  S. Cha, C. M. Vest: Tomographic reconstruction of strongly refracting fields and its application to interferometric measurement of boundary layers, Appl. Opt. **20**, 2787–2794 (1981)

4.286  Y. P. Presnyakov: Calculating the refractive index by measuring the derivative of the optical path length, Opt. Spektrosk. **40**, 514–517 (1976)

4.287  J. Radon: Über die Bestimmung von Funktionen durch ihre Integralwerte längs gewisser Mannigfaltigkeiten, Berichte Saechs. Akad. Wissen. (Leipzig) **69**, 262–277 (1917)

4.288  M. V. Berry, D. F. Gibbs: The interpretation of optical projections, Proc. R. Soc. London A **314**, 143–152 (1970)

4.289  R. A. Crowther, D. J. DeRosier, A. Klug: The reconstruction of a three-dimensional structure from projections and its application to electron microscopy, Proc. R. Soc. London A **317**, 319–340 (1970)

4.290  R. South: An extension to existing methods of determining refractive indices from axisymmetric interferograms, AIAA J. **8**, 2057–2059 (1970)

4.291  H. H. Chau, O. Zucker: Holographic thin-beam reconstruction technique for the study of 3-D refractive-index field, Opt. Commun. **8**, 336–339 (1973)

4.292  V. V. Pikalov, N. G. Preobrazhenskii: Abel transformation in the interferometric holography of a point explosion, Comb. Expl. Shock Waves **10**, 827–833 (1974)

4.293  T. F. Zien, W. C. Ragsdale, W. C. Spring: Quantitative determination of three-dimensional density field by holographic interferometry, AIAA J. **13**, 841–842 (1975)

4.294  D. W. Oldenburg, J. C. Samson: Inversion of interferometric data from cylindrically symmetric, refractionless plasmas, J. Opt. Soc. Am. **69**, 927–942 (1979)

4.295  H. G. Junginger, W. Van Haeringen: Calculation of three-dimensional refractive-index field using phase integrals, Opt. Commun. **5**, 1–4 (1972)

4.296  K. Murata, N. Baba, K. Kunugi: Holographic interferometry with a wide field angle of view and its application to reconstruction of refractive index fields, Optik **53**, 285–294 (1979)

4.297  G. P. Montgomery, D. L. Reuss: Effects of refraction on axisymmetric flame temperatures measured by holographic interferometry, Appl. Opt. **21**, 1373–1380 (1982)

4.298  M. Pavelek, M. Liška: Interferogram evaluation of axially symmetric phase objects, Opt. Acta **30**, 943–954 (1983)

4.299  M. Defrise, C. De Mol: A regularized iterative algorithm for limited-angle inverse Radon transform, Opt. Acta **30**, 403–408 (1983)

4.300  S. Helgason: *The Radon Transform* (Birkhäuser, Boston 1980)

4.301  S. Walles: Visibility and localization of fringes in holographic interferometry of diffusely reflecting surfaces, Ark. Fys. **40**, 299–403 (1970)

4.302  S. Walles: On the concept of homologous rays in holographic interferometry of diffusely reflecting surfaces, Opt. Acta **17**, 899–913 (1970)

4.303  C. Froehly, J. Monneret, J. Pasteur, J. C. Viénot: Etude des faibles déplacements d'objects opaques et de la distortion optique dans les lasers à solide par interférométrie holographique, Opt. Acta **16**, 343–362 (1969)

4.304  J. Monneret: Exploitation des systèmes d'interférence observables en interférométrie holographique par mesure des déplacements angulaires de l'onde diffractée par l'objet, Opt. Commun. **2**, 159–162 (1970)

4.305  J. Monneret: „Etude théorique et expérimentale des phénomènes observables en interférométrie holographique interprétation des interférogrammes et applications à la métrologie des microdéplacements", Thèse, Université de Besançon, Besançon (1973)

4.306  P. Jacquot: „Analyse de l'information contenue dans un interférogramme en double exposition: étude de quelques procédés sur des exemples concrets", Thèse, Université de Besançon, Besançon (1973)

4.307  W. T. Welford: Fringe visibility and localization in hologram interferometry, Opt. Commun. **1**, 123–125 (1969)

4.308  W. T. Welford: Fringe visibility and localization in hologram interferometry with parallel displacement, Opt. Commun. **1**, 311–314 (1970)

4.309  C. H. F. Velzel: Fringe contrast and fringe localization in holographic interferometry, J. Opt. Soc. Am. **60**, 419–420 (1970)

4.310  A. F. Fercher, R. Torge: Apertureinflüsse in der Hologramm-Interferometrie, Optik **30**, 521–526 (1970)

4.311  W. H. Steel: Fringe localization and visibility in classical hologram interferometers, Opt. Acta **17**, 873–881 (1970)

4.312  M. A. Machado Gama: Fringe localization and visibility in hologram and classical broad source interferometry, Opt. Commun. **8**, 362–365 (1973)

4.313  I. Přikryl: Localization of interference fringes in holographic interferometry, Opt. Acta **21**, 675–681 (1974)

4.314  J. Leroy: Localisation des franges en interférometrie holographique, Nouv. Rev. Optique **6**, 329–337 (1975)

4.315  T. Tsuruta, N. Shiotake, Y. Itoh: Formation and localization of holographically produced interference fringes, Opt. Acta **16**, 723–733 (1969)

4.316  I. Yamaguchi: Fringe loci and visibility in holographic interferometry with diffuse objects. I. Fringes of equal inclination, Opt. Acta **24**, 1011–1025 (1977)

4.317  I. Yamaguchi: Fringe loci and visibility in holographic interferometry with diffuse objects II. Fringes of equal thickness, Opt. Acta **25**, 299–314 (1978)

4.318  K. A. Stetson: A rigorous treatment of the fringes of hologram interferometry, Optik **29**, 386–400 (1969)

4.319  N. E. Molin, K. A. Stetson: Measurement of fringe loci and localization in hologram interferometry for pivot motion, in-plane rotation and in-plane translation, Optik **31**, 157–177, 281–291 (1970)

4.320  K. A. Stetson: the argument of the fringe function in hologram interferometry of general deformations, Optik **31**, 576–591 (1970)

4.321  N. E. Molin, K. A. Stetson: Fringe localization in hologram interferometry of mutually independent and dependent rotations around orthogonal, non-intersecting axes, Optik **33**, 399–422 (1971)

4.322  K. A. Stetson: Use of fringe vectors in hologram interferometry to determine fringe localization, J. Opt. Soc. Am. **66**, 626–627 (1976)

4.323  K. A. Stetson: Holographic strain analysis by fringe-localization planes, J. Opt. Soc. Am. **66**, 627 (1976)

4.324  K. A. Stetson: Problem of defocusing in speckle photography, its connection to hologram interferometry, and its solutions, J. Opt. Soc. Am. **66**, 1267–1271 (1976)

4.325  H. Kreitlow: „Untersuchung quantitativer Zusammenhänge in der holographischen Interferometrie insbesondere im Hinblick auf eine Auswertung holographischer Interferenzmuster", Thesis, Techn. Universität Hannover (1976)

4.326  M. Dubas, W. Schumann: Sur la détermination holographique de l'état de déformation à la surface d'un corps non-transparent, Opt. Acta 21, 547–562 (1974)

4.327  M. Dubas, W. Schumann: On direct measurements of strain and rotation in holographic interferometry using the line of complete localization, Opt. Acta 22, 807–819 (1975)

4.328  M. Dubas: „Sur l'analyse expérimentale de l'état de déformation à la surface d'un corps opaque par interférométrie holographique, en particulier à l'aide de la localisation des franges", Thèse (Juris-Verlag, Zürich 1976)

4.329  J. C. Charmet: „Holographie appliquée à l'analyse des structures sous contrainte: détermination de l'état de déformation, en particulier par l'étude du contraste de l'interférogramme, Thèse, Université P. et M. Curie, Paris (1977)

4.330  J. C. Charmet, F. Montel: Interférométrie holographique sur objets diffusants: application de la mesure du contraste à la détermination des gradients de déplacement, Rev. Phys. Appl. 12, 603–610 (1977)

4.331  J. Ebbeni, J. C. Charmet: Strain components obtained from contrast measurement of holographic fringe patterns, Appl. Opt. 16, 2543–2545 (1977)

4.332  K. Iwata, R. Nagata: Analysis on fringe formation in holographic interferometry, Opt. Acta 25, 19–39 (1978)

4.333  K. Iwata, R. Nagata: Fringe formation in multiple-exposure holographic interferometry, Opt. Acta 26, 995–1007 (1979)

4.334  M. Miler: Relation between geometry of localization of interference fringes in classical and holographic interferometry, Opt. Commun. 28, 156–158 (1979)

4.335  S. K. Chawla, C. A. Sciammarella: Localization of fringes produced by rotation of the recording plate in focused-image holography, Exp. Mech. 20, 240–244 (1980)

4.336  P. Dunn, B. J. Thompson: Object shape, fringe visibility, and resolution in far-field holography, Opt. Eng. 21, 327–332 (1982)

4.337  C. S. Vikram, M. L. Billet: Gaussian beam effects in far-field in-line holography, Appl. Opt. 22, 2830–2835 (1983)

4.338  R. L. Kurtz: Analysis of localized fringes in the holographic optical Schlieren system, NASA CR-161619 (1980)

4.339  I. S. Zeilikovich, V. A. Komissaruk: Localization of fringes in holographic interferometry of phase objects with diffuse scatterers, Opt. Spektrosk. 39, 985–987 (1975)

4.340  A. J. Decker: Holographic flow visualization of time-varying shock waves, Appl. Opt. 20, 3120–3127 (1981)

4.341  I. Prikryl, C. M. Vest: Holographic interferometry of transparent media using light scattered by embedded test objects, Appl. Opt. 21, 2554–2557 (1982)

4.342  W. Schumann, D. Cuche: Fringe visibility and homology in holographic interferometry of transparent isotropic media, SPIE 370, 30–37 (1983)

4.343  P. Beckmann, A. Spizzichino: The scattering of electromagnetic waves from rough surfaces (Pergamon, Oxford 1963)

4.344  L. I. Goldfischer: Autocorrelation function and power spectral density of laser-produced speckle patterns, J. Opt. Soc. Am. 55, 247–253 (1965)

4.345  S. Lowenthal, H. Arsenault: Image formation for coherent diffuse objects: statistical properties, J. Opt. Soc. Am. 60, 1478–1483 (1970)

4.346  R. B. Owen, H. K. Liu: Optical correlation of surface displacement, Opt. Eng. 18, 266–272 (1979)

4.347  K. N. Petrov, Y. P. Presnyakov: Holographic interferometry of the corrosion process, Opt. Spektrosk. 44, 309–311 (1978)

4.348  V. P. Shchepinov, B. A. Morozov, S. A. Novikov, V. S. Aistov: Determination of contact surface by holographic interferometry, Zh. Tekh. Fiz. 50, 1926–1928 (1980)

4.349  J. T. Atkinson, M. J. Lalor: The effect of surface roughness on fringe visibility in optical interferometry, Opt. Lasers Eng. 1, 131–146 (1980)

4.350  P. Jacquot, P. K. Rastogi: Speckle motions induced by rigid-body movements in free-space geometry: an explicit investigation and extension to new cases, Appl. Opt. **18**, 2022–2032 (1979)

4.351  L. H. Tanner: A study of fringe clarity in laser interferometry and holography, J. Phys. E **1**, 517–522 (1968)

4.352  J. De Jong: A small aperture-dependent error in quantitative holographic interferometry, SPIE **370**, 195–205 (1983)

4.353  N. E. Molin, K. A. Stetson: Measuring combination mode vibration patterns by hologram interferometry, J. Phys. E **2**, 609–612 (1969)

4.354  K. A. Stetson, P. A. Taylor: The use of normal mode theory in holographic vibration analysis with application to an asymmetrical circular disk, J. Phys. E **4**, 1009–1015 (1971)

4.355  K. A. Stetson, P. A. Taylor: Analysis of static deflections by holographically recorded vibration modes, J. Phys. E **5**, 923–926 (1972)

4.356  C. H. Ågren, K. A. Stetson: Measuring the resonances of treble viol plates by hologram interferometry and designing an improved instrument, J. Acoust. Soc. Am. **51**, 1971–1983 (1972)

4.357  J. A. Levitt, K. A. Stetson: Mechanical vibrations: mapping their phase with hologram interferometry, Appl. Opt. **15**, 195–199 (1976)

4.358  B. Breuckmann, H. Marcher, W. Thieme: Holographische Schwingungsanalyse, in *Optoelektronik in der Technik,* ed. by W. Waidelich (Springer, Berlin, Heidelberg 1982) pp. 95–99

4.359  C. H. Hansen, D. A. Bies: Optical holography for the study of sound radiation from vibrating surfaces, J. Acoust. Soc. Am. **60**, 543–555 (1976)

4.360  K. J. Schmidt, W. Kreitlow: Schwingungsanalyse durch kombinierte Anwendung der holografischen Interferometrie und der Berechnung mittels finiter Elemente, Mech. Res. Commun. **4**, 427–434 (1977)

4.361  O. Rusu, D. Borza: Holographic study of flexural vibrations of uniform circular plates, Rev. Roumaine Sci. Tech. Série Méc. Appl. **22**, 267–286 (1977)

4.362  R. Röhler, C. Sieger: Analysis of asymmetrical membrane vibrations by holographic interferometry, Opt. Commun. **25**, 297–300 (1978)

4.363  J. Janta, M. Miler, R. Vrabec: Radial vibrations of piezoceramic resonators investigated by holographic interferometry, Opt. Appl. **8**, 59–63 (1978)

4.364  J. C. MacBain: Displacement and strain of vibrating structures using time-average holography, Exp. Mech. **18**, 361–372 (1978)

4.365  B. Ineichen, W. Schneider: Comparison of holographic measurements and theoretical calculations for vibration studies, SPIE **210**, 203–206 (1979)

4.366  J. Politch: Combining holography with speckling for vibration analysis, Isr. J. Techn. **18**, 275–280 (1980)

4.367  A. Rosenberg, J. Politch: Investigation of a vibration piezoelectric ceramic disk by synthesis of optical coherent methods, Exp. Mech. **20**, 140–144 (1980)

4.368  Y. Xu, C. M. Vest, J. D. Murray: Holographic interferometry used to demonstrate a theory of pattern formation in animal coats, Appl. Opt. **22**, 3479–3483 (1983)

4.369  M. Murata, M. Kuroda: Application of holographic interferometry to practical vibration study, SPIE **398**, 74–81 (1983)

4.370  G. M. Brown, R. R. Wales: Vibration analysis of automotive structures using holographic interferometry, SPIE **398**, 82–89 (1983)

4.371  H. G. Leis: Vibration analysis of an 8-cylinder V-engine by time-averaged holographic interferometry, SPIE **398**, 90–94 (1983)

4.372  J. P. Sikora, F. T. Mendenhall: Holographic vibration study of a rotating propeller blade, Exp. Mech. **14**, 230–232 (1974)

4.373  A. L. Popov, V. E. Solodilov, G. N. Chernyshev: Holographic interferometry in problems of resonant vibration of shells of revolution, Mech. Solids **12**, 108–115 (1977)

4.374  M. D. Olson, C. R. Hazell: Vibration studies on some integral rib-stiffened plates, J. Sound Vibr. **50**, 43–61 (1977)

4.375  C. R. Hazell: Recording receptance of flat plates by holographic interferometry, J. Sound Vibr. **75**, 275–283 (1981)

4.376  M. M. Butusov, V. Y. Demchenko, Y. G. Turkevich: Holographic stroboscope based on a ruby laser having a passive shutter, Prib. Tekh. Eksper. **2**, 203–204 (1971)

4.377  F. Albe, P. Smigielski, H. Fagot: Application effective de l'interférométrie holographique par double exposition à l'étude des déformations de céramiques, dues à l'impact d'un projectile, Opt. Commun. **8**, 369–371 (1973)

4.378  W. T. Armstrong, P. R. Forman: Double-pulsed time differential holographic interferometry, Appl. Opt. **16**, 229–232 (1977)

4.379  A. Felske, A. Happe: Vibration analysis by double pulsed laser holography, SAE Transactions **86**, 88–104 (1977)

4.380  B. S. Hockley, R. A. J. Ford, C. A. Foord: Measurement of fan vibration using double pulse holography, J. Eng. Power **100**, 655–663 (1978)

4.381  B. Ineichen, J. Mastner: Vibration analysis by stroboscopic, two-reference-beam heterodyne holographic interferometry, SPIE **210**, 207–212 (1979)

4.382  N. Abramson: Light-in-flight recording: High-speed holographic motion pictures of ultrafast phenomena, Appl. Opt. **22**, 215–232 (1983)

4.383  A. J. Decker: Holographic cinematography of time-varying reflecting and time-varying phase objects using a Nd : YAG laser, Opt. Lett. **7**, 122–123 (1982)

4.384  R. Pawluczyk, Z. Kraska, Z. Pawlowski: Holographic investigations of skin vibrations, Appl. Opt. **21**, 759–765 (1982)

4.385  S. C. Gustafson: Pulsed holographic interferometry of objects subject to both uniform and vibrational motion, Opt. Eng. **19**, 849–852 (1980)

4.386  S. Kawase, T. Honda, J. Tsujiuchi: Measurement of elastic deformation of rotating objects by using holographic interferometry, Opt. Commun. **16**, 96–98 (1976)

4.387  K. A. Stetson: The use of an image derotator in hologram interferometry and speckle photography of rotating objects, Exp. Mech. **18**, 67–73 (1978)

4.388  W. F. Fagan, M. A. Beeck, H. Kreitlow: The holographic vibration analysis of rotating objects using a reflective image derotator, Opt. Lasers Eng. **2**, 21–32 (1981)

4.389  J. Geldmacher, H. Kreitlow, P. Steinlein, G. Sepold: Comparison of vibration mode measurements on rotating objects by different holographic methods, SPIE **398**, 101–110 (1983)

4.390  W. F. Fagan: Industrial applications of image derotation, SPIE **398**, 193–198 (1983)

4.391  J. C. MacBain, J. E. Horner, W. A. Stange, J. S. Ogg: Vibration analysis of a spinning disk using image-derotated holographic interferometry, Exp. mech. **19**, 17–22 (1979)

4.392  J. C. MacBain, W. A. Stange, K. G. Harding: Real-time response of a rotating disk using image-derotated holographic interferometry, Exp. Mech. **21**, 34–40 (1981)

4.393  J. C. MacBain, W. A. Stange, K. G. Harding: Analysis of rotating structures using image derotation with multiple-pulsed lasers and moire techniques, Opt. Eng. **21**, 474–477 (1982)

4.394  N. V. Morozov, K. P. Alim, Y. I. Ostrovskii: Holographic interferometry of rotating objects in oppositely directed beams, Zh. Tekh. Fiz. **51**, 355–360 (1981)

4.395  N. V. Morozov, Y. I. Ostrovskii: Holographic interferometry of rotating objects using a pulsed ruby laser, Zh. Tekh. Fiz. **52**, 577–578 (1982)

4.396  K. A. Stetson: Vibration measurement by holography, in *The Engineering Uses of Holography*, Proc. Symp. Glasgow 1968, ed. By E. R. Robertson, J. M. Harvey (University press, Cambridge 1970) pp. 307–331

4.397  K. A. Stetson: Holographic vibration analysis, in [ref 3.17, pp. 181–220]

4.398  M. Zambuto, M. Lurie: Holographic measurement of general forms of motion, Appl. Opt. **9**, 2066–2072 (1970)

4.399  C. S. Vikram: Holographic interferometry of superposition of motions with different time functions, Optik **45**, 55–64 (1976)

4.400  C. S. Vikram, K. Vedam: Unified approach for time-average holography of separable fine motions, Optik **57**, 385–389 (1980)

4.401  C. S. Vikram: Time-average holography of objects vibrating sinusoidally and moving with constant acceleration, Opt. Commun. **8**, 355–357 (1973)

4.402  C. S. Vikram: Stroboscopic holographic interferometry of vibration simultaneously in two sinusoidal modes, Opt. Commun. **11**, 360–364 (1974)

4.403  C. S. Vikram, G. Bose: Holographic interferometry of damped oscillations with two frequencies, Optik **43**, 253–258 (1975)

4.404  C. S. Vikram: Pulsed holography for interferometric analysis of superposition of three sinusoidal vibrations: A theoretical analysis, Pramana **8**, 541–544 (1977)

4.405  P. C. Gupta, K. Singh: Holographic interferometry of non-sinusoidal vibrations, Opto-electr. **6**, 305–311 (1974)

4.406  P. C. Gupta, A. K. Aggarwal: Effects of periodic, non-sinusoidal vibrations on the holographic process, Opt. Commun. **15**, 366–369 (1975)

4.407  P. C. Gupta, K. Singh: Time-average hologram interferometry of periodic, non-cosinusoidal vibrations, Appl. Phys. **6**, 233–240 (1975)

4.408  P. C. Gupta, K. Singh: Hologram interferometry of vibrations represented by the square of a Jacobian elliptic function, Nouv. Rev. Opt. **7**, 95–100 (1976)

4.409  P. S. Derus, I. B. Ekimov, V. N. Kudryavtsev: Vibration analysis by time-averaged hologram interferometry, Zh. Tekh. Fiz. **49**, 1692–1696 (1979)

4.410  K. A. Stetson: Hologram interferometry of nonsinusoidal vibrations analyzed by density functions, J. Opt. Soc. Am. **61**, 1359–1362 (1971)

4.411  R. Tonin, D. A. Bies: Analysis of 3-D vibrations from time-averaged holograms, Appl. Opt. **17**, 3713–3721 (1978)

4.412  R. Tonin, D. A. Bies: Time-averaged holography for the study of three-dimensional vibrations, J. Sound Vibr. **52**, 315–323 (1977)

4.413  R. Tonin, D. A. Bies: General theory of time-averaged holography for the study of three-dimensional vibrations at a single frequency, J. Opt. Soc. Am. **68**, 924–931 (1978)

4.414  C. S. Vikram, R. K. Sood: Triple-exposure stroboscopic holographic interferometry with thin phase recording materials, Nouv. Rev. Optique **4**, 109–110 (1973)

4.415  K. N. Chopra, G. S. Bhatnagar: Quadruple-exposure technique in stroboscopic holographic interferometry, Appl. Opt. **13**, 2468–2470 (1974)

4.416  C. S. Vikram, G. Bose, J. N. Maggo: Combination of time-average and stroboscopic techniques in holographic interferometry of sinusoidal vibration, Nouv. Rev. Optique **6**, 55–59 (1975)

4.417  M. Murata: A new method of holographic interferometry for practical vibration studies, Proc. ICO Conf. Opt. Methods in Sci. and Ind. Meas. Tokyo (1974) pp. 271–275

4.418  K. A. Stetson: Effects of beam modulation on fringe loci and localization in time-average hologram interferometry, J. Opt. Soc. Am. **60**, 1378–1388 (1970)

4.419  K. A. Stetson: Envelope factors due to laser modulation in time-average, holographic, vibration analysis, J. Opt. Soc. Am. **62**, 698–700 (1972)

4.420  N. Takai, M. Yamada, T. Idogawa: Holographic interferometry using a reference wave with a sinusoidally modulated amplitude, Opt. Laser Techn. **8**, 21–23 (1976)

4.421  I. Wojciechowska, A. Śliwiński: Examination of vibration amplitude distribution of ultrasonic transducers using optical holography with a modulated reference beam, Ultrasonics **19**, 115–119 (1981)

4.422  K. A. Stetson, K. Singh: Measurement of signal-to-noise ratio in hologram reconstructions, by vibration interferograms, Opt. Laser Techn. **3**, 104–108 (1971)

4.423  A. V. Vyshemirskii: Upper limit on the amplitude range for interference measurements of vibrations of diffuse surfaces, Zh. Tekh. Fiz. **52**, 67–73 (1982)

4.424  I. Péntek: Possibility of extension of holographic interferometry for analysis of high amplitude vibration, SPIE **398**, 95–100 (1983)

4.425  L. D. Siebert: Holographic coherence length of a pulse laser, Appl. Opt. **10**, 632–637 (1971)

4.426  S. Mallick: Pulse holography of uniformly moving objects, Appl. Opt. **14**, 602–605 (1975)

4.427  H. Fujiwara, A. Tomita: A holographic measurement of the pulse width of a Q-switched ruby laser, Opt. Commun. **14**, 17–20 (1975)

4.428  H. Fujiwara, M. Yasutake, K. Murata: A new holographic measurement of frequency sweeping of the output from a Q-switched ruby laser, Opt. Commun. **14**, 21–23 (1975)

4.429  H. Fujiwara, M. Yasutake: Holographic investigation of the temporal-coherence of the output of a Q-switched ruby laser, Opt. Commun. **14**, 318–321 (1975)

4.430  T. Sato, H. Fujiwara, K. Murata: Holography of a moving object using frequency chirp of a Q-switched ruby laser, Appl. Opt. **17**, 3096–3100 (1978)

4.431  H. Fujiwara, T. Sato, K. Murata: Upper-limit speed of a moving object in front-surface holography using a Q-switched ruby laser, Appl. Opt. **21**, 721–724 (1982)

4.432  K. A. Stetson: Method of vibration measurements in heterodyne interferometry, Opt. Lett. **7**, 233–234 (1982)

4.433  A. S. Bogomolov, N. G. Vlasov, E. G. Solovev: A method for time averaging in holographic interferometry of nonperiodically moving bodies, Opt. Spektrosk. **31**, 481–482 (1971)

4.434  C. S. Vikram, G. S. Bhatnagar: Application of holographic addition to time-average hologram interferometry of constant velocity motion, Appl. Opt. **13**, 720–721 (1974)

4.435  M. Miler: Investigation of ramp motion by means of multiple-exposure holography, Opt. Quant. Electr. **7**, 329–330 (1975)

4.436  M. Miler: Stroboscopic holography in ramp approximation, Opt. Commun. **14**, 406–408 (1975)

4.437  M. Miler: Linear approximation for strobe-hologram vibrational analysis, Zh. Tekh. Fiz. **47**, 396–404 (1977)

4.438  E. Müller: Holographie d'objets en mouvement, in [ref. 3.39, pp. 121–148]

4.439  J. G. Kelly, L. P. Mix: Measurements of high-current relativistic electron diode plasma properties with holographic interferometry, J. Appl. Phys. **46**, 1084–1090 (1975)

## Chapter 5

5.1  E. Müller: „Auswertung holographischer Interferogramme unter Berücksichtigung der durch Rekonstruktion mit geänderter Lichtwellenlänge erzeugten Bildmodifikationen", Thesis No. 7246, ETH Zürich (1983)

5.2  K. A. Stetson: Fringe interpretation for hologram interferometry of rigid-body motions and homogeneous deformations, J. Opt. Soc. Am. **64**, 1–10 (1974)

5.3  K. A. Haines, B. P. Hildebrand: Contour generation by wavefront reconstruction, Phys. Lett. **19**, 10–11 (1965)

5.4  B. P. Hildebrand, K. A. Haines: Multiple-wavelength and multiple-source holography applied to contour generation, J. Opt. Soc. Am. **57**, 155–162 (1967)

5.5  J. R. Varner: Simplified multiple-frequency holographic contouring, Appl. Opt. **10**, 212–213 (1971)

5.6  E. Menzel: Comment to the methods of contour holography, Optik **40**, 557–559 (1974)

5.7  A. A. Friesem, U. Levy: Fringe formation in two-wavelength contour holography, Appl. Opt. **16**, 3009–3020 (1976)

5.8  F. M. Küchel, H. J. Tiziani: Real-time contour holography using BSO crystals, Opt. Commun. **38**, 17–20 (1981)

5.9  B. P. Hildebrand, K. A. Haines: The generation of three-dimensional contour maps by wavefront reconstruction, Phys. Lett. **21**, 422–423 (1966)

5.10  N. Abramson: Sandwich hologram interferometry. 3: contouring, Appl. Opt. **15**, 200–205 (1976)

5.11  G. L. Rogers, L. C. G. Rogers: The interrelations between moiré patterns, contour fringes, optical surfaces and their sum and difference effects, Opt. Acta **24**, 15–22 (1977)

5.12  P. DeMattia, V. Fossati-Bellani: Holographic contouring by displacing the object and the illumination beam, Opt. Commun. **26**, 17–21 (1978)

5.13    A. S. Bogomolov: Use of reflection holograms to investigate the relief of diffuse sur-
        faces, Opt. Spektrosk. **51**, 337–341 (1981)
5.14    A. S. Bogomolov, E. S. Romashev, V. G. Seleznev: Use of thick-layer holograms in
        holographic interferometry, Opt. Spektrosk. **38**, 999–1000 (1975)
5.15    M. Yonemura: Holographic contour generation by spatial frequency modulation, Appl.
        Opt. **21**, 3652–3658 (1982)
5.16    N. Shiotake, T. Tsuruta, Y. Itoh, J. Tsujiuchi, N. Takeya, K. Matsuda: Holographic
        generation of contour map of diffusely reflecting surface by using immersion method,
        Japan. J. Appl. Phys. **7**, 904–909 (1968)
5.17    Y. Fainman, E. Lenz, J. Shamir: Contouring by phase conjugation, Appl. Opt. **20**,
        158–163 (1981)
5.18    K. A. Stetson: Holographic surface contouring by limited depth of focus, Appl. Opt. **7**,
        987–989 (1968)
5.19    N. Abramson: Light-in-flight recording: High-speed holographic motion pictures of
        ultrafast phenomena, Appl. Opt. **22**, 215–232 (1983)
5.20    N. Abramson: Holography using picosecond lightpulses, SPIE **370**, 170–174 (1983)
5.21    C. A. Sciammarella: Holographic moiré, an optical tool for the determination of dis-
        placements, strains, contours, and slopes of surfaces, Opt. Eng. **21**, 447–457 (1982)
5.22    C. M. Vest: *Holographic Interferometry* (Wiley, New York 1979)
5.23    W. Schumann, M. Dubas: *Holographic Interferometry. From the Scope of Deformation
        Analysis of Opaque Bodies,* Springer Ser. Opt. Sci. Vol. 16 (Springer, Berlin, Heidelberg
        1979)
5.24    P. Hariharan, B. F. Oreb, N. Brown: A digital system for realtime holographic stress
        analysis, SPIE **370**, 189–194 (1983)
5.25    P. Hariharan, B. F. Oreb, N. Brown: Real-time holographic interferometry: a mi-
        crocomputer system for the measurement of vector displacements, Appl. Opt. **22**,
        876–880 (1983)
5.26    T. R. Hsu: Large-deformation measurements by real-time holographic interferometry,
        Exp. Mech. **14**, 408–411 (1974)
5.27    T. Uyemura, Y. Yamamoto, K. Tenjimbayashi, N. Yokoyama: Real-time holographic
        interferometry with pulsed laser, SPIE **192**, 190–195 (1979)
5.28    H. Krepelková: The application of holographic interferometry to the analysis of compos-
        ite materials structure, Opt. Appl. **10**, 91–97 (1980)
5.29    K. M. Leung, T. C. Lee, E. Bernal, J. C. Wyant: Two-wavelength contouring with the
        automated thermoplastic holographic camera, SPIE **192**, 184–189 (1979)
5.30    B. Fischer, M. Cronin-Golomb, J. O. White, A. Yariv: Amplified reflection, transmis-
        sion, and self-oscillation in real-time holography, Opt. Lett. **6**, 519–521 (1981)
5.31    H. J. Tiziani: Real-time metrology with BSO crystals, Opt. Acta **29**, 463–470 (1982)
5.32    C. A. Sciammarella, P. K. Rastogi, P. Jacquot, R. Narayanan: Holographic moiré in real
        time, Exp. Mech. **22**, 52–63 (1982)
5.33    J. Politch: Real-time imaging and strain distribution of an angularly vibrating diffused
        plate, Opt. Acta **29**, 485–492 (1982)
5.34    N. V. Morozov, Y. I. Ostrovskii, L. M. Boeva: Real-time holographic interferometry of
        moving objects in oppositely directed beams, Zh. Tekh. Fiz. **52**, 1854–1858 (1982)
5.35    P. M. De Larminat, R. P. Wei: A fringe-compensation technique for stress analysis by
        reflection holographic interferometry, Exp. Mech. **16**, 241–248 (1976)
5.36    T. D. Dudderar, E. M. Doerries: Application of holographic interferometry to real-time
        studies of heat effects in multilayer circuit boards, Mat. Evaluation **37**, 41–50 (1979)
5.37    O. D. D. Soares: Hologram repositioning by an interferometric technique, Appl. Opt.
        **18**, 3838–3840 (1979)
5.38    R. Dändliker, B. Ineichen, F. M. Mottier: High resolution hologram interferometry by
        electronic phase measurement, Opt. Commun. **9**, 412–416 (1973)
5.39    R. Dändliker, J. F. Willemin: Measuring microvibrations by heterodyne speckle inter-
        ferometry, Opt. Lett. **6**, 165–167 (1981)

5.40    R. Dändliker: Measuring displacement, velocity and vibration by laser interferometry, in *Optoelectronics in Engineering,* ed. by W. Waidelich (Springer, Berlin, Heidelberg 1982) pp. 51–58

5.41    T. M. Hoffer, W. Fischer: Abnahme von Werkzeugmaschinen mit einem Laser-Meß-system, Feinwerktechnik & Meßtechnik **85**, (I) 229–235, (II) 343–359 (1977)

5.42    G. E. Sommargren: Optical heterodyne profilometry, Appl. Opt. **20**, 610–618 (1981)

5.43    J. H. Churnside, H. T. Yura: Laser vector velocimetry: a 3-D measurement technique, Appl. Opt. **21**, 845–850 (1982)

5.44    J. H. Shapiro, B. A. Capron, R. C. Harney: Imaging and target detection with a heterodyne-reception optical radar, Appl. Opt. **20**, 3292–3313 (1981)

5.45    R. Dändliker: Heterodyne holographic interferometry, in *Progress in Optics,* **17**, Chap. 1 (North-Holland, Amsterdam 1980)

5.46    P. V. Farrell, G. S. Springer, C. M. Vest: Heterodyne holographic interferometry: concentration and temperature measurements in gas mixtures, Appl. Opt. **21**, 1624–1627 (1982)

5.47    R. Dändliker, B. Ineichen, F. M. Mottier: Electronic processing of holographic inter-ferograms, in Digest of Papers, Int. Opt. Computing Conf. Zürich 1974 (IEEE, New York 1974) pp. 69–72

5.48    B. Ineichen. R. Dändliker, J. Mastner: Accuracy and reproducibility of heterodyne holographic interferometry, in *Applications of Holography and Optical Data Processing,* ed. by E. Marom, A. A. Friesem, E. Wiener-Avnear (Pergamon, Oxford 1977) pp. 207–212

5.49    R. Dändliker, E. Marom, F. M. Mottier: Two-reference-beam holographic inter-ferometry, J. Opt. Soc. Am. **66**, 23–30 (1976)

5.50    A. J. Decker, Y. H. Pao, P. C. Claspy: Electronic heterodyne recording and processing of optical holograms using phase modulated reference waves, Appl. Opt. **17**, 917–921 (1978)

5.51    R. Dändliker, B. Ineichen: Nonlinear cross-talk in two-reference-beam holographic interferometry, Opt. Commun. **19**, 365–369 (1976)

5.52    J. Katz, E. Marom: Effects of nonlinear recording in holographic interferometry, J. Opt. Soc. Am. **69**, 696–705 (1979)

5.53    J. Politch, J. Shamir, J. Ben Uri: Some characteristics of multibeam holography, Opt. Laser Techn. **3**, 226–228 (1971)

5.54    R. Dändliker, R. Thalmann, J. F. Willemin: Fringe interpolation by two-reference-beam holographic interferometry: reducing sensitivity to hologram misalignment, Opt. Com-mun. **42**, 301–306 (1982)

5.55    V. G. Kulkarni, P. N. Puntambekar: Holographic multiplexing using multiple reference beams, Opt. Acta **28**, 1611–1617 (1981)

5.56    T. Tsuruta, N. Shiotake, Y. Itoh: Hologram interferometry using two reference beams, Japan. J. Appl. Phys. **7**, 1092–1100 (1968)

5.57    D. Cuche, W. Schumann: Fringe modification with amplification in holographic inter-ferometry and application of this to determine strain and rotation. SPIE **398**, 35–45 (1983)

5.58    N. Abramson: *Making and Evaluation of Holograms* (Academic, London 1981)

5.59    N. Abramson: Sandwich hologram interferometry: a new dimension in holographic com-parison, Appl. Opt. **13**, 2019–2025 (1974)

5.60    N. Abramson: Sandwich hologram interferometry 2: some practical calculations, Appl. Opt. **14**, 981–984 (1975)

5.61    N. Abramson: Sandwich hologram interferometry 4: holographic studies of two milling machines, Appl. Opt. **16**, 2521–2531 (1977)

5.62    N. Abramson, H. Bjelkhagen: Sandwich hologram interferometry 5: Measurement of in-plane displacement and compensation for rigid body motion, Appl. Opt. **18**, 2870–2880 (1979)

5.63    H. Bjelkhagen: Pulsed sandwich holography, Appl. Opt. **16**, 1727–1731 (1977)

5.64   N. Abramson, H. Bjelkhagen: Pulsed sandwich holography 2: practical application, Appl. Opt. **17**, 187–191 (1978)

5.65   P. Hariharan, Z. S. Hegedus: Two-hologram interferometry: a simplified sandwich technique, Appl. Opt. **15**, 848–849 (1976)

5.66   P. Hariharan: Hologram interferometry: identification of the sign of surface displacements, Opt. Acta **24**, 989–990 (1977)

5.67   N. Abramson, H. Bjelkhagen, P. Skande: Sandwich holography for storing information interferometrically with a high degree of security, Appl. Opt. **18**, 2017–2021 (1979)

5.68   N. Abramson, H. Bjelkhagen: Deformation, displacement and vibration investigations in manufacturing applications using a new hologram interferometry technique, Opt. Lasers Eng. **1**, 51–68 (1980)

5.69   I. Dirtoft, N. Abramson, U. Sandström: Holographic measuring of deformations in complete upper dentures, SPIE **211**, 106–110 (1979)

5.70   S. Amadesi, A. D'Altorio, D. Paoletti: Sandwich holography for painting diagnostics, Appl. Opt. **21**, 1889–1890 (1982)

5.71   J. L. Doty, B. P. Hildebrand: The use of sandwich hologram interferometry for nondestructive testing of nuclear reactor components, Opt. Eng. **21**, 542–547 (1982)

5.72   S. Vukičević, I. Vinter, D. Vukičević: Sandwich hologram interferometry for determination of sacroiliac joint movements, SPIE **370**, 129–132 (1983)

5.73   M. Dubas, W. Schumann: Contribution à l'étude théorique des images et des franges produites par deux hologrammes en sandwich, Opt. Acta **24**, 1193–1209 (1977)

5.74   W. Schumann, M. Dubas: On the use of holographic interferometry in deformation analysis with additional degrees of freedom during the reconstruction, VDI-Berichte Nr. 313, 175–180 (1978)

5.75   G. S. Ballard: Double-exposure holographic interferometry with separate reference beams, J. Appl. Phys. **39**, 4846–4848 (1968)

5.76   L. A. Kersch: Advanced concepts of holographic nondestructive testing, Mat. Evaluation **29**, 125–129, 140 (1971)

5.77   E. B. Champagne: Quantitative data reduction with the use of fringe control techniques in conjunction with holographic interferometry, in *Eng. Appl. of Holography* (Proceedings) Los Angeles, (1972) pp. 133–145

5.78   J. Surget: Schéma d'holographie à deux sources de référence, Nouv. Rev. Optique **5**, 201–217 (1974)

5.79   J. Surget: Two reference beam holographic interferometry for aerodynamic flow studies, in *Applications of Holography and Optical Data Processing,* ed. by E. Marom, A. A. Friesem, E. Wiener-Avnear (Pergamon, Oxford 1977) pp. 183–192

5.80   J. D. Trolinger: Application of generalized phase control during reconstruction to flow visualization holography, Appl. Opt. **18**, 766–774 (1979)

5.81   G. Ferrano, G. Häusler: Kompensation von Ganzkörperbewegungen bei der holographischen Interferometrie, Optik **54**, 115–134 (1979)

5.82   N. G. Vlasov, S. G. Galkin, E. G. Semenov: Three-beam holographic interferometry of diffusely reflecting objects, Zh. Tekh. Fiz. **50**, 2019–2020 (1980)

5.83   D. Garvanska: Polarization holography applied to detection of shape deviations of metal surfaces, J. Opt. **12**, 201–206 (1981)

5.84   J. D. Hovanesian, Y. Y. Hung, P. D. Plotowski: Fringe compensation in hologram interferometry, Exp. Techn. **5**, 16–18 (1981)

5.85   J. W. C. Gates: Holographic phase recording by interference between reconstructed wavefronts from separate holograms, Nature **220**, 473–474 (1968)

5.86   R. J. Radley, A. G. Havener: Application of dual hologram interferometry to wind-tunnel testing, AIAA J. **11**, 1332–1333 (1973)

5.87   K. Matsuda: Holographic interferometry (I): Analysis of interference fringes produced by displacement of hologram, J. Mec. Eng. L. **30**, 267–279 (1976)

5.88   K. Matsuda: Lateral shear interferometer using twin three-beam holograms, Appl. Opt. **19**, 2643–2646 (1980)

220     References

5.89   I. S. Zeilikovich, N. M. Spornik, A. P. Ovechkin: Adjustment of an interference pattern
       using two separate holograms, Opt. Spektrosk. **42**, 969–972 (1977)
5.90   D. Cuche: „Modification des franges d'interférence en interférométrie holographique
       appliquée à la détermination des dilatations et des rotations", Thesis ETH, Zürich
       No. 7459 (1984)
5.91   K. A. Stetson: Homogeneous deformations: determination by fringe vectors in hologram
       interferometry, Appl. Opt. **14**, 2256–2259 (1975)
5.92   P. M. Boone: Surface deformation measurements using deformation following holo-
       grams, in *Applications de l'Holographie,* Proc. Symp. Besançon 1970, ed. by
       J. C. Viénot, J. Bulabois, J. Pasteur. Paper 5.1
5.93   P. M. Boone: Use of reflection holograms in holographic interferometry and speckle
       correlation for measurement of surface-displacement, Opt. Acta **22**, 579–589 (1975)
5.94   D. B. Neumann, R. C. Penn: Off-table holography, Exp. Mech. **15**, 241–244 (1975)
5.95   R. L. Van Renesse, J. W. Burgmeijer: Application of Denisyuk pulsed holography to
       material testing, SPIE **398**, 138–148 (1983)

# Author Index

Reference numbers appear between slashes, followed by page numbers indicating the specific location of each citation

Abraham, R. /2.1/4
Abramowitz, I.A. /3.86/52; /3.87/52
Abramson, N. /3.36/48,51; /4.21/111;
 /4.105/114; /4.382/144; /5.10/156;
 /5.19/156; /5.20/156; /5.58/171;
 /5.59-62/171; /5.64/171;
 /5.67-69/171
Abrukov, S.A. /4.250/118
Abrukov, V.S. /4.250/118
Ade, G. /3.94/52
Adkins, J.E. /2.20/37
Afanaseva, V.L. /3.110/54
Aggarwal, A.K. /4.406/144
Agren, C.H. /4.356/144
Aistov, V.S. /4.348/133
Albe, F. /4.377/144
Aleksandrov, E.B. /4.124/114
Alekseev-Popov, A.V. /3.72/52;
 /3.196-198/65
Alferness, R. /3.179/65
Alim, K.P. /4.394/144
Allano, D. /4.219/118
Altamirano, J.H. /3.92/52
Alum, K.P. /4.267/118
Amadesi, S. /4.28/113; /5.70/171
Amelin, A.D. /4.211/118
Andrews, C.L. /2.13/21
Antonini, G. /4.217,218/118
Antonov, V.M. /3.327/102
Appa Rao, T.V.S.R. /4.51/113
Archbold, E. /4.35/113
Aristov, V.V. /3.98/52
Arkhipov, V.I. /4.31/113
Armstrong, J.A. /3.265/75
Armstrong, W.T. /4.378/144
Arsenault, H.H. /3.134/58; /4.345/133
Arseneva, T.M. /3.289/84
Atkinson, J.T. /4.69/113; /4.349/133
Aung, W. /4.212/118
Auth, D.C. /3.114/54
Axelrad, D.R. /4.81/113


Baba, N. /4.296/127
Bach, D.R. /4.262/118
Bachalo, W.D. /4.207/118
Bacon, J. /4.245/118
Bailey, E.E. /4.196/118
Baily, P.K. /4.264/118
Ballantyne, J.M. /3.86/52

Ballard, G.S. /5.75/171
Bally, G. von /3.26/48
Beckmann, P. /4.343/133
Beeck, M.A. /4.388/144
Beketova, A.K. /4.205/188
Bellmann, A. /4.249/118
Belvaux, Y. /3.173/64
Belyaev, A.G. /4.204/118
Benlardi, B. /3.207/65; /3.240/68
Benser, W.A. /4.196/118
Benton, S.A. /3.309-310/102; /3.311/102
Ben Uri, J. /3.175/64,68; /5.53/167
Beranek, W.J. /4.156/115
Berdichevskii, V.L. /2.35/47
Bernal, E. /5.29/160
Bernard, A. /4.252/118
Berry, D.H. /3.306/102
Berry, M.V. /4.288/127,129
Bertho, P. /4.29/113
Bhan, C. /4.53/113
Bhatnagar, G. /4.415/144; /4.434/146
Biedermann, K. /3.19/48; /4.101/114;
 /4.141/114
Bies, D.A. /4.359/144; /4.411-413/144
Bijl, D. /4.92/114; /4.169/117
Billet, M.L. /4.337/131
Bitetto, D.J. De /3.302/102
Bjelkhagen, H. /5.62-64,67,68/171
Bjorklund, G.C. /4.246,247/118
Blaschke, W. /2.5/4
Bobrov, S.T. /3.269,270/75
Bochner, N. /4.232/118
Boehnlein, A.J. /4.62/113
Boerner, W.M. /3.76/52
Boeva, L.M. /5.34/160
Bogomolov, A.S. /4.433/146; /5.13,14/
 156
Bolognini, N. /4.174/118
Bonch-Bruevich, A.M. /4.124/114
Bonjour, C. /3.39/48; /4.22/111
Boone, P.M. /5.92,93/183
Born, M. /2.12/20,23,24,26,28,35;
 /3.102/52,75,79,96
Borza, D. /4.361/144
Bose, G. /4.403/144; /4.416/144
Bowley, W.W. /4.121/114
Bräuchle, C. /4.246,247/118
Bremmer, H. /2.10/20
Breuckmann, B. /4.358/144
Brillouin, L. /2.3/4

# Subject Index

Abel transform 129
Aberration
  astigmatism 59, 63, 79, 80, 89, 106, 164
  coma 79, 83
  dual 105, 160
  lateral 105, 164
  of a point 83, 92, 103, 105
  Seidel 75, 78, 83, 101, 108
  spherical 79, 81
  theory 52
  transverse ray 56, 82, 86, 94, 95, 97
  wave 55, 77, 86, 91, 95, 97
Absorption coefficient 68
Acousto-optical modulator 165
Affine connection 15, 16, 56, 116, 123, 127, 139, 170
Amplification effect 165
Amplitude
  complex 22, 23, 48, 50, 52, 112, 133, 136, 145
  distribution 68
  total 133, 136
  transmittance 49, 165
Analogy 124, 136
Angular fringe spacing 117, 123, 170
Aperture 138, 159
  circular 139, 149
Astigmatism 59, 63, 79, 80, 89, 106
  tensor 164
Asymptotic analysis 47
  behavior 104

Beat frequency 168
Bouguer's formula 28
Bragg condition 63, 64, 68

Caustic 97
Circularly polarized wave 22
Collimated object beam 117
Collimated reference beam 108
"Collinear image" 152
Collineation center 10, 78, 83, 87, 99, 118, 126
Coma 79, 83

Composite holography 101
Conductivity
  specific 20
Configuration at recording 152
  modified 152
Conjugate-image wave field 51, 166
Connections between surfaces
  affine 15, 16, 56, 116, 123, 127, 139
Constitutive
  relations 20
Contour lines 159
Contouring 156
Correlation
  auto 133, 137
  cross 137
Coupled-wave theory 63, 65
Coupling parameter 68
Cross-image 167, 170
Curl 9
Curvature
  change 80
  Gaussian 14, 51
  geodesic 14, 61, 163
  mean 14, 59
  normal 14
  of a fringe 164
  of a hologram 61
  of an interface 30
  of a surface 44
  of a wavefront 32, 57, 62
  "reduced" 46
  "tensor" 14
curved light ray 8

Debye-type integral 142
Decomposition
  additive 41
  of a tensor 12, 13
  of a vector 12
  polar 39, 40, 41
Deformation
  apparent 84, 89, 90
  gradient 37, 41, 60, 115, 178
  of a hologram 169

# Springer Series in Optical Sciences

Editorial Board: J.M. Enoch  D.L. MacAdam  A.L. Schawlow  K. Shimoda  T. Tamir